Linux Yourself

Linux Yourself
Concept and Programming

Sunil K. Singh

CRC Press
Taylor & Francis Group
Boca Raton London New York

CRC Press is an imprint of the
Taylor & Francis Group, an **informa** business

A CHAPMAN & HALL BOOK

First edition published 2022
by CRC Press
6000 Broken Sound Parkway NW, Suite 300, Boca Raton, FL 33487-2742

and by CRC Press
2 Park Square, Milton Park, Abingdon, Oxon, OX14 4RN

© 2022 Taylor & Francis Group, LLC

CRC Press is an imprint of Taylor & Francis Group, LLC

ISBN: 978-1-138-33328-4 (hbk)
ISBN: 978-1-032-03707-3 (pbk)
ISBN: 978-0-429-44604-7 (ebk)

Typeset in Palatino
by codeMantra

This book is dedicated to

To my mother, Smt. Sita Singh,

For unmatched love and encouragement.

You have always been my source of support and inspiration:

"Thank you, Mom for always teaching me that the greatest

endeavour happens outside from your comfort region."

Contents

Section II   Linux Programming

Section III   Case Studies

Preface

Presently, the information about Unix and Linux is extensively available to start working with Linux operating system (OS). But we still feel that an easy approach is required to understand the basic concept and programming skill of Linux to get proficiency in Linux programming. This book aims to give you a strong foundation with the help of various fundamental and advanced topics along with suitable examples in Linux.

Linux OS is now considered to be one of the most important OSs in use. Not only does it bring the power and flexibility of a workstation to a personal computer (PC), but it also enables to run a complete set of Internet applications along with a fully functional graphical user interface (GUI). Linux is an OS program referred to as a *kernel*, which was originally developed by Linus Torvalds at the University of Helsinki in the late 1990s. A kernel helps in the functioning of an OS, but it is not enough to form a complete OS by itself. Therefore, to build a complete OS, the kernel needs some additional software programs, application programs, and development tools. These programs are consolidated in modules as applications and tools for every kind of task and operation. Hence, a complete Linux OS which includes kernel and all other free and open-source software's ranging from user applications to development tools, such as networking tools, security application or others. Such tools and applications including kernel upgrade their features in regular interval of times. Linux is developed as a part of the open-source software project in which developers work together to provide free and highly efficient software to the users. Being an open-source OS, Linux has turned into a premier platform for open-source software development community mainly developing under the Free Software Foundation's GNU Project. Numerous applications are provided as part of standard Linux distributions.

Linux comes in many flavors, known as Linux distributions, but most of their operations and features are fundamentally the same. They use the same desktop, shell, application programs, file system, support tools, network configuration, and other features. Many Linux distributions offer their own GUI tools and application programs, but these differ only on the front end. GNOME and the Korn Desktop Environment (KDE) have become standard desktop GUIs for Linux because of their flexibility, control, and accessibility. Linux is also known as fully functional PC edition of Unix OS that brings the speed, efficiency, scalability, and flexibility of Unix OS to a PC. It has all the standard features of a robust Unix system, including a complete set of Unix shells such as Bourne shell (sh), Bourne Again Shell (bash), tcsh, csh, and zsh. Those familiar with the Unix interface and commands may also find that some key Unix commands, filters, and configuration features are also available in Linux. But, having prior expertise of Unix is not required for anyone to get proficiency in Linux. This book is not envisioned for a particular Linux distribution; rather, it provides a brief and comprehensive description of various commands/features common to all Linux-based OSs. Hence, this book may help readers to improve their expertise in Linux OS, even if you don't have any basic knowledge of Linux.

Linux is highly portable and powerful; the users may easily change or customize Linux to meet their requirements as it comes with the complete source code. A large number of applications ranging from wristwatches to huge data centers run on Linux-based systems. Developers and companies are normally releasing several desktop features and applications, for both users and systems, in the latest release of Linux distributions. Finally, we can say that technological growth may not be dreamed without Linux due to its outstanding features.

Why People Like Linux

There is no prerequisite knowledge to study this book, except the basic computer skills and understanding, such as how to work with a computer system. With this opinion, all topics are presented in such a way that they should be simple and have a clear explanation with suitable examples and figures. Linux is distinguished by its own power, flexibility, and open-source accessibility unlike other famous OSs. Linux is much more secure and has the least or nearly zero virus threats. It may also support heavy CPU loads. That is why Linux is used everywhere including in the modern embedded systems such as mobile phones, tablet computers, various networking devices, facility automation controls, modern televisions, and video game consoles. Android is another widely used OS for mobile devices which is built on Linux, as you are aware that Android is known even to the less technically sound users. In present time simply, we can say that Linux is touching every person life as a technology developer or as a user.

Target Audience for This Book

Numerous people still believe that It is not easy to learn and acquire expertise in Linux and only a professional can learn and understand how a Linux system works. But due to the advancements in development, Linux has gained much popularity both at home and at the workplace. In order to understand the popularity of Linux, this book will help and guide the people of all ages to gain knowledge of Linux in an easy approach which can be used for all kinds of purposes. This book is designed to serve as a textbook for undergraduate/postgraduate students of computer science and its related disciplines or any user who has a passion to learn Linux. This book will offer them a complete orientation on Linux concepts along with clear and detailed explanations of Linux programming features which help them to become a good Linux programmer, developer, or network or system administrator in the field of computer science/information technology.

Why Linux Yourself?

This book can be used as a starting point to learn basic concept of Linux, and then move towards learn Linux programming skill. It will help and guide you through the process of mastering the skills needed to build and run a powerful and productive computation-intensive Linux system and workstation. This book helps you overcome technical barriers by explaining complex concepts in simple language and highlighting precise ways to make your learning outstanding with an easier explanation of terms, topics, and technical concepts of Linux OS.

The word *"Yourself"* in the book title refers to the fact that the content of this book is designed in such a way to educate yourself as well as offer a personalized feed of Linux learning and programming. This book works as a bridge between you and the beautiful world of Linux. Each chapter of this book gives in-depth knowledge and examples that may motivate you to create programs for your own problems.

Organization of this Book

The basic goal of this book is to guide you that how to enhance your programming skills effectively in Linux. Linux programming is a wide area, and this book tries to cover conceptual topics from a wide range of related topics to give you a good start. It guides you from the basic of Linux installation, the command line concept, the vi editor, and shells to more advanced topics such as creating shell script, sed, awk, IPC, and X Window programming. This book is divided into three parts, namely Linux Concept, Linux Programming and Case Studies, which are comprised of fifteen chapters, each covering a unique aspect of Linux OS.

Each chapter aims to introduce the theory and then demonstrate it with suitable examples. The Linux commands associated with each topic are covered and explained well to make your learning fast. **Chapter 1** presents details about the fundamental concepts, features, and development history of Unix and Linux systems. Further, you learn about the GNU Project, Linux distributions, installation procedure, architecture, and shell features, which offer an enough background to start working with a Linux system. You also learn about the key component of a Linux system such as kernel, GUI (GNOME/KDE), boot loader, login/logout, and start-up scripts along with various run levels. The key Linux commands using the bash are described in **Chapter 2**, in which you find the way to categorize a Linux command based on whether it is a built-in command or not. Generally, all the commands available in Linux may be divided into four major parts according to their utilities and usage: user commands, universal commands, system commands, and networking commands. Furthermore, you will also know the basics of Linux directory structure and how to get the path of each directory and some other information by using some frequently used special commands, such as *find, locate, type, which,* and *whereis.*

A shell is a very important feature or interface of Linux which allows you to interact with a Linux system. **Chapter 3** describes shells, specifically sh and bash, and their features. Further, it presents shells' command line structure, interpretive cycle, and variables, including environment variables. You also find out metacharacters and how to use them as well as how to combine commands, namely sequence commands, group commands, chained commands, and condition commands, to execute together. You will also see how to redirect the input/output and connect commands through the pipe, tee, and xargs commands. At the end of this chapter, the programming language feature of shells is briefed by introducing shell scripts and how to build your own shell command. **Chapter 4** presents the vi editor, which is perhaps the most popular pervasive text editor available in Linux or any other operating environments. It describes vi commands, operating modes, how to insert and delete text, operation on regular expressions, command combinations, how to do the coding of various programming languages and its compilation, how to get help, and other relevant information regarding the vi editor. You will also find some useful information about vim, which is an improved version of the vi editor, along with some other popular editors such as nvi, gedit, and emacs. The vi editor provides a rich feature for coding in various programming languages to the Linux programmers.

A regular expression, referred to as *regex* or *regexp,* is represented by a set of characters that is used to search and match patterns in a string or to simplify the solution of text manipulation problems. **Chapter 5** briefs regular expressions, an elaborate pattern-matching concept, along with POSIX Basic Regular Expression (BRE) and POSIX Extended Regular Expression (ERE) metacharacters. Further, you also get to know about the grep family which handles the pattern search, edit, and replace operations. At this point, some basic regular filter commands are also described, such as comm, cut, expand, head, tail, more, less, split, tee, and zcat. Further, **Chapter 6** covers the stream editor (sed), one of the key pattern-matching languages on GNU/Linux. It works with a stream of data, based on commands or a set of rules that use a simple programming language to solve complex text-processing tasks with a few lines of code. Therefore, this chapter starts with sed overview, syntax, and addressing mechanism that help to write a sed script. Further, the sed command sets are divided into basic and advanced commands. The basic uses of sed commands are editing one or more files automatically, streamlining tedious edits to multiple files, and writing alteration programs. You will also see the examples of sed commands to perform various operations such as substitute, replace, append, insert, change, delete, and transform. Further, the sed advanced command section covers how to read and write data with a file along with commands to change the execution flow of sed script. You will also find out about sed internal storage space, namely pattern space and hold space, along with sed merits and limitations. Another powerful advanced filter, awk, is explained in **Chapter 7**. awk is a convenient and expressive programming language that allows easy manipulation of structured data and creation of formatted reports. Chapter 7 briefs you about the awk programming language and how you can use it effectively, such as gawk features, awk program structure, how to write an awk program, and its execution. Further, you will see the various tools to generate formatted output, widely known as fancier output, after performing the desired computation over given inputs. This chapter also provides many standard commands and built-in functions such as text handling, arithmetic functions, string function, and in-time function. You will find some key examples to explain the awk features that make your learning very interesting and informative. It is to note that Chapter 3 only covers the basic features of Linux shell and command line execution of various commands, but **Chapter 8** explains the basics of the shell scripting concept and its features using bash (*.sh* or *.bash*) including how to create a shell script and an interactive script, how to define variables, arithmetic function, control structure, comparison operation, and string operation. Additionally, Chapter 8 also covers functions, arrays, and test commands with respect to shell scripts and describes how to use them in shell scripts with suitable examples.

The discussion of Chapter 1 extends more to **Chapter 9**, which presents more useful tips and techniques for Linux administration. Linux administration is a specialized work and covers all the essential things that you must do to keep a computer system in the best operational condition. Therefore, every Linux learner or user must understand the important concepts related to Linux administration including disk usage, kernel administration, boot loader (GRUB, LILO), root user privilege, user interface (GNOME, KDE), RPM package management and network management commands, and many more information related to system administration as well as maintenance. Everything in Linux is based on the concepts of files, or we can simply say that everything in Linux is a file. **Chapter 10** describes the organization and terminology of the Linux file system. It starts with Linux file system architecture, the inode concept, and how to create a file and directory along with other related key operations. Some file operations such as file listing, how to display, copy, print, search, rename, and remove are also described. It also demonstrates how to find files, directories, and pathnames and how to locate a directory in a file system. Further, it explains how to archive and compress files and directories along with some key attributes as well as how to set and reset the various file access permissions. Further, **Chapter 11** stretches the discussion of Chapter 10 and describes the various disk partitioning methods, file system layout, how to manage a file system including creation, mounting, unmounting, checking, and repairing. Further, it also describes the error-handling function, /proc file system, LVM, VFS, and various Linux extended (ext) file systems such as ext, ext2, ext3, and ext4 and their upgradation procedure.

Chapter 12 describes the Linux system programming feature concept, along with the compilation procedure of C and C++ source code by using the gcc/g++ compiler, respectively, in the vim editor environment, GNU make, and GNU debugger. Further, it covers the details about the concept of processes and threads along with their associated details such as process creation using fork(), vfork(), PID, PPID, state, priorities (nice), termination (kill), synchronization, and zombie. It also describes device file, signals, system calls, and POSIX with respect to the programming concept. **Chapter 13** provides more details about system programming features with the description of inter-process communication (IPC) facilities such as shared memory, message queue, pipe, and sockets with their system call in the C language. **Chapter 14** gives an overview of the X Window System, X client, X server, X protocol, and X toolkit and covers the programming aspect. It describes how to create and manage windows as well as how to start and stop X. It also briefs X architecture, Xlib programming model, desktop environments, and how to upgrade X Window tools. The Case Studies section comprised with **Chapter 15** brief some key features of widely available Linux distributions, such as Red Hat Enterprise Linux, Fedora, Debian GNU/Linux, and Ubuntu Linux. It also highlights the ethical aspect of using Linux.

Feedback Information

Feedback and suggestion help to improve the quality content of this book in future editions. Therefore, kindly send your feedback, suggestions, comments, bug reports, code improvements, and other information to linuxyourself@gmail.com. We will try to answer every email in a time-bound manner as per the feedback or technical questions. But sometimes we could not reply in time due to the high volume of emails. Please be patient for such a cause. All constructive and positive feedbacks, technical suggestions, and updated features related to Linux must be included in a forthcoming edition of this book.

Warning and Disclaimer

- Linux is a trademark of Linus Torvalds.
- The Unix OS, trademark as Unix, was initially developed by AT&T (Bell Labs) in 1970. Thereafter in 1993, the Unix trademark was moved to The Open Group. Presently, UNIX® is a registered trademark of The Open Group.
- The logos, trademarks, and symbols used in this book are the properties of their respective owners.

The objective of this book is to offer information about the Linux concept and programming. The author has put all efforts and taken all precautions to make this book correct, complete, and feasible; however, no warranty of any kind and no liability for the contents of this book can be accepted. The provided information and content in this book is distributed on an "As-Is" source basis, but there may be a possibility of errors and inaccuracies in the information or programs provided in this book; in this case, the author is not taking any responsibility to any person or entity regarding any loss or system damages.

It is hoped that this book will help you to enjoy teaching yourself with "Linux Yourself"!!!

Acknowledgments

The success of writing this book, *Linux Yourself: Concept and Programming*, depends largely upon the encouragement, support, and guidance of many others. I take this opportunity to express my gratitude to the people who have been instrumental in the successful completion of this book. I thank all the developers, the industry professionals, and the members of the open source and academic communities, and others for their active support and work on LINUX/UNIX to make it as a highly adaptable and useful operating system accessible to everyone.

First and foremost, I offer my sincerest gratitude to the team at Taylor & Francis Group (CRC Press) that gave me this opportunity and provided me with outstanding support to work on this project. I especially must thank Ms. Aastha Sharma, the Senior Acquisitions Editor who has shown her faith in me and provided extraordinary support, encouragement and help throughout this project. Also, I thank Ms. Shikha Garg, Editorial Assistant, for handling and keeping things on track, as well as all communication, which has helped make this book more valuable to the readers. My thanks to codeMantra for their support and sincerity in copy-editing, typesetting, and indexing to make our work more readable and presentable. It is indeed a privilege to work under Taylor & Francis Group and I am extremely thankful to them for their priceless support.

I would also like to thank faculty members and students of my Department of Computer Sc. Engineering, CCET (Degree Wing), Chandigarh, for their moral and positive support. Especially, I must thank my colleague Mr. Sudhakar Kumar, Assistant Professor, for his valuable backing at each and every step during the writing of this book. Finally, the most important source of the strength in my life is my family. I would like to thank my parents, the families of my brother and sister, my lovely sons (Ritvik Singh and Sanvik Singh), and my wife Meenakshi Rathore, who always act as guiding force and inspiration in every moment of my life.

Ultimately, I thank everybody who has ever believed in me and provided an opportunity to understand the various aspects of my life to reach this point.

Dr. Sunil K. Singh
April 27, 2021

Author

Dr. Sunil K Singh is working as Professor and HOD, Department of Computer Science and Engineering (CSE), Chandigarh College of Engineering and Technology (Degree Wing), Affiliated to Panjab University, Chandigarh, Sector-26, Chandigarh, India. He did his graduation (Bachelor of Engineering), postgraduation (Master of Engineering), and Doctor of Philosophy (Ph.D.) in computer science and engineering, and he has a great passion for both teaching and research. His areas of expertise are high-performance computing, Linux/Unix, Data Mining, Internet of Things (IOT), Machine Learning, Computer Architecture, Embedded System and Computer Network. He has published more than 50 research papers in reputed international/national journals and conferences. He is a reviewer of several renowned national and international research journals, and a member of professional bodies such as ACM, IE, LMISTE, ACEEE, IACSIT, and IAENG. He has also received 01 patent granted and 02 patents published, and he is also on many other research and book projects. He is very active as an ACM professional member. He also has contributed to the Eminent Speaker Program (ESP) of ACM India.

Section I

Linux Concept

1

Getting Started

Many people still believe that learning, and acquiring expertise in, Linux is an uphill task or only professionals can understand how a Linux system works. However, due to the advancements in development, Linux has gained much popularity both at home and at the workplace. This book uses an easy approach and aims to help and guide people of all age-groups to gain knowledge of Linux which is useful for all purposes. The term Linux system refers to a system which runs on a Linux-based operating system (OS). There exist a large number of flavors or distributions of Linux. Before we can explore the details of Linux, we need to explain the basic concepts of an OS and its features in subsequent sections.

1.1 Getting Started with OS

An OS is an essential part of a computer system. Sometimes, users refer to OS as the heart of the computer. An OS serves as an interface between the user and computer hardware. The job of an OS is to offer an environment in which users can execute application programs in an interactive and efficient way. An OS is also defined as a set of software that supervises computer hardware resources, offers common resources, and controls the execution of various computer programs. If a computer is in running state, then the OS is the only program that is always executing and is referred to as the kernel, while all other programs are called application programs.

Before addressing the various features and operations of Linux OS, we must explore the underlying structure of a computer system. The fundamental modules of a computer system are OS, computer hardware, application programs, and users. All such modules are arranged as shown in Figure 1.1.

The whole computer system is broadly divided into three parts: *user*: end users and developers; *software*: application software and OS; and *hardware*: central processing unit (CPU), input/output (I/O) devices, and memory. According to Figure 1.1, the OS synchronizes and controls the utilization of computer hardware between the various application software packages for different users.

Further, to identify the various features of an OS, we must look at its facilities. An OS gives facilities equally to users and application programs to perform various operations such as input/output operations, program execution, file system manipulation, network communication, error detection and correction, resource distribution, safety, and security. Therefore, the size of an OS varies according to the numbers of services provided. OS's development is split into smaller modules because of its complexity and large size. The functionality of each module should be well defined so that it can effectively coordinate with other modules.

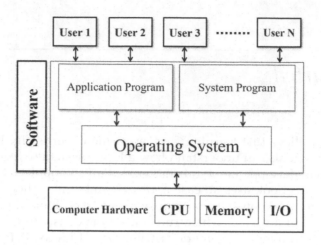

FIGURE 1.1
Arrangement of fundamental modules of a computer system.

1.2 A Brief Overview of Unix and Linux

Linux Yourself demonstrates how to build your concepts and programming skills to become an advanced Linux user or developer or both. Linux and Unix are other OSs such as Windows, DOS, and macOS. Linux, in many ways, is a clone of Unix. In recent years, Linux has become an observable fact and experience in the world of technology.

Unix is one of the most popular multitasking, secure, portable, and multiuser OSs. It was invented and developed by Ken Thompson and Dennis Ritchie at AT&T Bell Labs in the early 1970s. Initially, the basic objective was to create Unix for their personal use. Primarily, the code of Unix was written in B language and became operational on a PDP-7 in 1970. The first noteworthy achievement was changing the Unix system from the PDP-7 to the PDP-11. It gave the first intimation to everyone that Unix would act as an OS for all computer systems. In 1972, the subsequent major achievement was attained by altering (rewriting) the whole Unix programming code in C language. After converting, the C programming code of Unix confirmed the benefits of utilizing a high-level language, and presently, all Unix module implementations are written in C programming language.

Although Bell Labs was not in the computer business, even after 1974, Unix OS was first described in a technical journal. This prompted a great interest in various commercial institutions and universities. After this, Unix was distributed to commercial institutions as well as universities under AT&T (Bell Labs) licenses for their modified value-added version. After that, brilliant researchers started writing their own software and utility programs which were added to the Unix code. As a result, the most significant non-AT&T Unix system was developed at the University of California, Berkeley, known as Unix Berkeley Software Distribution (BSD). Subsequently, 1976 onward, like BSD, some other Unix flavors came into existence, such as IBM-AIX, SCO Unix, SunOS, HP-UX, FreeBSD, Linux, Mac OS X, SGI IRIX, and Sun Solaris. Unix is one of the utmost achievements in computer science, and for this supreme contribution, Dennis Ritchie and Ken Thompson received ACM Turing Award in 1983.

In 1990, 21-year-old Linus Torvalds, a computer science undergraduate, needed to purchase Unix OS for his own computer, but due to its high cost, he could not afford to buy it. Thereafter, he started writing his own Unix-like OS; later, it was announced as "Linux."

He posted the Linux on the Internet in 1991, so that anyone can use it without paying any money. Linux is a free and stable open-source OS because its source code is available publicly. The Linux kernel operates on several different platforms including Intel, IBM, and Alpha and even some handheld PDAs and others. Linux is available under the GNU General Public License (GPL) that comprises a complete set of development tools, graphical user interfaces (GUIs), hardware support, and application programs extending from office suites to multimedia applications. Linux has turned into one of the most important OSs and been operated for years on mainframes and minicomputers. The Linux OS is widely used for a broad range of systems such as personal computers (PCs), workstations, network servers, and high-performance computing systems.

Linux takes the Unix system as its inspiration, and due to this, Linux and Unix commands/programs are very similar in nature. In fact, almost all programs written for Unix may be compiled and executed on Linux also. Linux brings with it the speed, efficiency, scalability, and flexibility of Unix. Linus Torvalds originally developed the Linux kernel, which is the heart of the OS. However, Linux has continuously been circulated with a good amount of software applications, varying from network application to office applications including a wide variety of development tools. Linux is developed as a part of the open-source software initiative wherein individual programmers may collaborate to offer free and high-quality software along with other user applications. At present, Linux is a top-rated OS available to everyone. It is authored and maintained by thousands of developers from across the world, connected via the Internet to each other. Many companies have also extended their hands to provide support for Linux. Linux is widely used and provides support in TCP/IP networking, multiprocessor machines, scientific and parallel computing, grid computing, cloud computing, and much more.

The Institute of Electrical and Electronics Engineers (IEEE) has given a separate Unix criterion for the American National Standards Institute (ANSI). This new ANSI criterion for Unix has been widely known as Portable Operating System Interface for Computer Environments (POSIX). The given criterion identifies how a Unix-like OS needs to work and other classifying details such as system calls and interfaces with other modules. The POSIX classifies a complete universal standard which must be adhered to by all Unix versions and distributions [1]. On the other side, Linux was developed according to the POSIX standard from the beginning of its development. The development of Linux is currently handled by the Linux Foundation,* which is a union of Open Source Development Labs (OSDL) and Free Standards Group. The inventor of Linux, Linus Torvalds, works with this group to develop new Linux versions, i.e., new kernel versions. The latest Linux kernel version is regularly released after suitable upgradation and freely available to all of us. For more details, visit the kernel home page [2].

1.3 Unix/Linux History

To describe Unix/Linux, you can look at its history, as mentioned in Table 1.1. Unix and Linux are among the most accepted OSs worldwide because of their huge support base and a large number of distributions.

---

* http://www.linux-foundation.org.

TABLE 1.1

Timeline of Unix and Linux History

Year	Event	Description
1957	BESYS operating system (OS)	Bell Labs developed an OS, known as BESYS, for their own computer laboratories which was used to run several short batch jobs.
1964	Multics OS	Bell Labs was trying to be looking at a new generation of computers. That is why they agreed to work with General Electric and MIT to produce a new general-purpose, multiuser, time-sharing OS known as **Multics** (Multiplexed Information and Computing Service).
1969	**B** programming language	**B** is a programming language, developed at Bell Labs around 1969. It was created by Ken Thompson with the support of Dennis Ritchie.
April 1969	Multics OS called off by Bell Labs	Bell Labs decided to withdraw the Multics project due to differences with the other members. Thereafter, they purchased a new computer and started using GECOS as an OS. The GECOS was not practically as advanced as Multics. This situation generated the need among Bell Labs researchers to create something different and more powerful. Then, Ken Thompson, Dennis Ritchie, and others began to work using an old little-used DEC PDP-7 computer at Bell Labs. This led to the birth of Unix.
August 1969	First version of Unix written in assembly language	Ken Thompson wrote the first edition of an OS, yet unnamed, in combination with assembly language and B language for DEC PDP-7 minicomputer. This created OS was much smaller and simpler than Multics.
1969	Creator of Linux	1969 is the year of birth of Linus Torvalds, who developed Linux in 1991.
1970	Beginning of **Unix**	Ken Thompson named the newly written OS Unics (Uniplexed Information and Computing Service), but later, it was changed to **Unix**.
February 1971	First edition of Unix	The first edition of Unix was released.
November 1971	First Unix Programmer's Manual	The first edition of *Unix Programmer's Manual*, written by Ken Thompson and Dennis Ritchie, was released.
1972	Birth of C programming language	Dennis Ritchie developed the C programming language, which was an improved version of B language. B language was almost wiped out by C language.
1972	Second edition of Unix	The second edition of Unix was released on December 6, 1972.
February 1973	Third edition of Unix	The third edition of Unix was released.
November 1973	Fourth edition of Unix	The Unix code was rewritten again into C programming code. This step changed the history of OSs and made Unix portable.
January 1974	Unix moving out from Bell Labs	The Unix source code was moved out from Bell Labs, and the University of California at Berkeley received the first source code copy of Unix.
July 1974	First Unix article in a journal	The journal of Association for Computing Machinery (ACM) published an article "The UNIX Timesharing System," written by Dennis Ritchie and Ken Thompson. The authors explained that Unix is a general-purpose, multiuser, and interactive OS. This article created the first initial massive demand for Unix.
1975	Unix leaving home	Bell Labs makes Unix freeware (the code remained property of AT&T)
1976	Sixth edition of Unix	The sixth edition of Unix, which was broadly known as version 6 (V6), was released. This was the first version of Unix that was widely offered outside of Bell Labs. V6 became the beginning of the first version of Unix, commonly known as Berkeley Software Distribution (BSD), produced at the University of California, Berkeley.

(Continued)

TABLE 1.1 (*Continued*)

Timeline of Unix and Linux History

Year	Event	Description
March 9, 1978	1.0 BSD edition	The first Berkeley Unix 1.0 BSD was released. The first BSD version (1.x) was derived from V6.
1978	Unix adopted by commercial organization	Numerous commercial organizations started to receive Unix under legal license from Bell Labs (AT&T). The growth of a Unix-like system was increased, for example, System V, BSD, IBM-AIX, HP-UX, Sun Solaris, and SGI-IRIX.
1983	System V	AT&T released the first version of System V.
1983–1984	GNU (GNU's Not Unix).	GNU was founded by Richard Stallman (Open Software Foundation, GPL)
1985	4.2BSD	4.2BSD version was released by the University of California at Berkeley that included networking (TCP/IP), signals, and many other features.
1987	Minix	An open-source Unix clone OS was written by Andrew Tanenbaum, called as Minix.
1988	POSIX	The IEEE announced and published a set of standards for Unix interfaces, known Portable Operating System Interface (POSIX).
1989	X Window System (GUI) tool	MIT developed the X Window System (GUI tool)
1991	**Linux**	Linux Torvalds, a Finnish student, wrote Linux kernel as a personal project motivated from Minix. When Linux kernel is composed of GNU software packages to produce a free GNU/Linux OS, then such system may simply be referred to as "Linux." In the same year, Sun revealed Solaris, an OS built on SVR4.
October 5, 1991	Linux posted on the Internet	Torvalds sent a post to the newsgroup "comp.os.minix" for announcing the release of Linux kernel.
1992	Linux 0.99 kernel released using GNU GPL	Kernel version 0.99 was published using the GNU GPL by Linux Torvalds in the middle of December 1992. For this, Linux and GNU developers worked together to incorporate GNU components with Linux kernel to make a fully functional open-source OS.
1993	FreeBSD 1.0	FreeBSD 1.0 version was released in December 1993.
1994	Linux 1.0 kernel	Linux 1.0 kernel version was released.
1994	Red Hat Linux	Red Hat Linux came into existence.
1994	SUSE Linux	SUSE Linux came into existence.
1994	Beowulf computing	The Beowulf computing was developed and announced by NASA based on inexpensive clusters of computer systems running Unix or Linux on a TCP/IP LAN network.
1996	Linux 2.0 kernel	Linux 2.0 kernel version was released
1996	KDE	KDE development was started by Matthias Ettrich
1999	GNOME	GNOME, a desktop environment, came into existence. GNOME is defined as "*GNU Network Object Model Environment.*"
1999	Linux 2.2 kernel	Linux 2.2 kernel version was released.
2001	Linux 2.4 kernel	Linux 2.4 kernel version was released.
	Linux 2.6 kernel	Linux 2.6 kernel version was released.
October 2004	Ubuntu	The first version of **Ubuntu** was released.
2007	Ubuntu in Laptop	A leading computer system manufacturing company, Dell, started delivering laptops with preinstalled **Ubuntu**.
2011	Linux 3.0 kernel	Linux 3.0 kernel version was released.

(*Continued*)

TABLE 1.1 (*Continued*)

Timeline of Unix and Linux History

Year	Event	Description
2013	Android	Linux-based **Android mobile OS,** owned by Google, claimed to gain around 75% of the smartphone market share as per the total amount of phones distributed.
2014	Ubuntu users	**Ubuntu** claimed to have 22,000,000 users.
2014	Linux 3.12 kernel	Linux 3.12 kernel version was released.
2015	Linux 4.0 kernel	Linux 4.0 kernel version was released.
2015	**Android Things**	It is a Linux-based Android embedded OS targeting IoT devices.
2016	Linux 4.9 kernel	Linux 4.9 kernel version was released.
February 19, 2017	Linux 4.10 kernel	Linux 4.10 kernel version was released.
January 28, 2018	Linux 4.15 kernel	Linux 4.15 kernel version was released.

1.4 The GNU Project

The GNU (a recursive acronym for "GNU's Not Unix") Project is a cooperative project that gives computer users freedom and control with respect to the use of a software package. It was started on September 27, 1983, by Richard Stallman at MIT. The main objective of the GNU Project was re-coding the whole Unix OS, so that it could be freely circulated. The GNU Project page [3] recounts the story of how the idea of this project aimed at promoting the development, modification, and redistribution of free software to users, programmers, and developers came about in Stallman's mind. In 1985, as the GNU Project started to gain popularity, the Free Software Foundation (FSF) came into existence and got involved in the GNU Project to add other free software packages (both GNU and non-GNU).

Here, the term "free software" means freedom or liberty and has nothing to do with cost. In this context, if a particular user is using free software, then:

- The user has the privilege/freedom to execute the program for any purpose.
- The user has the privilege/freedom to amend the program as per the needs, i.e., modifying the source code.
- The user has the privilege/freedom to make any number of copies and redistribute them, either free of charge or with fee.
- The user has the privilege/freedom to circulate customized or upgrade versions of the program, so that anyone can get the benefit to use for their own improvements.

The FSF is a key contributor to the GNU Project that has written and produced many important tools and software packages. Linux is entirely dependent on free software, and this is the key reason for its popularity among users. Due to the significant contributions of the GNU Project, the distributions of Linux with such accompanying utilities are known as GNU/Linux. In due course, the free software terminology has been exchanged with the open-source software.

The GNU GPL is formed by the GNU software project that specifies how the open-source software should be used and controlled. Although there are a good number of licenses available to safeguard software's rights, offered as free of charge, the GPL is possibly the most well-known license. The Linux kernel itself comes under this license and, therefore, can be distributed freely as specified by the FSF. As a result, Linux is available to anyone who wishes to use it.

The GNU GPL is comprised of the following basic and key features:

- **Author rights**: The inventor or developer of the software has software rights.
- **Free distributions**: Anyone can use, alter, and redistribute the GNU software for their own purpose, but they must provide the source code with software during redistribution.
- **Copyright maintained**: The novel GNU agreement must be sustained during software repackaging and reselling in public domain. As a result, everyone has an opportunity to modify/alter the source code of GNU software.

For the protection of the authors and developers, the software covered by GPL does not have any warranty. In addition to this, GPL demands that revised versions should be noticeable as modified/upgraded version, so that any problem related to it may not be attributed to writers of the prior versions.

1.4.1 Open-Source Software

Open-source software can be freely used, modified, and redistributed (in modified or unmodified form) by anyone. It is often prepared by numerous people or groups and circulated under the license that protects freedom rights legally and complies with the meaning of open source. The open-source movements are well illustrated in the Open Source Initiative (OSI) webpage. According to the website [4], open-source software is described as follows:

> *In the open source software, programmers can read, redistribute and modify the source code for a portion of software. In the software growth, users may improve and fix bugs in it. This growth process may happen at a speed. This rapid evolutionary process manufactures improved software than the traditional closed model of development, mostly used by commercial organization, in which only a very few developers may see the source code and everybody else must blindly use.*

The main objective of open-source software is to make the source code in the public domain, i.e., available to all; other objectives are also highlighted by the OSI [4] in its Open Source Definition documents. Linux is a well-known open-source OS. The following rules secure the freedom and reliability features of open-source code:

- **Free distribution**: An open-source license, i.e., GPL, cannot demand any money/fee from anyone who sells or resells the software as open source.
- **Source code**: The complete source code should be integrated with software, without any restriction on its redistribution.
- **Derived works**: The license should permit all alterations/modifications along with its redistribution of source code with the same terms and conditions.
- **Truthfulness of the writer's source code**: If anyone uses the source code and eliminates original project's version or name by altering/modifying the source code, then a new license may be required for this project and to be marked as changed.

- **Equality among persons or groups**: The license allows that all people are equally entitled to utilize the source code.

- **Supplementary license**: No additional/extra licenses are required to use as well as redistribute the software.

- **No product-specific license**: The product license will never restrict the source code toward any specific software circulation.

- **License never controlling another software**: The license never stops open-source software to share a similar platform with other commercial software packages.

- **License never restricting a particular technology**: The license may not restrict any specific technology; it should be technology neutral in nature.

1.5 Features and Advantages of Linux

Linux is famous not only for its strength and flexibility but also for the accompanying professional-level development tools, the massive range of applications, fully functional GUIs, networking tools, and many others [5]. Some key features of Linux OS are highlighted below:

- **Portable**: A *portable* OS can be installed and run on different types of hardware platforms such as desktop, laptop, workstation, and computing-intensive server. The whole Linux OS is written in C programming code, which is a high-level and machine-independent language. Linux runs not only on Intel-based platforms but also on other platform systems such as PowerPC, Apple computers, Compaq (Digital Equipment Corporation (DEC)), Motorola-based machines, IBM, and others including 32- to 64-bit machines.

- **Multiuser**: Linux is a multiuser OS that allows multiple users to work and use all system resources simultaneously, such as memory, RAM, and application programs. The primary objective of a multiuser OS is to enhance the use of costly resources.

- **Multiprogramming**: Linux allows running multiple applications simultaneously while using common system resources.

- **Shell**: Linux offers a unique translator program that is used to execute commands and system calls. A good number of shells are present in a Linux system. Some popular shells are the Bourne shell (sh), the Bourne Again Shell (bash), TC shell (tcsh), C shell (csh), and Z shell (zsh).

- **Hierarchical file system**: Linux offers a single file structure in which all system/user files and directories are organized in a treelike structure.

- **Strong security model**: Linux provides a high level of user and system security. Generally, Linux systems do not require any antivirus software for security.

- **Multitasking**: Another important feature of Linux OS is multitasking, which permits every user to execute many tasks or jobs simultaneously. In this, all tasks may interact/communicate with each other but stay completely secured from one another, such as interaction of kernel with other processes.

- **Multiprocessor support**: Linux supports and may run on multiprocessor machines.

- **Graphical User Interface (X Window System; GNOME/KDE):** Still, it is an assumption of numerous people that Linux is a command line text-based OS. Initially, this was true. But now, Linux offers many packages such as GNOME and KDE to provide a complete graphics-based interface to the user, such as the Microsoft Windows OS.

- **Support of multiple file systems:** A file system is a structured way to organize all files on a disk drive/partition that basically monitors how the data is kept and recovered. Various forms of file systems presently exist. Each file system is differentiated with others according to its structure, logic, flexibility, security, size, etc. Linux is also supporting numerous file systems, such as **xia, ext, ext2, ext3, ext4, Jfs, xfs, msdos, vfat, UDF, iso9660, nfs, hpfs, sysv, FAT32, and NTFS.**

- **Rich set of networking protocol functions:** Linux offers a wide variety of network tools that support several networking functions and features and facilitate you to connect with other systems or with the Internet to access a variety of networks.

- **Availability of valuable services:** Linux offers a huge list of utility programs to cater for the user requirements; often, utility programs are mentioned as *commands,* required by users.

- **Software development:** One of the most impressive features of Linux is to provide the widest software development environment. Linux provides various development tools such as compilers, debuggers, and interpreters to configure and build a system. Apart from C and C++ programming languages, Linux also gives support to other famous programming languages such as Ada, Fortran, Java, Lisp, Pascal, perl, shell scripting, and Python for system configuration and software development.

1.5.1 Advantages

- **Free code (open source):** The Linux source code is freely available to everyone; i.e., its code is open source and circulated under the GNU GPL. Since its inception, various developer communities have been working in cooperation with others to improve the facilities of Linux OS. Due to this, the improvement/addition in Linux feature is a nonstop process.

- **Cost:** The open-source software packages including Linux are accessible with almost no cost and do not impose any licensing fees on their users.

- **Linux distribution:** The availability of various flavors of Linux OS provides a much greater freedom of choice. These include distributions such as Red Hat, Ubuntu, CentOS, and Fedora.

- **Live CD/USB:** Nearly, all Linux distributions come up with the live CD/USB feature that permits the users to run the OS, even without installing it on the system.

- **Open-source applications:** Linus has a rich availability of open-source applications and a wide variety of tools to accomplish tasks such as graphics processing, word processing, administration security networking, and web server.

- **High security:** Linux provides high security levels and mechanism to protect data from unauthorized access. Linux is also generally safer than other OSs and does not require a normal end user to use an antivirus tool.

- **Excellent developer community:** Linux OS has a very strong developer community that provides all technical support to any user or programmer across the globe totally free of cost.

1.6 Linux Distributions

In the original spirit of Linux, it is only a kernel, not a complete OS. That is why you can access and download all standard versions of Linux kernel from its official home page [2]. Therefore in true sense, Linux OS is a Linux distribution, and more exactly a GNU/Linux distribution. Presently, a large number of Linux distributions are available and being actively in use. Every organization or group combines, as package of Linux kernel and other software utility, including open-source software packages, in a different way as per their need and use. This package is released from time to time and widely referred to as Linux distribution, usually available through disk drive, such as DVD or CD-ROM, and also by downloading from the inventor website. Thereafter, all such releases may contain updated versions of package or new software packages.

Presently, the existing prominent Linux distributions are Manjaro, Red Hat, Ubuntu, Mint, SUSE, Fedora, Debian, and many others. But all Linux distributions have the same copy of kernel, but may vary in kernel version. The kernel versions are available at the kernel website [2]. The aim of most distributions is to offer a complete package with full support of all tasks. Such distributions may differ with other distributions in terms of provided software utility support and other features. For instance, CentOS is based on Red Hat Enterprise, and similarly, Ubuntu is derived from Debian Linux. Some distributions may be designed and developed to cater more specific tasks or offer precise technical support features. For example, the Debian Linux offers critical-edge developments, while other distributions provide more commercial versions, typically combined with commercial applications, such as secure servers and databases.

Further, some organizations such as Red Hat and Novell offer both the versions of Linux, i.e., commercial distribution and free or open-source distribution. The open-source distribution is widely supported by open-source community to adding new features, such as the Fedora project of Red Hat Enterprise. Other distributions, for example, Knoppix and Ubuntu, are available with dedicated features in live-CD form. It means that the complete Linux OS is available on a single CD and you can access such features directly from the CD without installing the OS on the machine. You can visit *distrowatch* web home page for further details and also see the list of numerous Linux distributions with their complete particulars and features [6]. But here, Table 1.2 provides the list of some more popular Linux distributions with their weblinks. Each Linux distribution has used its own applications and programs; you can refer to their distribution documentation for complete details of installed packages. There are many Linux distributions available, approximately 600, but presently, around 400–500 are active in development and use [7]. (See Section III: Case Studies for more details.)

1.7 Installation Procedure and Issues

1.7.1 Linux Installation

This section briefs the required initial process of installing Linux. Preparation is one of the most important factors in the successful installation of any OS. Linux installation is not always a relaxed and easy procedure if you are installing it as a first-time user. Therefore,

TABLE 1.2

Linux Distributions and Their Weblinks [6]

S. No.	Name of Linux Distribution	Weblink
1	Manjaro	https://manjaro.org/
2	Red Hat Linux	https://redhat.com
3	Mint	https://linuxmint.com/
4	Ubuntu	https://ubuntu.com
5	Debian	https://debian.org
6	Elementary	https://elementary.io
9	Solus	https://solus-project.com
10	Fedora	https://getfedora.org
11	openSUSE	https://opensuse.org
13	TrueOS	https://trueos.org
14	CentOS	https://centos.org
15	Knoppix Linux	https://knoppix.net/
16	Kali	https://www.kali.org
17	antiX	https://antix.mepis.org
18	ReactOS	https://reactos.org
19	SUSE	http://suse.com

in this case, it is highly recommended to use a bootable live CD of Linux OS and install desktop version of Linux system later.

- **Run Linux through a bootable live CD**: The main benefit of bootable Linux is that you may have the experience of Linux without installing it on system hard drive. The bootable Linux usually comes with CDs or DVDs known as live CDs and DVDs that you can try directly. Some of the live CDs also provide the feature of installation to your system. Despite their advantages, live CDs tend to run slower as compared to the OSs installed on the system. The changes made when using a live CD are not persistent and are lost when the system reboots.
- **Install a Linux on desktop system**: The first step of installing Linux on a desktop is to choose right distribution as per your need and requirement. Installing Linux on a desktop gives more flexibility in terms of data handling, software installation, or permanently customizing the system, as necessary.

1.7.2 Key Issues to Install Linux

It is necessary to emphasize the process to obtain the Linux software, i.e., various prepackaged distributions and how to prepare your desktop system for Linux installation [8]. There are some key concerns that need to be described regarding the disk partitioning before Linux installation, so that Linux may coexist with Microsoft Windows or other OSs. Numerous Linux distributions presently exist. Therefore, you may choose the one that suits your purpose. Linux distributions were already discussed in Section 1.6. These distributions are available via an anonymous FTP from the Internet, CD-ROM, and DVD, as well as in retail stores.

If you want to install Linux on a system that already has another OS, e.g., Microsoft Windows OS, and you want to keep both OSs on the same hardware platform, then both OSs should be present on an entirely distinct respective partition on the same hard disk. Before using any specific OS, first you may choose that OS to boot through installed boot loader (GRand Unified Bootloader (GRUB) or Linux Loader (LILO)). The subsequent sections highlight the basic procedure to find the required disk space or partition for Linux installation.

- **Adding or repartitioning hard drive**: Instead of changing the arrangement of current hard drive partition, i.e., Microsoft Windows partitions, it is suggested to create a new partition on the existing drive or simply attach a new hard disk and dedicate it entirely to Linux OS. If this is not possible, you may also resize your Microsoft Windows partitions to make a dedicated partition before the Linux installation. There are various commercial tools available to resize your disk partitions. These tools include Symantec PartitionMagic and Acronis Disk Director. In some Linux distributions (particularly in bootable Linux CDs), disk partition tools are provided during the installation process.

- **Creating Linux partitions**: The fdisk utility, available in most Linux distributions, can be used for creating a new partition with the required space dedicated to the Linux OS. You are required to create at least two partitions: **root (/) partition and swap partition**. The root partition will mount with main root (/) directory and have a Linux native file system (ext2 or ext3 or preferably ext4). The Linux swap partition does not have any mount point, but instead of this, it has a swap file system. A mount point is used to enter into the partition. It is important to note that the complete data of mounted Linux partition is seen under the top-most directory (/) and then divided subdirectories replicating a tree structure. Partitions never share a file system, and each partition on the hard drive has a separate file system.

- **Creating swap partition**: Many Linux distributions require creating an active swap space before the installation. If the amount of physical RAM is not enough, the installation procedure may not succeed unless you have some amount of swap space. There are two ways to create swap space. The first way is to create the swap file that is available in system and use it as a virtual RAM. The second way is to create a separate swap space partition. Most people use a swap partition instead of a swap file. The size of your swap partition depends on how much virtual RAM you require. Generally, a swap partition size is twice the size of your physical memory (RAM) size. For example, if the system has 512 MB of physical RAM, then the desirable swap partition size should be 1024 MB, which is enough for smooth working of Linux.

1.7.3 Key Steps of Linux Installation

The booting step of Linux is the first phase in the installation process and is usually performed using the bootable installation media. Boot prompt is automatically displayed when the bootable installation media is inserted at the starting of the computer (Basic Input/Output System (BIOS) uploading). This will be followed by a set of procedures for installation of Linux on the system. If you are going to install Fedora, then insert your Fedora Linux USB/DVD/CD or any other media comprised of ISO image of

Fedora Linux and restart your computer [9]. Thereafter, the following steps are required to follow:

- Firstly, the Linux installer will ask for the mode of the installation procedure. The mode can either be text-based or GUI-based. In the text mode, keyboard instructions will guide the procedure, and in the graphical mode, GUI will take input from the mouse to select various requirements in the procedure. Using graphical mode is recommended.

- Next, it will ask to select preferred geographical language which is being used during the installation process, such as English or Spanish or any other geographical language.

- Next, the user will have to select the time zone for the system using the time zone selection wizard. To ensure the correct date and time of system, click on the "Date & Time" icon wizard.

- Then, it will ask for the layout type of the keyboard (e.g., US English) that the user will use with the system.

- Disk partitioning setup will follow with automatic partitioning selected by default. You can also select the manual partitioning for appropriate partitions as per your requirement, do the partition, and install the file system. Manual partitioning helps you to use particular disk volumes for your Linux system.

- This will be followed by boot loader configuration setup in which one has to select the boot loader for the system. Generally, GRUB is set as default boot loader in most Linux distributions. However, LILO can also be select as boot loader.

- Next, in the network configuration setup box, the user has to configure the network devices of the system and go to ipv4/ipv6 settings tab and select the method to provide IP address, Netmask, gateway, and DNS information. For this, you may use DHCP for dynamic configuration of all network settings or may provide all these details manually.

- The firewall setup will follow. The firewall prevents any unauthorized access to the computer. In this setup, the user must select the services that are allowed under the firewall.

- The next page is package or software selection including development tools and others, where you can select required software/packages as per your requirement and then begin installation procedure.

- After the installation process is completed, a root user can log in as the administrator and create local user accounts for a general user.

- After this, all the procedural requirements will be over and now user can interact or log in with system via the graphical mode (CUI) or text-based mode (CUI).

1.8 Linux Architecture

Linux is distinguished by its simple, articulate, and graceful design with the remarkable feature that influences millions of people to install it on their system. Linux is used by a wide variety of people from a home user to the technological development community due to its well-structured architecture. Linux architecture comprises four major layers [10]. These are as follows:

- **Application program**: It provides major functionalities of an OS to the users.
- **Shell**: It is an interface between user programs and kernel which hides the complexity of the kernel's functions from the user. The shell receives commands as input from the user and executes it with the help of kernel functions.
- **Kernel**: The kernel is central part of OS and widely referred to as supervisor, core, or central part of the OS. The key components of a kernel are interrupt handlers (handling various requests), scheduler (process scheduling among multiple processes), memory management system (best utilization of various memory spaces), and system services such as networking and inter-process communication (IPC). The kernel is also responsible for managing available hardware resources.
- **Hardware layer**: Hardware peripheral interface is comprised of the hardware ports to connect all peripheral devices along with system hardware resources.

The basic view and developer view of Linux architecture are shown in Figures 1.2 and 1.3, respectively.

1.9 Shell and Its Features

The shell is a command language interpreter that provides a textual environment which behaves as an interface between the OS and the user. The shell offers a path to produce an executable script file, compile program code, execute code, work with various file systems, and manage and operate the computer system. There are two main ways to communicate or interact with the system: GUI and command user interface (CUI) or text-based interface. GNOME and KDE are two popular GUI tools that provide the desktop UI tool environment, whereas the shell provides a CUI text-based prompt in Linux. In most Linux OSs, the command line interface can also be accessed through the terminal application of GNOME/KDE desktop environment or virtual terminal on the remote system.

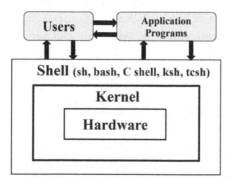

FIGURE 1.2
User view of a Linux system.

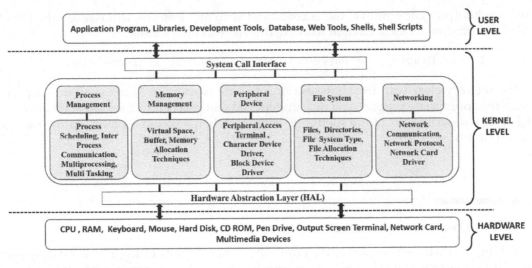

FIGURE 1.3
Developer/professional view of a Linux system.

Numerous Linux professionals believe that the shell is a significantly more effective tool than GUIs, due to its various advanced features, which are inbuilt in the shell. A number of shells have been developed and currently being used for Linux [13]. Some more prominent shells are listed here:

- **The Bourne Again Shell (bash):** It is an improved version of the original Bourne shell. The bash is written by Brian Fox in 1989 for the GNU Project. That is why it is also known as GNU bash, similar to the original Bourne shell which was created as the first Unix shell. The Bourne shell was developed at Bell Labs in 1979 by Stephen Bourne and its symbolized by '*sh*'. Presently, the GNU bash is available with all prominent Linux distributions and to be considered as one of the standard shell for all Linux OSs.
- **C shell (csh):** This is popular and widely accepted among BSD Unix users.
- **Korn shell (ksh):** This is well known for Unix system V.
- **TC shell (tcsh):** This is an improved version of the csh, created as part of BSD Unix.
- **Z shell (zsh):** This shell comprises the features of several other shells, including ksh.

Generally, several users might prefer to use a distinct shell. The preference of shells validates one of the benefits of the Linux distributions that provide a customized interface to the end user. This book uses bash command line interface to illustrate the concepts of Linux. The Linux command line interface comprises entered commands, its options, and arguments. By default, the prompt of bash is dollar sign ($), which is the prompt of a normal user. However, every individual shell has its own prompt; for example, csh has "%" sign as shell prompt. On the other hand, the shell prompt for root user is "#" sign. A shell prompt is given as follows:

```
$    (Normal or Local user)
#    (Root or Super user)
```

For any command execution on shell prompt, you can write a command along with command options and arguments. The command option is optional and depends upon the required output. For example, the **ls** command with an –l option will display the list of files and subdirectories of current working directory in long format.

```
$ ls -l    // Display the current directory content in long format
```

It's not very clear with the above example how to work with Linux shell initially, but with the appropriate support, you may understand and practice various key shell features. Chapter 3 describes the Linux shell in detail, which helps you to understand its features and usability.

1.10 An Overview of Kernels

The Linux kernel, the core part of the Linux OS, is accountable for the allocation of system resources and scheduling of various executing processes and other user jobs. This allows each job to get required part of system resources, such as getting CPU time, memory space, network connections, and peripheral devices (such as hard disk, DVD, and printers). The application programs work together with kernel via specific functions, referred to as system calls. The kernel is also viewed as an intermediary layer between software and hardware parts that makes your system functional. The kernel is responsible for managing all the peripheral hardware resources through its device drivers. The kernel is updated regularly to provide multitasking and multiuser support of newer technologies such as SCSI controller, USB, CD/DVD ROM, video card, and other peripheral devices that may not have been appropriately supported by the earlier versions of kernel.

The kernel may be divided into two main categories based on their design: *monolithic kernels* and *microkernels*:

1. **Monolithic kernels**: The design of monolithic kernels is very simple, and before the 1980s, all kernels were following this design concept. In a monolithic kernel, the entire core OS code is implemented as a single file, running in a single address space, i.e., kernel space. All kernel services exist and execute together with all its subsystems such as memory management, file systems, IPC, networking, and device drivers. The entire code of the kernel is packed into a single file that makes kernel bulky. The supporters of this design model highlight the simplicity and performance of the monolithic approach. Most Unix/Linux systems are following the monolithic design approach (Figure 1.4).

2. **Microkernels**: A microkernel is not implemented as a single large file. Instead, the kernel comprises only basic elementary functions and services such as file system, scheduling, and messaging (IPC), and networking and memory management functions (but the actual memory allocation function is implemented in user space). All functionalities of the kernel are broken down into separate processes which are assigned and loaded in user mode and run as daemons/servers. This keeps the kernel size small. Theoretically, this is a very graceful method because the individual parts are clearly separated from each other and communicate with kernel via communication interfaces such as message passing, shared memory, and other IPC mechanisms. The most prominent examples of microkernels are Minix 3, QNX, GNU Hurd, Escape, and Mach (Figure 1.5).

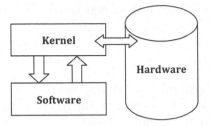

FIGURE 1.4
Monolithic kernels.

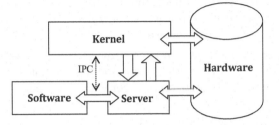

FIGURE 1.5
Microkernels.

The performance of monolithic kernels is better than that of microkernels. However, the key innovation has been introduced in the form of kernel modules that help to add functionality. The kernel module is basically a set of codes that can be inserted or removed while the system is up and running to support the dynamic addition of functions to the kernel. This feature tries to reduce some of the disadvantages of monolithic kernels. Linux 3.0 kernel includes over 15 million lines of code.

1.10.1 Kernel Module

After successful installation of Linux in your system, the kernel is properly arranged/configured with various attached devices of the system. Further, the kernel uses various modules to add the functionality or feature of a new device or enhance the technical support feature of present devices. This involves rearranging the current kernel to help the functioning of the newly added device, known as building or compiling the kernel. For smooth functions of the system, the kernel requires regular update of its modules to provide improved support with new or enhanced features to the attached devices.

Kernel modules are pieces of code. If you want to write a kernel module, you must have the proficiency in C language because the whole Linux kernel is written in C language. Further, C++ language is preferred over C in many environments due to its object-oriented feature. The GNU Project C compiler supports both C and C++. The key advantage of kernel module is that it can be loaded and unloaded into the kernel on functionality demand without the need to reboot the system or reinstallation of Linux kernel. Various websites provide the source of numerous kernel modules and help manual, comprising the complete procedure of how to load and unload kernel module and other details. From such Web sources, you can download and install the various kernel modules on a requirement basis. For example, device driver is one form of module that permits the kernel to gain access of hardware attached with the system. In the absence of module concept, the developer would have to construct monolithic kernels and insert any new functionality directly into the kernel image file.

To see the list of loaded modules in installed kernel, use command **lsmod**, which gets the details from the file */proc/modules*.

1.10.2 Linux Versions

The numbering of kernel version is always referred to as an important topic. The kernel version is to identify rank, feature, and revision part of kernel release. The kernel version number is comprised of four parts (Figure 1.6):

1. **Major version number**: This denotes the major release version of the kernel.
2. **Minor version number**: This reflects the stability of the particular released kernel; for example, an even number indicates stable release and an odd number indicates a development (experimental) release.
3. **Revision number**: This number indicates the revision sequence of overall release of a particular kernel; for example, 3.4.5 is the fifth revision sequence of 3.4.0 kernel version. Generally, this revision number is used for adding or supporting new characteristics.
4. **Bug fix and security patch number**: This indicates the bug fix and security patch version of any kernel.

As described under point 3 regarding revision number, a pre-release version (pre-patch of "RC") of any kernel has a name like 3.6.18-rc that includes a new feature and must be tested before releasing its stable version. Pre-patch kernels are maintained and released by Linus Torvalds.

A system can install more than one version of the kernel. By using the command **uname** with the **–r** option (the **–a** option gives more complete information), you can see the kernel version of currently running kernel, as given below:

```
$uname -r  // Display kernel version
```

The Linux kernel is available under the GNU GPL licensing manner which permits users to download the kernel in all forms from the Internet, but the latest and stable kernels are

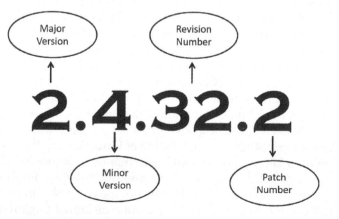

FIGURE 1.6
Details of kernel version numbers.

available at the kernel official home page [2]. Distributions may include different kernel versions as per the requirement of Linux OS. Broadly, This book is based on the Linux kernel 3.10 series and also cover previous kernel versions according to the topic, relevancy.

1.11 The GNOME and KDE Desktops

Before starting your work on Linux OS, you must be familiar with how to interact with or access a Linux system. Basically, there are two ways to access the system: desktop/GUI and CUI. Once you start working with the Linux system, then you have two ways to interact, either through GUI or through command line interface. The X Window System [10,11], a GUI tool, is used to facilitate the graphical interface. This GUI tool was developed by researchers working on a distributed computing project, called project Athena, at Massachusetts Institute of Technology lab with the help of DEC in 1984. GNOME and KDE are two widely known desktop GUI tools that operate under X Window System. They are used in most of the Linux- and Unix-like OSs. It is significant to highlight here that GNOME and KDE are the desktop terminology interfaces that appear almost the same. However, they are two different wholly integrated desktop interfaces with individual tools for choosing user preferences. These two GUIs coordinate with Linux kernel via window manager.

1.11.1 The Window Manager for GUI

A window manager is a set program that regulates feel and appearance of a window (a rectangular area on screen where images may be drawn/displayed) on the desktop. It operates under a desktop manager, such as GNOME or KDE, to monitor all operations and position aspects of windows such as open, close, move, resize, and minimize in X Window System environment. It also provides absolute application features such as desktop icons, fonts, toolbars, wallpapers, and desktop widgets for better interaction and familiarity to the user.

- **Enter GUI mode from command line mode**
  If you log into the system in command line mode interface, then you will have the option to start X Window System (GUI) interface by using GNOME or KDE desktop tools. In Fedora Linux, you can use the **startx** command to begin GNOME desktop by default. Once you exit from GNOME desktop, you will return to your logged login in command line mode.

```
# startx      // Enter GUI mode
```

1.11.2 GNOME Desktop

GNOME stands for GNU Network Object Model Environment and is a desktop environment comprised of free and open-source software that runs on most of the Linux/Unix distributions. GNOME is developed under the name of the GNOME Project, which is a part of the GNU Project. For more details, you can visit the GNOME home page [12]. It was developed to give an efficient, stable, and user-friendly free desktop environment for Linux. GNOME offered several features along with default desktop manager, such as file, folder, and display manager. It also supplies development platform, a framework for building applications that integrate with the desktop. Default window managers for GNOME are Mutter, Metacity, and Compiz.

The GNOME Project has its own release schedule and version number. Generally, a full GNOME stable version is released roughly in a six-month schedule along with its essential libraries such as GIMP Toolkit (GTK) and GLib. The GTK is used for drawing widgets and having a unique set of libraries developed particularly for the GUI Image Manipulation Program (GIMP). The GTK is written in C programming language; however, bindings for C++ and other languages are also available. GNOME 3.28 is the latest version of GNOME 3 that has numerous new features and enhancements for GNOME 3.28. Several prominent Linux distributions, such as Fedora, Debian, Ubuntu, Red Hat Enterprise Linux, CentOS, Oracle Linux, and Kali Linux, use GNOME as a default desktop environment on many major Linux distributions; for example, GNOME is also a default desktop environment for Solaris Unix.

1.11.3 KDE Desktop

K Desktop Environment (KDE) is an open-source and user-friendly graphical desktop environment to provide GUI. KDE was founded by Matthias Ettrich in 1996, now supported by KDE developer community [13]. The KDE GUI is equipped with the user's basic needs, including window manager, file manager, system configuration tool, and other tools to facilitate customization of desktop working environments. The famous application products of KDE are Plasma Desktop, Plasma Mobile, KDE frameworks, KDE neon, and other multiplatform programs to run on various OSs such as Linux/Unix distributions, Android, and Microsoft Windows.

The KDE framework is a combination of KDE libraries and Qt toolkit. KDE is mainly written in C++ programming language on top of the Qt framework and also includes bindings for other programming languages. The Qt toolkit framework, comprised of development tool, cross-platform APIs, libraries, and classes, was originally developed by Trolltech developers. The Qt license was not compatible with the GPL when KDE development was started. That is why the FSF agreed to endorse a different project, which was the GNU Network Object Model Environment (GNOME), instead of KDE. However, presently Qt is being developed by both Qt Company (formerly Trolltech) [14], a publicly listed software company based in Finland, and Qt Project under the open-source concept which provides the opportunity to involve individual developers and firms working to advance Qt. Now, Qt is available in both forms, i.e., proprietary and open-source (GPL) licenses.

The first KDE 1.0 version of desktop environment was released on July 12, 1998, for Unix OS. The latest announcement of KDE is Plasma Desktop, i.e., KDE Ships Plasma 5.12.5 version on May 1, 2018. Plasma Desktop is one of the KDE's flagship products that offer a unified environment for managing applications and running on varied appearance aspects such as desktops, netbooks, smartphones, and tablets. The Plasma Desktop is offered as a default desktop environment on various Linux distributions, such as openSUSE, Chakra, Mageia, Kubuntu, and PCLinuxOS.

1.12 Boot Loaders

To prepare and set up a computer system for Linux or some other OS, you must have the familiarity with the devices connected with the system and their configuration along with start-up or booting procedures. Booting is a procedure where system installation involves various steps to load system software parts including firmware initialization, boot loader execution, loading and running Linux kernel image, execution of various start-up scripts, and daemon process to set up the whole Linux system for work. When the system starts, power-on self-test (**POST**) is used to perform system hardware initialization and configuration check by the PC firmware. The firmware, stored in PC motherboard ROM chip, is commonly known as BIOS or ROM BIOS or PC BIOS. It is the first software that runs when the system is switched on.

1.12.1 BIOS

BIOS is stored in EEPROM chip on PC's motherboard that takes the control of computer system while turning on or resetting it. It is the responsibility of BIOS to get all the system hardware to a condition at which they are ready to initialize or boot an OS. The key part of system initialization is locating the OS from among the several storage devices attached to the system. It could be hard disk, CD-ROM, DVD, USB, network card, video card, etc. The Unified Extensible Firmware Interface (UEFI) and Extensible Firmware Interface (EFI) are the latest firmware used in modern systems after the BIOS. As per the BIOS boot sequence setting, the first boot device will start the booting process. The selection of booting device can be updated in the BIOS settings, as per the availability of OS in that respective device. After testing the hardware, BIOS searches for an OS loader, known as boot loader, to load and configure Linux OS.

A boot loader is a very small program that resides on ROM and typically starts after the BIOS works. Boot loader fetches the Linux kernel from specified boot device into the physical memory to bring computer into a completely operational state. Normally, the boot loader exists in the starting sectors of a hard drive, referred to as master boot record (MBR). The MBR file is a very tiny program that comprises the information of hard disk partitions and OS as shown in Figure 1.7. Due to any reason, if this file becomes corrupted or not accessible, then it's not possible to boot or install the OS in a computer system. Hence, it is suggested to keep the backup of MBR, so that in the case of MBR file damage, it may reinstate or update with MBR backup copy file.

The size of MBR may be 512 bytes or more, situated at the first sector of disk drive. MBR may comprise one or more files such as the following:

- **Partition table**: It illustrates the storage device partitions. In this view, the boot sector may also be known as partition sector.
- **Bootstrap code**: It is a set of instructions to recognize the configured bootable partition and subsequently load and run its volume boot record (VBR).
- In early days, it was possible to boot/start Linux or any other OS from floppy disk, but now, it is done through hard disk, optical medial (CD/DVD) or USB drive. A

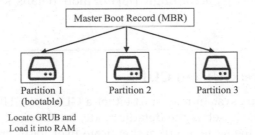

FIGURE 1.7
Details of MBR.

boot loader is also known as a boot manager or bootstrap program to find the selected kernel if the computer system has more than one OS installed on it, and to get it loaded into the physical memory along with user-selected option. The role of boot loader is over when it starts the selected kernel; then, the control of system hardware is placed under running kernel.

- Both Linux and Unix OSs support dual boot installation by having some boot loader program or manager. That is why they are referred to as multiboot OSs. The Linux boot managers may configure simply to boot Linux along with other OSs such as Microsoft Windows, Mac, and other Linux variants on the same system. A Linux system supports generally two most common boot loaders that are widely used by Linux distribution: **GRUB** and **LILO**. The */boot* folder in Linux contains the information of installed boot loader.

```
/boot        —files used by the bootstrap loader (LILO or GRUB)
```

- **GRUB**: It is a multiboot loader and is available under the GNU Project. Therefore, the GRUB boot loader is also referred to as GNU GRUB boot loader. GRUB is used for most Linux distributions and was originally designed and implemented by Erich Stefan Boleyn. When you turn on or restart your system, the GRUB boot loader helps you to select an OS from menu-driven interface to load and setup. By using keyboard arrow keys, you can move up and down the control to select a particular OS and press the Enter key to start the chosen OS. You can see the GRUB Man page [10] for various GRUB options. Two versions of GRUB are available: GRUB Legacy and GRUB2. You can see all details about GRUB at its GNU GRUB home page [15].

  You can modify the GRUB configuration file to support multibooting feature in OS from time to time. In Debian and Ubuntu, the */boot/grub/menu.lst* file keeps the GRUB configuration setting details, but in the case of Red Hat, Fedora, and other similar distributions, the */boot/grub/grub.conf* file holds the GRUB configuration setting [1]. You just make some essential entries in the GRUB configuration file, and when you reboot the system, GRUB will automatically read all such entries from its configuration file (for more details, see Section 9.6).

- **LILO**: LILO is another popular boot loader program or boot manager for Linux. In the initial years of its inception and popularity, it was the default boot loader for many Linux distributions. It performs similar tasks as GRUB. But now, most Linux distributions are using GRUB as the default boot loader manager. You can also modify your LILO configuration file to support multibooting feature in the OS. You can access LILO configuration file */etc/lilo.conf* directly and modify accordingly [16] (for more details, see Section 9.6).

1.13 Linux Interface: GUI and CUI

After booting, the Linux system may start either a GUI or a CUI. Most of the Linux distributions have GUI mode set as the default mode, but it may be changed as and when necessary. Thereafter, the system will authenticate user login credential (username and password) to provide access.

- **GUI**: In most of the OSs' cases, graphical mode is set as the default mode of GUI. GNOME or KDE desktops are two most popular ways to provide GUI mode to manage your desktop. We have already mentioned about their description in the earlier sections. You can also start text mode within the desktop GUI through a terminal window application.

  The majority of Linux distributions usually provide a simple way to start text mode via shell from GUI. You can start a terminal window application from Linux desktop in two simple ways which are given as follows:

  - **Right-click on the desktop**: In Fedora Linux, you can make right click on the desktop and then select Open Terminal followed by click to open.
  - **Click on the panel menu**: Numerous Linux desktops incorporate a panel at the bottom of the screen from where you can also start terminal application. For GNOME desktop, select Applications, then click Accessories, and then click on Terminal to open terminal window for CUI mode within GUI.

- **CUI**: Another way to access the system is text-based or command-based user interface, also referred to as character console. When Linux system starts in text mode, the whole display screen is black, showing a blinking white character as input cursor. The text-based login is different from a graphical login in which first you need to provide username followed by pressing the Enter key to type the password. It is worth mentioning here that while typing the password, there is no movement of cursor. It is for security reasons. This is the normal way to authenticate a user in text-based login interface. After successfully authenticating the user credentials, a user can view shell prompt as regular user or root user (depending on user rights). Now you can type commands on shell prompt to work with the Linux system.

By default, the dollar ($) sign is appeared on shell prompt to recognize logged user as regular user:

```
$, but it may be customized.
```

The default shell prompt for root user is always hash (#) sign:

```
# (super user)
```

In most of the Linux systems, the prompt sign may be **$** or **#** and is headed by your username, system name, and current directory name. This book is following the same notation of shell prompts to highlight the incredible amount of shell features in the coming sections. In the majority of Linux distributions, the bash is available by default. To know the current login shell, type the following command:

```
echo $SHELL
/bin/bash (Output in most Linux system)
```

One of the most important features of a Linux system is that you can start multiple shell sessions (virtual text-based consoles) along with graphics console (GUI-based desktop). By default, generally, Linux has six text-based consoles/terminals (CUI) and one graphical console/terminal (GUI). You can switch between these terminals by pressing Ctrl+Alt+F_n, where n=1–7. You can start any of the six virtual console terminals (tty) in a text-based environment by using the following procedures, as shown in Table 1.3:

TABLE 1.3

List of Virtual Consoles

Ctrl+Alt+F1	Virtual console 1 (tty1)
Ctrl+Alt+F2	Virtual console 2 (tty2)
Ctrl+Alt+F3	Virtual console 3 (tty3)
Ctrl+Alt+F4	Virtual console 4 (tty4)
Ctrl+Alt+F5	Virtual console 5 (tty5)
Ctrl+Alt+F6	Virtual console 6 (tty6)

The graphical console environment can be started with Ctrl+Alt+F7. You can switch between virtual terminals as similar as you would switch between workspaces or applications in GUI environment. For example, after your login in CUI mode, you are on virtual console (tty1) and you want to start another virtual console. To do so, you can simply press Ctrl+Alt+F2 (or F3, F4, and so on up to F6) to start one of the virtual terminals and subsequently provide the username and password of an existing user to access the system from that newly created console. You can switch among these six virtual text consoles by pressing Ctrl+Alt+F1…F6. As shown in Figure 1.8, at any point of time, you can switch from any text console to GUI terminal by pressing Ctrl+Alt+F7 and vice versa.

1.14 Login and Logout

Whenever you are going to use command line mode, it actually seems to work with shell. A shell is a program or interface that takes input from the user, in the form of commands, and passes it to the kernel for further processing. Every Linux distribution provides a shell program, which is GNU bash in most Linux distributions as the default shell. We are going to refer to this shell in all examples and command descriptions.

1.14.1 Login

You can access and use a Linux system in a number of ways. Linux is a multiuser OS. If you want to use text-based login mode, then you use the *login* command. In many Linux distributions, the functionality of login is unseen by the user. At the time of first user login, this command is run automatically.

```
GNU/Linux
Kernel 3.10.0-514.16.1.el7.x86_64
```

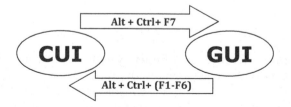

FIGURE 1.8
Switching between two modes of interface.

```
Login: sanrit // Login as sanrit (normal user)
Password:          // Provide password, and then, press the Enter key.
[sanrit@Linux~]$
```

NOTE: While typing the password, there is no movement of cursor shown on screen for security reasons.

The shell prompt appears after the successful login and indicates that the shell is ready to accept input. It may vary in appearance, depending on the Linux distribution. Generally, it will include your username along with system name, such as *username@systemname*, followed by the current working directory and then a dollar sign. The home directory of a user is denoted by the character ~.

In Linux, after successfully logging into a user account, command line prompt is headed by the hostname and home directory name which are surrounded by a set of brackets, as given in the below example. The username and password are case-sensitive in Linux.

To know the home directory path, use the pwd command:

```
[sanrit@Linux~]$pwd
/home/sanrit
[sanrit@Linux~]$
```

1.14.2 Logout

If you want to end your login session on Linux, you need to log out. This can be accomplished using the *exit* command or the *logout* command. Before logging out, you must save your work and close all presently running or active processes. After that, you use the following way to end your session:

- To exit or end (log out) the shell terminal session, you may apply the following command on shell prompt:

  ```
  [sanrit@Linux~]$ exit
  OR
  [sanrit@Linux~]$logout
  ```

After you logged out or exited from your current session, then the login prompt will appear again, and the system is available for other users.

- **Managing user accounts**

  Linux is a multiuser OS that allows more than one user to work with the system simultaneously. As discussed earlier, there are two broad categories of users: *root users* and *normal users*. A root user, also known as superuser, has power to access all system files and control over the system. On the other hand, a normal user has limited privileges toward accessing files or system resources. A superuser, also known as administrator, has the right to add, delete, and modify various user accounts of a Linux system.

  Every user of the system has a unique login name (username) and a numeric user ID (UID). The complete user account information of each user is stored in the */etc/passwd* file. The password record is stored in an encrypted form on a separate file, known as shadow file. This file is located at */etc/shadow*.

The */etc/passwd* file includes the following additional information:

- **Home directory**: The user home directory into which the user is placed after logging in.
- **Group ID**: The numeric group ID of which the user is a member.
- **Login shell**: The name of the shell program to be executed to interpret user commands.
  - **Groups**: For a better system management and control over the access to files and other system resources, it is helpful to organize users into groups. Each group is recognized by system group file, located at /etc/group. This file comprises information such as the unique group name, group ID (GID), and the list of users belonging to the group.
  - **Superuser**: The root user or superuser has complete control over the system. The superuser account has a fixed unique ID (UID 0 or 1), and root is the login name. The superuser can access any file in the system. Further, the Linux system administrator uses the root user account to carry out the various administrative tasks to manage and control the system functionalities.

The subsequent section explains some operations on a user account.

- **Creating a new user account**
  To add a new user account, you can run the *adduser* or *useradd* command, as root user:

```
#adduser [User_Name]         // Create new user

#password [User_name]  // Assign the password to the mentioned username,
but be very careful while using this command. If you forget to give the
username as argument for this command, then the password command will
assign or overwrite the password of root user.

#password<enter key>  // Reset the password of root user.
```

- **Adding a group**
  To add a new group, use the below-mentioned command, as root user:

```
#groupadd [group_name]   // Add a new group
```

- **Deleting a group**
  To remove or delete a group, use the below-mentioned command, as root user:

```
#groupdel [group_name]   // Delete the group
```

- **Deleting user accounts**
  You can delete an account (along with its home directory and all the files residing therein, and also the mail spool) by using the *userdel* command with the -r option, i.e., recursively remove.

```
# userdel  <option>  [user_name]
# userdel  user_name
# userdel  -r user_name
```

User creation/deletion is always done by the superuser.

1.14.3 Switching Users

The *su* (switch user) command is used to switch the user from one another.

1. Normal user to root user:

```
$ su            // No username; by default, switch to root user from normal
user.
Password:    // Type the root password and then press Enter. If you are a
valid root user, then # prompt will display on prompt.
#
```

2. Normal user to another user:

```
$ su <user_name>
Password:          // Type the password of the given username.
$
```

3. Root user to any user:
   Superuser can enter in any normal user account by using the *su* command, but in this case, the system will not prompt to enter the password for normal user from superuser.

```
# su <user_name> // Type the name of normal user and directly enter the
                 normal user account.
 $pwd           // Display the present working directory of user
 $exit          // Exit from normal user and back to root user because you
    enter normal user account from root user account.
#
```

1.14.4 Shutdown

To shut down the system, use the shutdown command in superuser (root user) mode:

```
[root@Linux ~]# shutdown
To shut down the system right now, then use the following option with the
shutdown command:
# shutdown -h now      // -h is the option for time in minutes
OR
# shutdown -h +0
```

To shut down with a warning message to all the users from root user account:

```
# shutdown -h +15 "Type_message"    // To shut down the system in 15
                                    minutes with the given message
```

For example:

```
# shutdown -h +15 "System will shut down in 15 min, please save your work"
```

It broadcasts the message to the entire users from root user account.

• **Showing the log of system shutdown/reboot**

The *last* command will show a log of all reboot and shutdown details, because the log file of shutdown and reboots was created under Linux OS:

```
# last reboot
OR
# last shutdown
```

NOTE: Almost every Linux command is associated with various command options. For the list of command options, see the manual page of respective command by using the **man** command followed by the name of command, such as #man <command_name>.

1.15 Start-Up Scripts and Run Levels

1.15.1 Start-Up Script

The steps of booting sequence from power-on to setup of a user prompt are generally known as start-up or setup scripts of a Linux system. The following steps are required to set up a Linux OS:

1. System start-up/hardware initialization (BIOS)
2. Boot loader stage 1 (loading of MBR)
3. Boot loader stage 2 (LILO or GRUB boot loader loading)
4. Kernel initialization (loading the kernel into physical memory and starting the init process)
5. The init process (initialization and checking of all system resource)
6. Starting the display manager (GNOME or KDE) and displaying the login screen and prompt for credential to login into the system

1.15.2 Run Levels (init, inittab, and rc Files)

Once the kernel is loaded in the physical memory and device drivers are initialized, the kernel initiates the *init* program during booting of the computer system [17]. The term *init* stands for initialization. It is a daemon that keeps on running until the system is shut down. It is located in /etc, /bin, or /sbin or /sbin/init. It generates new processes and restarts certain programs as per the requirements of the system. For example, each virtual console has a getty process running on it, and after successful login, the getty process is replaced by the shell process. Subsequently, after logging out, *init* restarts a new getty process, permitting a user to log in again.

The details of *init* program are mentioned in the *inittab* file, located at */etc/inittab*. In order to understand this file, first you must be familiar with the concept of run levels. When a Linux system boots, the */sbin/init* program reads the */etc/inittab* file to regulate the actions of each run level by executing the script */etc/rc.d/rc.sysinit*. For each run level, the rc (run commands) script files are stored in the directory */etc/rc.d/rcN.d*, where *N* is the run level number. The convention path of this directory may vary among different Linux distributions.

The basic responsibility of the *init* process is to place the system in the default run level mode, which is generally 2, 3, or 5, depending upon the Linux distribution, and to prepare the machine for use. The basic description of various run levels is given in Table 1.4.

TABLE 1.4

Description of init Run Levels

ID (N)	Run Level	Description
0	Halt	Shut down or stop the system.
1	Single-user text mode	Start the system in single-user mode for administrative tasks.
2	Multiuser mode	Start the system in multiuser mode without configuring the network interfaces, the same as 3 if no networking.
3	Multiuser mode with networking (CUI)	Start the system in multiuser mode with network interface configuration, but in text mode (CUI).
4	Not used (user-definable)	Reserved for special purposes.
5	Multiuser mode with networking (GUI)	Start the system with GUI-based login screen with network interface configuration.
6	Reboot	Reboot/restart the system.

1.16 Summary

Currently, Linux is the most influential and stable open-source OS that is widely available to home users, workplaces, and development environment. Linux was created in the early 1990s by Linus Torvalds. Linux systems are being used almost everywhere due to its features and vast range of available applications. The continuous development over the years has made the Linux OS an excellent alternative for the desktop systems as well. You can easily find a wide variety of open-source applications for any task including word processing applications, games, applications for playing music, and email applications. As compared to other desktop computing systems, Linux offers a much more secure computing environment due to its strong networking and built-in security features. Since it is open source, Linux offers you the liberty to customize the computer system according to your needs and solve problems that may be unique to you.

In this chapter, we surveyed a range of fundamental concepts, features, and development history related to Unix and Linux systems. An understanding of the GNU Project, Linux distribution, installation, architecture, and shell feature must offer an enough background to readers to start working with a Linux system. We also briefly covered the key components of a Linux system such as kernel, GUI (GNOME/KDE), boot loader, login/logout, and start-up scripts along with various run levels.

1.17 Review Exercises

1. Write down the various features and services of Linux operating systems and classify the following terminologies:
   i. Multiprogramming systems
   ii. Multitasking systems
   iii. Multiprocessor systems.
2. Discuss the design principles of Linux operating systems? Write the name of interface (a, b, and c) as shown in the figure given below with proper justification:

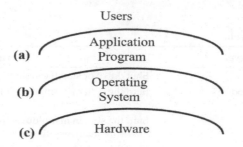

Users

(a) Application Program

(b) Operating System

(c) Hardware

3. List out the key reasons why AT&T Labs decides to rewrite the whole Unix code into C language.

4. Kernel brings hardware close to the user. Discuss how the kernel does that since many components of hardware have their own specifications and design.

5. Linux can be successfully installed on various platforms such as Intel, PowerPC, DEC, and IBM. Discuss what makes Linux so portable.

6. Why AT&T distributed the Unix code to commercial institutions and universities?

7. Justify why more than 90% of the code which was initially written in B language was converted into the C language. Justify with some key reasons.

8. Discuss PDP-7 and PDP-11 minicomputers.

9. Discuss the terms of GPL, BSD, open source, POSIX, and OSDL.

10. Which process is initiated, just after the system power-on?

11. Discuss virtual memory. How kernel creates a virtual memory?

12. How open source helps in the growth and popularity of the software community?

13. How BIOS helps while installing Linux or any operating system?

14. Highlight the installation procedure of the Linux operating system?

15. List down some Unix-like operating systems announced by commercial institutions in the late "1970s."

16. What are the various key rules of GNU projects to preserve the freedom and reliability of open-source software?

17. Highlight six key Linux features which differentiate it from other operating systems.

18. Define Linux distribution and mention some key distributions along with their source.

19. What are the key differences in Linux operating system architecture and other operating systems?

20. Discuss the role of shell and its features in a Linux operating system.

21. Discuss the various parts of Linux kernel version. How to know the current kernel version?

22. Discuss the key difference between monolithic kernels and microkernels.

23. List some popular Linux boot loaders along with their function.

24. Write the procedure to switch from one desktop environment to another, like switching from GNOME to KDE.

25. How to create and switch among text terminal and graphical console environments?
26. How to distinguish between superusers and local users? Why local user does not have permission to create another user?
27. Define the importance of the various run levels along with their description in Linux.

References

1. Richard Petersen. 2008. *Linux: The Complete Reference*. The McGraw-Hill.
2. Linux Kernel Organization, Inc., The Linux kernel archives in the World Wide Web. https://www.kernel.org/.
3. Richard Stallman, The GNU project in the World Wide Web. https://www.gnu.org/gnu/thegnuproject.en.html.
4. The Open Source Initiative, The open source definition in the World Wide Web. https://opensource.org/osd.
5. Machtelt Garrels. 2008. *Introduction to Linux: A Hands on Guide*. Fultus Corporation.
6. Linux Distributions, The Major Distributions: An overview of major Linux distributions and FreeBSD in the World Wide Web. https://distrowatch.com/dwres.php?resource=major.
7. Jose Dieguez Castro. 2016. *Introducing Linux Distros*. Apress.
8. Werner Heuser. 2011. *Linux on the Road*. Open Source under GNU Free Documentation License.
9. Wale Soyinka. 2012. *Linux Administration: A Beginner's Guide*. The McGraw-Hill.
10. Richard Blum and Christine Bresnahan. 2015. *Command Line and Shell Scripting Bible*. John Wiley & Sons, Inc.
11. Ian Darwin, Valerie Quercia and Tim O'Reilly, 1995. *X Window System User's Guide*. OPEN LOOK Edition.
12. The GNOME Project, The GNOME's technologies in the World Wide Web. https://www.gnome.org/technologies/.
13. The K Desktop Environment (KDE), Software open communities, The KDE Announcements in the World Wide Web. https://kde.org/announcements/.
14. The Qt Company, The Qt for Developers in the World Wide Web. https://www.qt.io/developers.
15. Bruce Dubbs, GNU GRUB home page in the World Wide Web. www.gnu.org/software/grub.
16. Adam Haeder, Stephen Addison Schneiter, Bruno Gomes Pessanha and James Stanger. 2010. *LPI Linux Certification in a Nutshell*. O'Reilly Media.
17. Evi Nemeth, Garth Snyder, Trent R. Hein and Ben Whaley. 2011. *Unix and Linux System Administration Handbook*. Pearson Education, Inc.

2

Linux Commands

The human interaction module is a key required component of each computer system that makes interacting or working with the system easier. Today, the computer users are more familiar with graphical user interface (GUI) than with text-based interface (CUI). But in terms of communication performance with system, text-based interface or command line interface (CLI) is a much more effective way of communication with computer. Possibly, by using the graphical interface, you can manage and execute the task very easily, whereas the use of CLI helps you to handle complex tasks very straightforwardly. There are no shortcuts when it comes to gaining expertise in Linux. Learning and practicing various commands through the CLI is extremely rewarding and will help you become a powerful Linux professional, as user or developer. That is why every hardcore programmer and developer prefers to work in the command line mode rather than the graphics mode.

Linux is a clone of Unix operating system. Due to this, it shares a pool of commands with Unix. Whenever the CLI is taken into consideration, then it simply refers to Linux shell or shell. The shell acts as an intermediary layer between the Linux kernel and the user which takes input from the user and passes it to the kernel and vice versa. A good number of Linux shells have been created to use in Linux distribution. However, the Bourne Again Shell (bash) is the most popular among all available shells. The bash is created as a part of the GNU Project and accessible under GNU General Public License (GPL). The bash is an enhancement of the original Bourne shell (*sh*), and it was written by Steve Bourne at Bell Labs. You can run shell independently on a physical terminal or within a terminal window of GUI mode. But in both the cases, the functionality and feature of the shell remain the same. The command **man bash** displays the help manual of bash.

This chapter has tried to include and explain the essential features of primitive Linux commands using the bash. After successful installation of Linux, for example, you boot your system in command user interface (CUI) or CLI to interact or work with the system. For the CLI, the login prompt appears on your screen with the hostname that you assigned or given to your system, as shown below:

```
Linux release
Kernel 3.10 on x86_64

Linux login//
```

This is your initial login prompt where you have to provide the username and password to log into the system. After a successful login, the system displays a prompt which is preceded by the hostname and the current directory, both of which are bounded by a set of brackets and followed by a login prompt sign, either # or $. The dollar sign ($) is the shell prompt for regular users, whereas the hash sign (#) is the shell prompt exclusively assigned to the root user or administrator in all Linux distributions. In CUI, when shell prompt appears on the screen, the system is prepared to take user inputs, in the form of Linux command. The command is normally comprised of a set of text lines, i.e., one or more lines of text. Such interface is referred to as CLI for end user. For example, if a regular user, named *sanvik*, logged into a Linux system, called *Linuxhost*, then the login prompt would appear as shown below:

```
[sanvik@Linuxhost /home/sanvik]$
OR
[sanvik@Linuxhost ~]$
(Note: ~ denotes the home directory path of logged user, i.e., /home/
sanvik.)
```

Once you finish your work, you may **log out** or **exit,** so that the Linux system, called *Linuxhost*, can be available to another user for login into the system for work.

```
[sanvik@Linuxhost /home/sanvik]$ logout      // Press the Enter key
```

2.1 Command Syntax, Options, and Arguments

It is extremely important to understand command line syntax, while entering a specific command on shell prompt. The basic syntax of a user input command on shell prompt is comprised of the name of the command followed by various associated options with the command and then a list of arguments. The command syntax is given as follows:

```
$<command_name> <-options> <List_of_arguments>
```

The mentioned command line syntax prescribes sequence and separation of various options and arguments with respect to the given command name on command prompt that decides the execution behavior of the mentioned command. In addition to this, the Linux system normally consists of four components:

- **Command_name**: A valid Linux command which may be a built-in shell script or system call or user program that stored in some directory which is registered in PATH environment variable,* or some already mentioned program directories of PATH environment variable in Linux file system. In command line syntax, *command_name* is must and other parts are optional which depends on command operation behaviors. For example:

```
$ ls  // To see the content of the present working directory, then only
command name (ls) is required and it does not require any options as well
as arguments.
```

- **Options**: Most of the commands have one or more options that help to categorize various operations of a particular command. Each distinct operation of a command is represented with a character or word, headed by one (-) or two hyphens (--), respectively. Option is also referred to as one-letter code that changes the style

---

* See Sections 3.8.1 and 3.8.3 for more details.

of command action. The options are added after *command_name* and separated with space. It is optional and totally depends on the type of action being performed by the command. For example:

```
$ ls -l  // The -l option of ls command changes the output behavior of
ls, by providing the outputs in a longer detailed format.
```

In the majority of cases, single-dash options, i.e., one-letter code, may be merged or mentioned separately along with input command, but in both the instances, the output behavior of command is unchanged. The association of option part with command line input is optional. For example:

```
$ ls -l -t   // The two options, -l and -t, separated by space, are given
with the ls command. To get output in long format, use the l option.
Similarly, to get the sorted output, based on file modification time, use
the t option.
```

```
$ ls -lt // Combining the l and t options with a single hyphen provides
the same output as above.
```

```
$ ls -lt -r // It provides a similar output as above by reversing the
order of sort.
```

- **Arguments**: Arguments may be described as input to the given command on shell prompt or command line after command options (this may be optional, totally depending upon operation behavior). An argument can be in the form of a directory or a file's name that can be passed as an input to the entered command on shell terminal or console. Several commands accept arguments after options or toward the end of the entire command line [1]. In a Linux system, all files' names and commands are case-sensitive. Hence, File1 and file1 are referring to two separate files, for example:

```
$ ls -l *.c // Display only the C program file details in long format of
current working directory. Here, *.c is an argument that permits to match
and fetch all files having .c extension.
```

```
$ ls -l /etc  // Display the content details in long format of /etc
directory. Here, the /etc directory acts as an argument or input to the
ls command.
```

```
$cat /home/abc  // Show the contents of the /home/abc file. Here, abc is
a file that acts as an input to the cat command. This file resides in
home directory.
```

The options and arguments part in command syntax format may or may not be optional, subject to the command.

- **Metacharacters**: Usually, by using any characters, you can make any variable, which may be used to store some value. But in shell, you must be careful to create a variable, because some characters are used as operators in shell operation. Every shell has some specific set of characters that have a special meaning to the shell command line, known as metacharacters.

Some key characters, referred to as metacharacters, are ?, *, ., [,], $, >, <, &, |, etc. For more details, see Table 3.2. Every metacharacter has a special meaning, when it is used in the processing of a given input on command line. For example, asterisk (*), question mark (?), and brackets ([ ]) are used to produce the lists of filenames; dollar sign ($) is used to get value from a variable; the greater than (>) and less than (<) signs are used as redirection operators; the ampersand sign (&) is used to run a process or command in background; the pipe sign (|) is used to transfer the output of one command as the input to another command; and the dot sign (.) is used to signify the current directory and many others [2]. It is worth mentioning here that if you wish to use any of these characters as variable, then before using them on command line, you must disable their special meaning interpreted by shell as metacharacters, so that these cannot be executed as metacharacters by shell. You may use single quotes, double quotes, and backslashes to disable the special meaning of such metacharacters, so they can be treated as normal characters by shell during their processing. Metacharacters are also discussed in detail in Chapter 3 (Section 3.15) and Chapter 5 (Section 5.1). Some examples of metacharacters are as follows:

```
$ls -l > xyz // Redirect the output of ls -l into file, named as xyz.
Here, > is the metacharacter, that is, output redirection operator.

$ls | more // Transfer the ls command output as input to the more
command. The more command is a filter for paging through text.
```

2.2 Internal and External Commands

All the Linux commands may be categorized into **internal** and **external** commands.

- **Internal commands**: All built-in commands are referred to as internal commands that are executed directly within the shell itself. These commands are loaded into the shell at the time of booting. They are a part of the shell and don't have a path since they are not coded in separate files. Therefore, the shell does not require a separate process to execute these commands. This is the main reason why the internal commands do not have any process ID during their execution. Some examples of internal commands are cd, pwd, and echo.

- **External commands**: All commands that are not built into the shell are referred to as external commands. Such commands are written as programs that have their own binary files. Most commands are external in nature, and the shell starts a separate sub-process to execute such commands due to the existence of a binary. These files mostly reside in /bin, /sbin, and /usr/sbin directory locations. The execution of an external command is possible only if the command directory location is present in the path specified by the $PATH variable. Otherwise, it will give an error. If the command binary file is available but the path of this file is not mentioned or included in the $PATH variable, then it will also give an error message: "command not found." Most external commands are stored in the /bin directory. The commands that are mainly used by the system administrator or superuser are found in /sbin and /usr/sbin and typically need root privileges to execute. $PATH is an environment variable, and if an external command is placed in a directory, which is listed or mentioned in the $PATH variable, then such a command is available to the user for execution. Some examples of external commands are ls, mv, and cat.

- **Identifying Commands**: One might wonder how a command is recognized as either internal or external in nature. The type utility may be used to check the nature of a given command as **internal** or **external**. All the commands that give the "**shell built-in**" message are internal commands, whereas the commands for which some directory path is shown are referred to as external commands [3]. If a command exists as both internal and external, then the internal command may get a higher priority. For example:

```
$type cd      // Specifying that cd is an internal type command with the
              message "cd is a shell built-in."

$type pwd     // Specifying that pwd is an internal type command  with the
              message "pwd is a shell built-in."

$type type    // Specifying that type is an internal type command  with
              the message "type is a shell built-in."

$type more    // Specifying that more is an external type command with the
              message "more is /bin/more."

$type cat     // Specifying that cat is an external type command with the
              message "cat is /bin/cat."
```

2.3 Command Location and User Commands

One of the important features of Linux-/Unix-like operating systems is that all the resources, tools, utilities, and APIs are stored in the form of files that are categorized into various file types. These files are further stored in directories. A directory is also a type of file. All these files and directories are organized in the form of a hierarchical directory structure that can be thought of as treelike structure of directories (sometimes called folders). The first directory in this hierarchical structure is referred to as *root* directory, represented as /. The root directory (/) further comprises various files and subdirectories, which may include more number of files and subdirectories. Everything on the Linux system is located under the root directory (/), as illustrated in Figure 2.1.

All data of the Linux system has been organized or kept under a single treelike structure that comprises all files and directories. It never takes into account how many storage devices or drives are connected to the computer system. On the other hand, if we take the example of Microsoft Windows operating system, it may have a separate file system structure for each drive or storage device or even partitions. For more details about the file system, see Chapters 10 and 11.

The Linux commands are sets of an executable stream of bytes which are stored in a file. Hence, besides typing the command on the command line, we must be aware of how to traverse the Linux file system to find the command's location. In the file system hierarchy, the directory in which the user is presently working is referred to as the present working

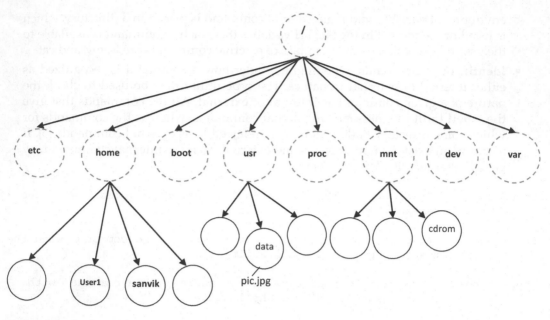

FIGURE 2.1
Linux hierarchical directory structure.

directory (pwd). By using the change directory (cd) command, we can move or traverse to reach the desired directory from the current place of directory.

The path of each directory may be categorized into two sections: **absolute pathnames** and **relative pathnames**.

- **Absolute pathnames**: An absolute path is defined as getting the location of a file or directory, moving from the root directory (/). In other words, start from the root directory (/) and follow the tree branch by branch (i.e., directory by directory), until the desired directory or file is found.
  For example (Figure 2.1):

```
/home/sanvik
/usr/data/pic.jpg
/mnt/cdrom
```

All are absolute paths. The first slash (/) is the root directory, and the intermediate slashes (/) are placed to separate the directory names.
- **Relative paths**: Relative path is described as the path of a file or directory from the present working directory. A relative pathname starts from the current working directory to the desired directory or file in downward direction. It **never starts from root directory, i.e., /.** It uses the following special symbols:
  - "." (dot)—it represents the current working directory
  - ".." (dot dot)—it represents the parent directory of the current working directory.

For example, suppose the present working directory is */home/user1*; then, there are two ways to change the directory: first, by using absolute path, and second, by using relative path. For example, in Figure 2.1, assume that there is a directory named Music under the sanvik directory; then,

```
$ pwd
    /home/sanvik
    $cd Music (using relative path)

    $pwd
    /home/sanvik/Music

    or

    $ pwd
    /home/sanvik
    $cd /home/sanvik/Music/ (using absolute path)
    $ pwd
    /home/sanvik/Music/
```

Before executing any command, we must be aware of its type and location. A command is basically an executable program, written in programming language such as C/C++, or in any scripting language such as bash, Ruby, Python, or perl. As discussed earlier, some commands are already available or built into the shell that are executed directly by the shell itself. Such commands are known as internal commands and don't have any path. All other commands have their own location in the hierarchical file structure and are known as external commands.

Many times we know that a particular file, directory, or command exists, but don't know where and how to find it. The following commands may be used to search or find the location path (absolute path) for it: **find**, **locate**, **type**, **which**, **whereis**, and **whatis**. These commands are also discussed in Section 2.3.1. You can also visit the manual page or man page to get the complete details of various options associated with such searching commands [4].

- **The find command—search for files in a directory hierarchy:**
    The *find* command helps to search for a file on a file system. It has many options that may be applied to customize the search operation. It is very useful to search recently built or renamed files where the *locate* command may not be suitable. The syntax of the *find* command is given as follows:

```
$ find [paths] [options] [expression]
```

The *find* command allows you to search a file in a variety of ways, such as by file type, date, size, permissions, and owner. To see the complete details of various options associated with the *find* command, you may run *info find* or see Section 10.5.4 of Chapter 10.

The *find* command seeks recursively through directory tree for files and directories that match the particular attribute. Further, it can print either the file or directory that fulfills the matches or executes other operations on the matches as defined. If no path is mentioned, then by default, the current directory path is applied, and similarly, if no expression is provided, then the expression "-print" is utilized. For example:

```
$ find /usr -type f -name "*.JPG"
```

It searches all regular files in the /usr directory that matches the wildcard pattern *.JPG; -type f, as mentioned in command line, it put limit on search to regular files only. But to search for directories only, use -type d. The "-*name*" option is case-sensitive, but the "-*iname*" option is used to search regardless of case sensitivity.

- **The locate command—a way to find files from the database:**
  The locate command searches for files on the local system (entire directory hierarchy) performing a rapid database search and prints absolute pathnames of every filename matching with the given patterns:

```
$ locate [options] "patterns"
```

To retrieve a file with the help of the *locate* command is considerably quicker than with the *find* command. It is because the *locate* command will move directly to the database file. It is required that the *locate* command must get the latest or updated database before searching any file. By using *updatedb* utility, it updates the database frequently to incorporate changes in a file system.

For example, to locate all files that have the given search pattern in their name from the entire file system, i.e., from the whole directory hierarchy, use the following:

```
$ locate tcs        // Search pattern is "tcs"
/home/ritvik/tcshrc.c
/root/tcsc.cpp
/root/ xyz.tcs
/usr/bin/sun-message.tcs
/user/src/xyz/tcs.txt
...
```

The following would list the absolute paths of all files, named *file1*, and all directories, named *dir1* //:

```
$locate file1 dir1
/home/XYZ/file1345 or /usr/local/mydir145/home.html
...
```

THE LOCATE DATABASE CREATION

In few Linux distributions, sometimes the *locate* command does not work just after the Linux installation, but after some time or the next date, it will start doing well. It is because the *locate* database is generated by another package/utility, called *updatedb*. Normally, the *cron* daemon or job executes it at a regular interval. Most of the systems consist of *locate*, which executes the *updatedb* program one time in a day or more depending on the size of database. If the database is not updated regularly, then it is suggested to run the *updatedb* program manually with superuser privilege.

- **The type command:**
  The *type* command is a shell built-in command that shows command type, i.e., internal command or external command. The syntax of the type command is given as follows:

```
$type Name_of_command
```

- **The whatis command**:
  The aim of the *whatis* command is to offer help/description about the command functionality in one line. For example:

```
$ whatis whereis
whereis(1)- locate the binary, source, and manual page files for a
command                            //output

$whatis cal
date(1)—Display a calendar          //output

$whatis ls
ls(1)—list directory contents       //output
```

- **The whereis command**:
  It discovers the binary, source, and manual page of a specific program (command), and it handles only exact matches. If you know the command name incompletely or partially, then take the help of the *slocate* (secure locate) command.

```
Syntax: whereis [options] filename
```

For example, to know the directory location of the tar command as well as its man page (i.e., help page), type the following command:

```
$ whereis tar
tar(1) - /bin/tar /usr/share/man/man1/tar.1.gz
```

It shows that the directory location of the tar command is /bin/tar directory, and its man page directory location is /usr/share/man/man1/tar.1.gz directory.

- **The which command**:
  The *which* command is a very useful command to locate the exact location of executable file within the system, related to the given or input command. It searches only in $PATH (environment variable) of user's shell. If more than one version of an executable program (command) is installed in the system, then by using the -*a* option, it displays the list of all occurrences of a particular command in the PATH environment variable. The syntax of the *which* command is given as follows:

```
$which Command_Name
```

For example:

```
$ which ls
/bin/ls
```

The *which* command operates only for executable programs, not shell built-in or alias programs that are replacements for actual executable programs. If you attempt to apply the *which command* on a shell built-in or internal command, such as *cd* or *echo*, then you will get either an error message or no response:

```
$ which cd
Error// command not found
```

- **System call**

Fundamentally, the system call is an interface between computer program and kernel. Basically, the system tells how a computer program interacts with kernel to execute some task. By using the system call, programs execute various functions, such as opening files, open(); reading files, read(); creating processes, fork(); process communication, pipe(); writing files, write(); and file protection, chmod(). Here, fork(), open(), write(), pipe(), chmod(), and exit() are some examples of system call [5].

When a program requires a system call, the arguments are wrapped up and given to the kernel, which gets over the execution of program until the call finishes. A system call is not a simple function call, but rather, it requires a special technique to shift control to the kernel. There are numerous distinct approaches for computer programs to make system calls, but the low-level instructions or program for creating a system call differs among numerous CPU architectures. You can obtain the complete list of system calls by visiting the man page for syscalls().

2.3.1 User Commands

The major benefit of both Linux and Unix operating systems is that they provide a rich set of utilities or functions through various commands [6]. You can use all these utilities through command line of Linux. You can invoke them immediately by simply typing their name on command line, i.e., in CUI environment, or indirectly access them from a menu-driven option or icon in graphical mode. The subsequent section describes some essential and frequently used Linux commands/utilities that can be executed from the command line.

2.3.2 Universal Commands

You are already aware that the Linux operating system comes in several flavors, known as Linux distributions. Presently, more than 400 Linux distributions are being actively in development and use [7]. Some of the more popular distributions are Red Hat, Ubuntu, CentOS, SUSE, Fedora, and Debian. Every distribution has its own features and properties. This section is now going to discuss some commands that are generally available in all flavors, preferably referred to as universal commands. Table 2.2 describes some universal commands apart from those already discussed in Section 2.3.1.

2.3.3 System Commands

A rich number of commands associated with the Linux kernel can be used to start and manage the various system resources and processes. Some of the key system commands are mentioned in Table 2.3.

2.4 Communication and Other Commands

Networking is a way to connect more than two computers to make a computer network that enables exchanging information or resources among all systems, widely referred to as network of computer network, i.e., Internet. When it comes to networking, Linux plays a vital role in building all sorts of networking systems, machines, or devices such as servers, switches, routers, firewalls, cloud storage boxes, and gateways [8]. Table 2.4 highlights some of the most frequently used commands that are regularly used to monitor, manage, and transfer files in a network.

TABLE 2.1

List of Frequently Used User Commands

Command Name	Description
who -option	It displays a list of currently logged users into the system.
whoami	It displays the username of the present/current user.
ls [options] [path_names]	It lists the contents of a given directory. • *ls* // Display the list of filenames from the current working directory • *ls -l* // Display the list of filenames in long format, i.e., with detailed information of the current working directory • *ls -l /etc* // Display the list of filenames in long format (with detailed information) of the /etc directory. The most useful options are -F, -R, -l, and -s.
clear	It clears the terminal display/screen.
cal [-jy] [month][year]	It displays the calendar. Some examples are given below: • *cal* // *Display the calendar of current month* • *cal 3* // *Display the calendar of March month of current year* • *cal 5 2017* // *Display the calendar of May of 2017* • *cal 2017* // *Display the whole year calendar of 2017.*
date	It displays the system's current date and time.
mkdir directories_names	It creates one or more directories, if they do not already exist. • *mkdir xyz* // Make a single directory with the name xyz. • *mkdir xyz1 xyz2 xyz3* // Make three directories with names *xyz1*, *xyz2*, and xyz3.
rmdir directories_names	It is used to delete the empty directory from file system. For a non-empty directory, use the –r option with the rm command. • rmdir dir1 dir2 // Remove dir1 and dir2, but both directories must be empty.
uname -option	It displays the system hardware and operating system details. Some examples are given below: • uname // Print the name of the operating system installed in the system, i.e., Linux • uname –a // Print all the information of the system • uname –r // Print the current kernel version of Linux system • uname –o // Print the name of the operating system • uname –p // Print the processor type of the system
echo -option string echo $val	*echo* is a shell built-in command to display the given string input as well as the variable value, similar to the printf function of C language. For example, if a variable name, age, stores the value 10, i.e., age=10, then • echo "I am $age years old" The output is as follows: I am 10 years old.
cp -option <src> <dest>	It performs the copy of files and directories from one place to another (src—source; dest—destination). • *cp file_1 file_2* // Copy the file_1 content in file_2. If file_2 exists, then it is overwritten. Use the -i option to confirm before overwriting. • *cp file_1 file_2 dir_1* // Copy file_1 and file_2 into directory dir_1. The dir_1 should be available before performing the copy operation. • *cp dir_1/* dir_2* // Using a wildcard, all the files of dir_1 are pasted into dir_2. The *dir_2* should be available before the operation. • *cp -r dir_1 dir_2* // Copy the content of directory dir_1 into directory *dir_2*. If directory dir_2 does not exist, then first create a directory with the name *dir_2* and then copy the content of directory *dir_1* into *dir_2*.

(Continued)

TABLE 2.1 (*Continued*)

List of Frequently Used User Commands

Command Name	Description
mv -option <source> <destination>	It is used to rename the name of files and directories and also to move a set of files to a directory. The renaming and moving operation of mv is depending on how it is used. When it is used to rename a file, then simply use the new filename as the second argument. But in the case of moving a set of files, then the last argument should be the path of directory where you want to transfer the operation. • mv item1 item2 (in the same directory // rename) (*item1 is renamed item2*) • mv item1 item2 Dir1 (in other directory location // move) (move the files item1 and item2 into the directory named Dir1) • mv file_1 file_2 // Rename file_1 to file_2. If any file with the name file_2 exists, then it is overwritten by the contents of file_1, and you may accidentally delete the original file_2, so be careful while renaming the file. • mv -i file1 file2 // The same as above. Just make a confirmation before it is renamed or overwritten. • mv dir_1 dir_2 // It moves the content of directory dir_1 into directory dir_2. If directory dir_2 does not exist, then first create directory dir_2 and then move the content of directory dir_1 into dir_2 and thereafter delete dir_1.
rm -option item_name	This command is utilized to delete (remove) files and directories. • rm * .html // Remove all files with the .html extension. • rm file_1 // Delete file_1 silently. • rm -i file_1 // Remove file_1 after the user confirmation. • rm -r file_1 dir_1 // Delete file_1 and dir_1 along with their contents.
ln	This command creates links between files, and it may be either a hard or symbolic link file. • ln file_1 file_2 // Create a hard link file, named file_2, of file_1. The file_2 is a hard link with file_1. It means that if you are accessing the content of file_2, it shows the file_1 content. If there is any change in the content of file_1, then it will be automatically updated in file_2. If file_1 is deleted, then file_2 exists with the last updated content of file_1. • ln -s file_1 file_2 // Create a symbolic (soft) link, named file_2, of file_1. The file_2 is a soft link with file_1. It means that if you are accessing the content of file_2, it shows the file_1 content. If there is any change in the content of file_1, then it will be automatically updated in file_2. If file_1 is deleted, then file_2 exists, but with no content. It works similar to "shortcut icon of any file, created on desktop of Microsoft Windows."
more <filename>	It displays the content of given filename, one screen at the first time, i.e., if the content of given file covered the entire screen. Apart from this display of file content, you can easily see the filename and content percentage of the file that has been viewed, at the bottom of the screen. If the file content is longer than one screen, then press the Spacebar or the F key to forward one page, press the Enter key to forward one line—i.e., help to view file content line by line—and press the Q key to quit.
less <filename>	Every Linux OS provides both the more and less commands. The functions of both commands are almost similar, but the less command has added features as compared to the more command; for example, the less command can search for a pattern in backward direction also in contrast to the more command. The less command displays the content of a file screen by screen. Press the Spacebar to continue to the next screen, press the J key to go one line up, press the K key to go one line down, and press the P key to quit.

(Continued)

TABLE 2.1 (*Continued*)

List of Frequently Used User Commands

Command Name	Description
lpr -option <filename>	The lpr (line printer) command sends files for printing. • lpr -# <number> // Set the number of copies to be print for a particular file. • lpr -#4 <sample.txt> // Print four copies of the sample.txt file. • lpr -P <printer> // Give the printer name to be used for printing. If no printer is mentioned, use system's default printer for printing. • lpr -P <headroom> <report> // Select the printer with name (headroom) for printing the file, named report.
wc -option <filename>	It displays the number of lines, word count, byte count, and character count of the given files mentioned as command arguments. Some options of the *wc* command are as follows: • -c—to print byte count • -m—to print character count • -l—to print newline count • -L—to print the length of the longest line • -w—to print word count
tar [-option] <archive-file> <directory-and-file-names>	The command "*tar*" stands for tape archive and is used to generate compressed or uncompressed archive files. Some options of the *tar* command ID are given below: • -c—to create archive • -x—to extract an archive • -f—to create an archive with the given filename • -t—to display or list files in the archive file
gzip <filename>	It is used to compress one or more files. The compressed files are marked with the .gz extension. • *gzip <filename>* // Compress the input file, as filename.gz • *gzip -d <filename.gz>* // Decompress filename.gz. This is similar to the gunzip command.

TABLE 2.2

List of Frequently Used Universal Commands

Command Name	Description
cat -option <filename>	It shows the content of a filename. If the file does not exist in the current directory, then provide the absolute path of that filename. • cat /etc/password // Display the content of the password file It can also be used to combine the contents of multiple files into a single file. For example: *cat file1 file2 file3 >> sample.txt* // Copy the contents of file1, file2, and file3 into another file, named sample.txt.
chmod mode <directory-and-file-names>	It is used to modify/change the permissions of a given list of files and directories for owner, groups, and other users. Mode is an octal number used to set/reset the permission bits.
chown -useraccount filename	It is used to change the ownership of files and directories in a Linux file system. • *chown <myuser> <sample.txt>*
df	It stands for disk free (df) and shows the amount of free disk space available on each file system. If the filename is not given, then it shows the free space available on all presently mounted file systems.
du -option file	It stands for disk usage (du) and estimates and displays the disk space used by files. • du -s *.txt // To report the size of each file, having .txt extension, in the current directory.

(Continued)

TABLE 2.2 (*Continued*)

List of Frequently Used Universal Commands

Command Name	Description
env -option	It is used to print a list of the current environment variables or to execute a program in a custom environment without changing the current one. • *env* // With no options, it displays the current environment variables and their values.
touch <filename>	It creates an empty text file with the given filename.
exit	It is used to exit from the command prompt or the command shell and terminate the program.
export <variable_name>	It is used to convert a given local variable into global variable. Such converted variables are known as exported variables which are also being referred as environment variable. These exported variables are automatically broadcasted to the programs which may be executed later by the shell. • export -p // Display the whole list of names that are exported in the current shell.
find <pathnames> <conditions>	It traverses the directory and subdirectories starting from the specified pathname and searches for files with names that fulfill the mentioned conditions. If no path is mentioned, then the current directory is taken as the default pathname.
fdisk -option	It helps to manage and manipulate disk partitions. With the use of the fdisk command, you can do operations of various partitions on a hard drive, such as view, create, resize, change, copy, delete, and move. All these operations can be performed through text-based menu-driven interface. • fdisk -l // List all the available disk partitions on Linux.
file <name of file>	It prints the output information related to the input filename, such as file types and last modified details.
grep	The grep command stands for "**g**lobal **r**egular **e**xpression **p**rint." It is a pattern-matching command that matches the given search pattern or text line by line from a given filename and prints all lines that comprise the given pattern. The basic syntax of this command is as follows: • grep -option <search option> <filename> • grep -w "hope" sample.txt
man command_name	The *man* command is used to provide the complete information about a given command. • *man ls* // Display the complete information about the *ls* command.
pwd	It displays the absolute path of the current working directory.
sed -option	*sed* is an abbreviation for "stream editor" and performs complex pattern matching. It performs numerous complex operations/functions on a file, such as search, find and replace, and insertion or deletion. It also filters and transforms the text. • sed G myfile.txt>newfile.txt // Add double spaces between the contents of the file, named myfile.txt, and write the output to another file, named newfile.txt. • sed "s/unix/linux/" file.txt // Replace the word "unix" with inux" in the given file.txt
shutdown -option	The *shutdown* command may be used to halt, power off, or reboot the machine. • *shutdown -r now* // Restart the system immediately.
sudo -option	It permits a local user/normal user to run/execute a command as superuser privilege for some define moment, as define in the *sudoers* file.

TABLE 2.3

List of Some Key System Commands

Command Name	Description
init runlevel	The init process is considered to be the parent of all running processes. It is responsible for initializing the system in a particular way. The *init* command is executed only by the superuser. • # *init 0* // Shut down/halt • # *init 1* // Single-user mode # *init 6* // Restart
ps -option	The ps command stands for process status and is used to show the list of all currently running processes along with their other relevant information such as PID, terminal type, and the time for which the CPU has been executing that process. • *ps* // Print all processes associated with the current shell terminal • *ps -a*// Print all the running processes on the Linux system
fork	It is used to create a child process from the current running process.
getpid	It prints the process identification (PID) number of the specified executing process or program.
getppid	It prints the parent process identification (PPID) number of the specified executing process or program.
mount	It is used to attach or mount a file system which is found on an external storage device or a device partition. This external file system is attached with the main Linux file system to make it accessible. By default, the /mnt directory is used to attach or mount any external file system, but you can mount it at any place in the Linux file system. The syntax of the mount command is given below: • mount -t type <*device_path*> <*dir_path*> device_path // Path of external device through which Linux accesses the new file system. *dir_path* // Path of the directory (in main Linux file system) in which you want to attach/mount the external file system or new file system.
umount	It is used to unmount a file system, i.e., de-attach any file system from the main file system of Linux. You can give either a device path name or the directory path name (mounted place) as argument of the umount command to remove the file system from Linux. The syntax of the umount command is as follows: # umount <device path name> OR # umount <directory path name>
nice	It is used to change process priority. A higher nice value of a process indicates low priority of that process or program.
idle	This is an internal system call and is applied during bootstrap. The idle() system call may called by only process 0 and never returns for process 0, but always returns -1 for a user process.
kill	The kill command is used to terminate a running process. The syntax of the kill command is follows: • *kill* -option <pid> // To terminate a process, with bearing process ID (pid). • *kill -l* // Kill all the processes running in the system including the **kill** process also, but except the root process (PID 1).
lseek	The lseek() system function is used to relocate the read/write file offset of open file linked with the file descriptor (*fd*).
link [-option]	It is used to create a new link, also referred to as a hard link, to an active file. The syntax of the link command is given as follows: • link <file_1> <file_2> // Make a hard link file, named file_2, which has the same index node to an active file, named file_1. Both files have to point the same data space on the disk with the same permissions and ownership due to the same index.

(Continued)

TABLE 2.3 (*Continued*)

List of Some Key System Commands

Command Name	Description
setup	It is used to start all configuration functions to configure various devices and file systems with Linux kernel and also mount the root file system. This system call, setup(), is called once from *linux/init/main.c*. No user process may call this setup command.
umask	It is used to display the set default file permission in Linux system for all newly created files. You can also use this command to change the current file permission of any files. File permission is a four-digit octal number, also referred to as *umask value*. • *umask* // See the default file permission of the system. • *umask <octal_no>* // Set the default file permission of the system.
wait	This command is used to delay/suspend the execution of any process, mentioned as process ID (PID). If PID is not given, then delay is applied to all currently active processes to complete and return status is zero for all. • *wait <pid>* • *wait 3452* // Delay for process **3452** to terminate and return its exit status

TABLE 2.4

List of Some Communication Commands and Other Important Commands

Command Name	Description
hostname -options	This command is used to see the hostname of the Linux system or to set the hostname or domain name of the Linux system: • hostname -d // Provide the domain name of the Linux system • hostname -a // Provide the alias name of the Linux system
ping -options	The command ping stands for **P**acket **I**nternet **G**roper and is used to send an ICMP ECHO_REQUEST to the network hosts. It also helps to test the connectivity between two nodes. • *ping www.kernel.org* // Check the connectivity by sending the ICMP ECHO_REQUEST to kernel host. If there is connectivity, the ICMP ECHO_RESPONSE will come from kernel host.
ifconfig -option	It is used to either configure or display the configuration setting of all network interfaces that are currently in operation. • *ifconfig eth0* // Display the settings of the first Ethernet adapter installed in the system. • *ifconfig -a* // Display all interfaces that are currently available, even if down, i.e., not active .
ifup -configuration	This command is used to bring a network interface up. • *ifup eth0* // Enable or up the network interface *eth0*.
ifdown -configuration	This command brings a network interface down. • *ifdown eth0* // Disable or down the network interface *eth0*.
netstat -option	It is used to display/print the network connections, routing tables, interface statistics, masquerade connections, and multicast memberships. • netstat -at // Display all TCP ports • netstat -au // Display all UDP ports
ftp -option	It stands for **F**ile **T**ransfer **P**rotocol. This command permits a user to transfer files between the host and remote systems on network.
rlogin	It is used for logging into a remote system and performing various operations by starting a terminal session on remote system. The basic syntax is given as follows: • *rlogin -option <username> <domain_name>* • *rlogin -l sanrit domain.com* // To log in as username "sanrit" into remote system, having the address "domain.com."
telnet -option	It is used communicate with another remote system/host. Basically, the telnet command/protocol allows a user to access a system remotely.

(*Continued*)

TABLE 2.4 (*Continued*)

List of Some Communication Commands and Other Important Commands

Command Name	Description
fsck -option	It is widely known as file system checking command and used to check and repair one or more file systems. • fsck -A // Check all the configured file systems (through the /etc/fstab file) in a single run of this command. • fsck /dev/hda3 // You can check a specific file system installed on a particular partition for example, /dev/hda3.
mkfs -option	The "make file system" (mkfs) command is applied to create a file system on a formatted storage device or media, generally a partition on hard disk drive (HDD). • mkfs -t vfat /dev/hda2 // To create a vfat (Microsoft Windows-compatible file system) on the second partition of the first attached hard drive.
rusers	It gives output similar to the "*who*" command. The *who* command prints the list of users presently logged on the local system, but the rusers command shows the list of all machines/hostnames on the local network. The syntax of the rusers command is given as follows: *#rusers -option <hostname>* • rusers // To print the list of the users on a local network that are logged in remote machines. • rusers -h // To print the list of users sorted alphabetically by hostname. • rusers -h xyz // To print the list of users on an "*xyz*" host.
ssh	The Secure Shell (ssh) provides secure login into remote systems or remote transmission. For secure connection over network/Internet, use the ssh or Kerberos version of telnet. The basic syntax of ssh is given as follows: #ssh <remote_host> // The "remote_host" is defined with IP address or domain name of the remote system for secure connection.
traceroute	It is used to print the network route which is followed by packet to traverse from source to destination. The only parameter you need to provide with the *traceroute* command is the IP address of the destination host system/node. The basic syntax of the *traceroute* command is given as follows: # traceroute<hostname> // Print the route from source host to destination. The destination host details may be provided as its hostname or IP address.

2.7 Summary

Linux commands can be categorized based on various aspects and utilizations. One of the aspects can be divided based on whether the command is a built-in command or not. Built-in commands are executed directly in the shell itself and are called internal commands, but external commands are not built into the shell. These commands exist in binary files located in some directories. Initially, the shell locates all such binary files in a particular directory, specified by $PATH variable and executed by a separate sub-process. A user can recognize internal or external commands by using the "type" command. Generally, all the available commands in Linux may be divided into four major categories according to their utilities and usage: user commands, universal commands, system commands, and networking commands. In Linux, all resources, tools, utilities, and APIs are stored in the form of hierarchical directory structure, referred to as Linux file system, containing various directories and files of Linux system. Each file has a unique path from root, known as absolute path. Some special commands may be used to locate and provide some key information/details of various Linux commands, for example, *find, locate, type, which,* and *whereis.* A shell is a command-based interface between the user and the system. In the subsequent chapter, you can find some more details about shell and other important key features.

2.8 Review Exercises

1. Describe command syntax, options, and arguments of Linux command line.

2. Explain metacharacter and how it is important in a Linux environment.

3. By using the cd command, you go to /bin/xyx and then type cd ~ and press enter. Then, find out your current place.

4. How to distinguish between an internal command and an external command? Explain why internal commands do not have process ID whereas external commands have.

5. Assume that you type a command for execution and then see the message "command not found," but the command exists in its location. Explain the possible cause of such error message.

6. How the Linux file system differs with Microsoft Windows operating system? Highlight some key pros and cons.

7. Discuss "absolute pathname" and "relative path" mechanisms of searching the directory or filename, and list the absolute path of the directory where Linux commands are stored.

8. How to know the directory location of the *cat* command as well as its man page?

9. How to locate the exact location of executable file of the ls, date, who, and mkdir commands?

10. Discuss the main difference between the locate and find commands.

11. How to see the echo command functionality? Describe it in one line, and give an example.

12. How to display the complete details along with inode numbers of /bin directory?

13. How to display the current date as mm/dd/yy? Further, write the command to display the date and time of 30 seconds ago, 2 years ago, and the next day.

14. Write the commands to delete a non-empty directory and to see the processor detail and kernel version of a system.

15. If a file named ABC_1 is write-protected, then how to write the content of a file named ABC_2 into ABC_1?

16. How to create a soft link and hard link file of any given file? What is the use of such link files in Linux?

17. Write the command to print five copies of a file named *report.txt* on a color printer named *headoffice*.

18. What will be the function of the command **more +35 sample.txt**?

19. How to print the total number of lines, words, and characters of a file named sample.txt?

20. Write the command to concatenate the archive files. Further, also write commands for adding files, namely **test.txt** and **result.txt**, into an already existing **Question.tar** file.

21. How to see the details of installed file system in each partition of a given hard drive?

22. How to show the amount of free disk space available on each file system? Write the output of **du -s *.png.**

23. Describe the init process along with various run levels of a system with examples.

24. Execute the man command in your Linux system and write a small summary of your observation.

25. Describe environment variable. How to display the list of the current environment variables?

26. Write commands for the following operations:

    a. How to see the process ID and parent process ID of the currently running process?

    b. How to change the priority of a process?

    c. Write down the output of **df -i.**

    d. How to terminate a process immediately?

    e. What is the output of **umask 574**?

    f. When will you use the **sudo** command?

    g. How to display the complete details of all installed network devices of a system?

    h. How to activate the network connection of the installed network card **eth0**?

References

1. Gareth Anderson. 2006. *GNU/Linux Command–Line Tools Summary*. Open Source under GNU Free Documentation License.
2. Richard Blum. 2008. *Linux Command Line and Shell Scripting Bible*. Wiley Publishing, Inc.
3. Neil Matthew and Richard Stones. 2001. *Beginning Linux Programming*. Wrox Press Ltd.
4. Linux man pages in the World Wide Web. https://linux.die.net/man/.
5. The Linux Information Project, The Comprehensive Index in the World Wide Web. http://linfo.org/main_index.html.
6. J. Purcell. 1997. *Linux Complete: Command Reference*. Red Hat Software, Inc.
7. Jose Dieguez Castro. 2016. *Introducing Linux Distros*. Apress.
8. Wale Soyinka. 2012. *Linux Administration: A Beginner's Guide*. The McGraw-Hill.

3

The Shell

A text-based or command line user interface (CUI) communication with Linux/Unix is provided by a shell, which is normally used for improving system interaction and administration. A shell takes input in text form and gives output in the same way. It is also a programming language as well as a command interpreter. As you are aware that the shell is also provide a rich set of commands for Linux. With the support of these commands, users can do various tasks that are directly input by them and execute with the help of kernel. In the programming language features of the shell, it allows to place various commands in a requisite sequence and format into a file to complete a task. Such files are known as shell scripts or shell programs. The shell program is not compiled to create a separate executable file. Here, all shell programs are executed in interpretive mode. In this mode, the shell executes the script file line by line by searching and executing all the commands mentioned in each line on the system. The shell is always available with any Linux distribution. Therefore, shell scripts are relatively portable and have few dependencies such as inbuilt shell syntax in addition to the commands they call up. In this chapter, we have taken Bourne shell (*sh*) and its updated version, i.e., Bourne Again Shell (*bash*), as languages for scripting, along with regular expressions as a general notation. Due to the rich features and greater role in Linux, the shell, i.e., *bash*, has been discussed at many places in this book.

3.1 What is a Shell?

A shell is a command language interpreter that offers a command line user environment to execute various commands and perform operations. It helps all learners to learn Linux in a very efficient way. It works as an interface layer between the kernel and end user. After booting, if the user wants to start Linux in CUI mode, then first Linux verifies the username and password of the user and thereafter its setup and starts available shell, like bash. After successful start of the shell, the user is presented with a command prompt interface, i.e., text-based working environment. The prompt indicates that Linux system is ready to accept commands, written in text format or shell script for execution. The size of command or shell script format may usually occupy some set of characters. This interface is commonly referred to as command line interpreter or text-based interface. The key responsibility of shell is to provide the command prompt and then interpret commands. A command itself is a program, and after it is executed and the output is displayed on console, the shell again displays the command prompt and is ready to receive new commands for execution and other operations. The first shell was developed by Stephen Bourne for Unix and is known as Bourne shell and represented as "*sh*." Most of the Linux distributions used bash, which is an improved and GNU version of sh, as the default shell.

3.2 Why Use a Shell in Linux?

The shell provides a powerful programming environment allowing automation of almost any task that you can imagine and prefer on a Linux system. It is very important to mention here that shell is not an operating system, but it is a part of system through which the user can interact with Linux kernel to execute commands and various applications. There are numerous reasons to use shells in Linux. Some of them are as follows:

- Shells access system calls directly through the command name.
- Shells provide an environment where multiple users can log in into the Linux system and perform their respective work independently.
- Shells are used to avoid repetition of command, put all the commands in a file, and execute the file, i.e., the shell script.
- Each shell does the same job, but has its own syntax, semantics, and built-in functions to write shell script programs.
- The shell is not a part of the kernel, but the kernel helps to execute programs and applications. Therefore, users have an option to select, and install, the desired shell from various available shells for a Linux system.
- The shell can operate individually, as on a physical terminal, or as a window terminal in graphical mode. But in both cases, their functionality is the same.

3.3 The Login Shell (Shell Prompt)

This chapter primarily covers the conceptual features of the *bash* and an effective way to execute various Linux commands in a text-based environment. After successful validation of the login credential of a user, the default prompt of bash appears on console, i.e., $ for a regular user and # for a root user. So after successful login into a Linux machine, you first see a prompt (# or $). You may think that nothing is happening at the prompt, but actually a Unix command is running at the terminal. This command is special because it starts running just after the login into the system and remains active until you log out [1]. This command is your current shell, i.e., bash. If you run the **ps** command (which shows the list of currently active processes), then you will see the below output which comprised with currently running processes, as shown below:

```
$ ps
PID     TTY     TIME    CMD
357     pts/6   0:00    bash            // bash shell
```

When you type a command, it goes as input to the shell. The shell first checks the entire input command parameter for metacharacters, because metacharacters are special characters that have no significance to the command but represent something special to the shell. The details of such metacharacters are explained in the subsequent parts of this chapter. The shell provides a command line interface (CLI) on a Linux system where you can type/ submit commands using the keyboard. To know the name of currently active or running shell, then type the following command on command line:

```
#echo $SHELL
#ps $$
```

NOTE: The output could be shown as **/bin/sh** for sh; **/bin/csh** for C shell (csh); **/bin/ksh** for Korn shell (ksh); and **/bin/bash** for bash or others (as per the installed shell).

Some basic command line-editing short keys are given below. You can use the below key sequences or combinations to edit and recall commands:

- **Ctrl+L:** Clean up the whole screen, similar to the "clear" command.
- **Ctrl+W:** Remove the word before the cursor.
- **Ctrl+U:** Remove the line before the current cursor place. If the cursor is placed at the end of the line, remove the whole line.
- **Up and Down arrow keys**: Recollect the previous commands on command line (to see command history).
- **Ctrl+R:** Search through formerly used commands (see command history)
- **Ctrl+C:** Kill the currently running command.
- **Ctrl+A**: Move to the beginning of the line of the current cursor place.
- **Ctrl+E:** Move to the end of the line of the current cursor place.
- **Alt+F:** Move cursor in one-word forward direction in the current line.
- **Alt+B:** Move cursor in one-word backward direction in the current line.
- **Ctrl+H:** Similar as the Backspace key.
- **Ctrl+T:** Exchange the last two characters before the cursor place.
- **Esc+T:** Exchange the last two words before the cursor place.
- **Ctrl+K:** Remove the line after the cursor.
- **Tab**: Auto-complete files, directory, command names, and many more.
- **Ctrl+D:** Exit the current shell.

3.4 Command Line Structure of Shells

The shell has features of both a programming language and a command interpreter. The shell permits a user to run commands by typing them manually on command prompt or automatically by using a program, known as shell script. The behavior of a command is dependent upon various types of information provided along with command name. The various commands may be run by themselves or by giving additional information as option/arguments to make them do a more specific task [2]. The syntax of command line structure of the shell may be as follows:

```
#command_name [-options] [list of arguments]

  For example:
     $ls       // Command name but with no option and no argument
          Output: Display the list of files from the current directory
```

```
$ls -l    // Command name and option but no argument
    Output: Display the list of all files and subdirectories of the
        current working directory in long format

$ls -l /usr/bin // Command name, option, and argument
    Output: Display the list of all contents from bin directory in
        long format.

$cat /etc/passwd // Command name and argument but no option
    Output: Display the contents of the passwd file
```

3.5 sh Command

When Unix came into existence in 1971, it had a very basic shell, written by Ken Thompson, known as Thompson shell. Thompson was one of the inventors of Unix and C programming language at AT&T Bell Labs. But in 1979, Steve Bourne wrote a new scriptable Unix shell at AT&T Bell Labs, known as Bourne shell. Bourne Shell or bash was the default shell that came with the seventh version of Unix. The Bourne shell was not identified by bsh because it's only "shell," so it's denoted by sh and its path is represented by /bin/sh. Thereafter, many other shells have been written with more features, but they generally maintained their compatibility with the sh. However, sh had remained in high demand due to its better features to control jobs, input/output implementation, and other important tasks. But, the most significant feature provided with the sh is the concept of pipeline that allows one process to pass its output to the input of another process. Further, Steve Bourne also introduced the feature of changing the shell from being a very basic command interpreter into a flexible scripting language. Presently, various other shells are also available, but almost all of them are centered on either sh or csh.

When you login into a Unix/Linux system, you will see a prompt in text-based interface. The prompt could be a #, $, %, or any other symbol. This symbol depends upon the shell of the system. The default prompt of sh is $ for a regular user (or # for the superuser/root user), and for csh, it is %. Every shell provides a good number of its own built-in or native commands.

The *bash* is an *sh*-compatible command language interpreter that executes commands, which is given at command line terminal as a standard input or provided from a specified file [1,3].

```
sh, jsh—the standard sh command interpreter
jsh    —job control is enabled
bash   —GNU Bourne Again Shell
```

- **bash syntax:**

```
# bash [options] [file]
```

The *bash* has some commonly used built-in commands that are not required to call another program for its execution. However, all built-in commands are executed directly in the shell. Such built-in commands are dissimilar among the distinct shells. Some generally utilized built-in commands are as follows:

- **:** Null command or return an exit status of zero
- **.filename:** Load the file *filename* for execution
- **case:** Conditional loop to select a choice for multiple options
- **cd:** Change the present working directory
- **echo:** Display a string on screen or to print the value of a variable
- **eval:** Evaluate the given arguments and return back the result to shell
- **exec:** Execute a command that completely replaces the current shell process
- **exit:** Exit from shell
- **export:** Share the required environment variable with the subsequent child process
- **for:** A conditional loop
- **pwd:** Display the current working directory
- **read:** Read a line from a file or from standard input
- **set:** Set or update variables for shell
- **test:** Evaluate an expression for conditional statement as true or false
- **trap:** Indicate a signal trap, to be executed while receiving a certain signal
- **umask:** Set a default file permission, to be set for new files
- **unset:** Remove shell variables
- **wait:** Wait for a process to terminate or exit

3.6 Basics and Interpretive Cycle of Shells

Irrespective of which Linux operating system you are working, you always find a common feature, i.e., shell. Generally, all Linux distributions use GNU bash as the default shell. It gives a method to create shell script files, execute programs, compile programming code, work with various file systems, supervise computer resources, etc. Even though the shell is not as much of interactive as compared to graphical user interfaces (GUIs), a majority of Linux users, as developers, believe that the shell environment is much more powerful than a graphical environment. Currently, a rich number of Linux shells exist, and many advanced characteristics have been incorporated regularly into them to enable a better interaction with the system.

After successful login, you will see a prompt that is ready to receive a command for execution. You must provide commands to the shell prompt in a defined syntax/format, so that it understands and recognizes all parameters before execution. Every shell command comprises three main parts, namely command name, command options (if required), and command arguments (if required). All three parts are separated with a white/blank space. The basic operations performed by the shell, as shown in Figure 3.1, in its interpretive cycle are given below:

- **Accepting a command:** The shell presents a prompt (# or $, in bash) and waits for user input, as command.
- **Interpreting a command:** After a command is submitted, the shell checks the whole command line for metacharacters and expands abbreviations to reconstruct a simplified command line.

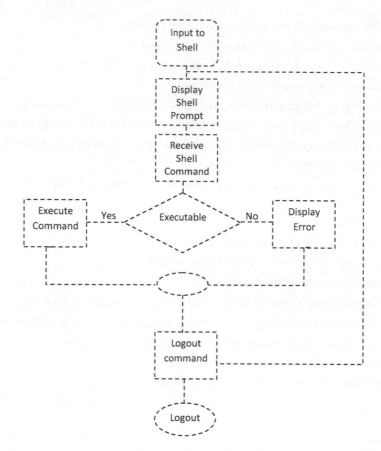

FIGURE 3.1
Shell's interpretive cycle.

- **Executing a command**: The shell submits the simplified command line to the kernel for execution.
- **Giving output**: The shell stays till command execution and gets the output from kernel to display on screen. Usually, the shell may not do any work during this process.
- **Waiting for the next command**: After displaying the output, shell resurfaces the prompt character (# or %) and waits for the next command input.

3.7 Starting a Terminal Shell

There are many methods to acquire a shell interface in the Linux system. Almost all Linux distributions provide an interactive access to a system via GNU bash. The most common ways to interface with a system are shell prompt, terminal window, and virtual terminal (remote access). We have already seen how to interface and start a shell prompt in CLI mode. On the other hand, if you are using a GUI, then you may execute a terminal emulator program to start a shell by opening a window-based terminal, commonly referred to as terminal shell.

Most Linux distributions provide the required feature to start a shell in GUI mode. The basic procedure to start a terminal window shell from a Linux desktop is given below:

- **From desktop**: On GUI desktop, click the right button of mouse device to open context menu, and see for shells, terminal window, or some other similar item, and after finding the appropriate item on context menu, then select it to begin shell. In Fedora Linux, click the right button on desktop to open context menu and click Open Terminal to start the shell.

- **Starting through the menu panel**: In many Linux distributions, GUI desktop comprises an information panel at the left side base of the desktop screen from where you can start any application. For a GNOME desktop system, you can click on the menu panel, select applications article, then select accessories, and then click terminal article to open a terminal window shell, but this procedure may vary among various Linux distributions.

3.8 Shell Variables: User-Defined and Predefined

Every shell has its own syntax and semantics for writing shell scripts. The shell script may get input data from a file or given as input by the user. With the help of shell script, you may create your own command resulting in a simplified workflow and an efficient use of time. Similar to other programming languages, bash also has the concept of variables, symbolic names, a portion of storage unit to which we can assign values and read data, and other operations for further processing. Generally, program data is kept in main memory (RAM) which is partitioned into smaller locations, and each location has a unique number called memory location/memory address. These memory addresses are utilized to hold program data. The coder/programmer may provide a unique name to these addresses, referred to as variable or memory variable. Such features permit you to store data under the defined variable name in memory that can be easily accessed through the same variable name by any other program or script running from the shell.

The subsequent section introduces the use of variables in shell and briefs the syntax of setting and reading various types of shell variables. The shell syntax is fairly exclusive in the note of using shell variables.

In shells, basically the variables are categorized into two types:

1. Global variables
2. Local variables

3.8.1 Global Variables

Global variables, also known as *environment variables*, are accessible and noticeable in all active shell sessions including various subshells of each shell. The environment variables are very useful in running applications/processes that make child processes which sometimes require some information from the parent process [2]. When you start your shell session, Linux system sets numerous global environment variables. To display all the available global environment variables, use the **env** or **printenv** command on shell prompt.

```
$ printenv          // Display the global environment variables
```

Basically, there are two ways to provide various environment variables: the first offered by system, i.e., inbuilt environment variables; and the second created by the user. The inbuilt environment variables are always offered in capital letters to distinguish them from user-created environment variables. Table 3.1 shows some common environment variables that are arranged by the system during the login process to start a shell.

TABLE 3.1

List of Common Shell Environment Variables

Variable	Description
BASH	Display the complete directory path of the bash command, which is usually */bin/bash*.
BASH_VERSION	Display the bash command version.
EUID	Effective user ID number of the current user that is allocated during the start of the shell, based on the user's entry in the */etc/passwd* file.
HISTFILE	The location of your history file.
HISTFILESIZE	Display history entries' size/limit on memory. When history entries reach the maximum number limit, then it starts discarding the history entry on an FIFO manner from the memory.
HISTCMD	Print the history number of the current command in the history list.
HOME	User home directory, after every login, a user logged into this directory and on that moment, it is current working directory of user. By simply typing the *cd* command, you can return from any place, in the file system, to your home directory.
HOSTTYPE	Print system processor/hardware architecture on which the Linux system is running, such as i686 and x86_64.
MAIL	Display the complete directory path of your mailbox file, i.e., */var/spool/mail/user_name*. The filename is usually assigned as your username.
OLDPWD	Print the old working directory before swapping to the present working directory.
OSTYPE	Display the name of the present operating system.
PATH	It is a list of colon-separated directories path of various commands. While going to execute a command on command line, if the path, absolute or relative, of your executing command is not listed in PATH, then you need to add the path of your executing command to the PATH variable before execution; otherwise, the command will not execute and print the message "command not found." For a user, the value of PATH is shown as /usr/local/bin:/bin:/usr/bin:/usr/local/sbin:/usr/sbin.
PPID	Display the parent process ID of the current executing process or command.
PSI	Using PS1, you can set up the value of your shell prompt which is used as the primary prompt or default prompt of Linux. This default prompt will appear every time in console after successful login by a user. The default value is [\u@\h:\w]\$, which displays username, hostname, the current working directory, and user privilege. Further, you can also use other available shell variables such as PS2, PS3, and PS4 to set up/change command prompt with more features such as color, display, and date time. PS stands for "Prompt Statement."
PWD	It stores the path of your current working directory. The value of PWD will be updated every time you switch from your current working directory to another directory by using the cd command.
RANDOM	Generate and print a random number between 0 and 99999.
SHELL	Complete the pathname to the shell that you are using.

(Continued)

TABLE 3.1 (*Continued*)

List of Common Shell Environment Variables

Variable	Description
SHLVL	After login into Linux, the bash command starts, and opens a new shell. The SHLVL environment variable stores the total number of shell levels that are linked with the present shell session, and increases the SHLVL value by one every time a new shell is started. After login, when you start the shell, then the initial value of SHLVL is 1.
TERM	Terminal type.
TMOUT	In bash, the TMOUT variable is used to automatically log out the logged users after a certain period of inactivity. This certain period is defined as N seconds, which is given as follows: *export TMOUT=N* // Set TMOUT to N seconds; it will terminate the shell if there is no activity for N seconds. For example, terminate the shell if there is no activity for 4 minutes. 　*export TMOUT=240* *export TMOUT=0* // To disable auto-logout, just set the TMOUT to zero. *unset TMOUT* // Reset the auto-logout This is one of the important security features of Linux that avoid the access of inactive shells by unauthorized people.
TMPDIR	The name of the directory where the bash makes temporary files.
USER	The name of the user; i.e., print the username of the current user
UID	Every username is assigned with a user identification number, referred to as UID, which is kept in the */etc/password* file.

By using the **printenv** command, you can display an individual environment variable's value, and the **env** command cannot be used for this:

```
$ printenv HOME
/home/Sanrit

$ env HOME
env: HOME: No such file or directory
$
```

The **echo** command can also be used to display the value of a variable. Now in the case of environment variable, you must put a dollar sign ($) before the environment variable name:

```
$ echo $HOME          // Parent bash
/home/Sanrit
```

As stated earlier, global environment variables are also visible and available to all child processes running under the same shell session. By using the bash command, you can always create a new subshell in the same shell session. For example, you can start a new subshell by using the bash command and then display the value of the HOME environment variable. You can find that it is the same as it was in the parent bash. Since the value of the HOME environment variable is set by the system, the value is the same in all shells and subshells including the main shell.

```
$ bash                // Start a new sub-bash
$ echo $HOME          // Display the value of HOME
/home/Sanrit          // The same value as in the parent bash
$
```

3.8.2 Local Variables

Local variables are not visible and available in all shells, and they are visible and accessible only in the current shell in which it is defined.

These variables may also be assumed to be user-defined variables. The built-in command **set**, without any option, is used to display all variables including user-defined variables, local environment variables, and global environment variables. No such specific command is available to display only the local variables.

```
$ set                      // Display all variables
BASH=/bin/bash
BASH_ARGC=()
BASH_ARGV=()
[...]
BASH_ALIASES=()
BASH_ARGC=()
[...]
HOME=/home/sanrit
[...]
my_variable='Hello My World'
[...]

$
```

The **set** command also sorts all the variables alphabetically and displays them. The **env** and **printenv** commands are used to display only global environment variables which are in unsorted form.

3.8.2.1 Variable "Type"

Every programming language has its own conventions and rules regarding the declaration of "type" of variables, whether it is integer, float, Boolean, character, string, array, or something else. In some languages, "data type" declarations are very strictly typed, which means that they will not allow a float to be compared with an integer, character, string, etc. They even discourage the reinterpretation of representations, while some languages give more relaxation in typecast and allow performing a number of implicit conversions between data types.

In the shell, there is no concept of a "type" or "data type" declaration of variables at all. But the shell treats everything as a string type. Another important point regarding shell variables is that, unlike other programming languages, there is no need to clearly declare them before using them. The shell will not show any errors for referring to an undefined variable. Such a variable is almost equivalent to a variable which contains the null string. So when you define a variable in the shell, it declares the variable at the same time.

3.8.2.2 Creating and Setting User-Defined Variables: =, $, set(export), unset

Apart from the predefined shell variables, you may also create your own variables immediately from the bash and avoid those variables that are already in use. Variables are case-sensitive, but in the case of an environment variable or a global variable, it is capitalized by default.

But generally, local variables are declared in lowercase. However, you are free to use any names, as desired. A variable's name may be defined with any set of alphabetic characters, including the underscore. Variable names may also contain a number, but the first character should not be a digit. You can use both upper- and lowercases or a mixed case to define a variable name. It is worth mentioning here that some characters such as an exclamation mark (!), an ampersand (&), or even a space may not be included while creating a variable name. Such symbols are reserved and used by the shell only. By using the bash help page, you can see the list of variables that are already in use.

```
$ man bash
```

After starting the bash, you can create your own user-defined variables, also referred to as local variables that are visible only within the current shell session. The created variable may be assigned a string, an integer, a constant, or an array data value using the assignment operator (=). While doing so, ensure that there are no blank spaces on either side of the assignment operator; otherwise, it may produce errors.

You can give any set of characters as value to a variable. In the following examples, the variable *my_variable* is assigned various values as shown below:

```
$ my_variable1=Hello
$ my_variable2="45"
$ my_variable3="Hello, my name is sanvik"
```

In the above example, we created variables with the names *my_variable1*, *my_variable2*, and *my_variable3* and assigned values respectively. If the value is containing any blank space, then to avoid the chance of any error, we may put quote on the either side of assigning values and declare the assigned value as string.

After assigning a value to the variable, then you can retrieve or print the value by using variable name. Generally, the command uses the values of variables as arguments. You can use the $ operator along with variable name to print its assigned value. The echo command is used to display the content of a variable. Wherever a $ is placed before the variable name, the variable name is replaced with the value of that variable. For example:

```
$ echo $my_variable1
Hello
$ echo $my_variable2
45
$ echo $my_variable3
Hello, my name is sanvik

$ my_variable=Hello World  // Putting quote on the either side of
   assigning string values is mandatory.
-bash: World: command not found
                      $ my_variable="Hello World"
$ echo $my_variable
Hello World

$My_Name="Ritvik"
$ echo "What is your name : Answer- My name is $My_Name"
What is your name : Answer- My name is Ritvik

$ my_variable2 ="45"        // Space is not allowed on the left side of
equal sign (=)
   bash: my_variable2: command not found
```

By default, all the variables that are created by users are set as local environment variables. They are available and accessible only for current shell processes and not for any other child or subshell processes, until you convert the user-defined variables into global environment variables.

```
$ echo $my_variable3              // Working in the current shell
     Hello, my name is sanvik


$ bash                 // Create a child shell
$ echo $my_variable3   // Not working in the child shell as it is a
                          local variable of the previous shell.
  bash: my_variable3: command not found


$
```

3.8.2.3 *How to Set a Local variable into a Global Environment Variable*

Global environment variables are accessible and visible to all active shells and subshells. The conversion of a local variable into a global variable is known as exporting, and the *export* built-in command is used for it. Setting and exporting a local variable into global variable is generally performed in one step, as shown in the below example:

```
$ export Local_variable_name="value"
```

For example:

```
$ echo $my_variable3       // Working in the current/parent shell.
Hello, my name is sanvik
$ export my_variable3      // Make it as an environment variable.
$ bash                     // Create a child shell.
$ echo $my_variable3       // Working in child shell also.
Hello, my name is sanvik
$ exit                     // Back to the parent shell
$
```

A subshell can change the value of a variable that will be visible only for the current shell. But the change made by the child shell does not affect the values set in the parent shell. This is verified in the below example:

```
$ my_status="I am globally known"// Value set in parent shell
$ export my_variable             // Make environment variable
$
$ echo $my_status
I am globally known
$ bash                     // Create a subshell
$ echo $my_status          // Display the same value in the subshell
I am globally known
$
$ my_status="My name id 25"       // Change the variable value in the
subshell
$
$ echo $my_variable // Update the value of the variable in the current
shell
```

```
My name id 25
$
$ exit // Exit from the subshell and enter into the parent shell
$ echo $my_variable    // Display the value in the parent shell.
I am globally known
$
```

3.8.2.4 *How to Unset a Local Variable from a Global Environment Variable*

If you can set any user-defined variable as a new environment variable, then you can also unset/remove an existing environment variable and convert it into its original form, i.e., local variable. The following steps can be used to unset an environment variable by using the *unset* command:

```
$test="this is my first script"   // Assign value to variable
$echo $test                        // Print variable value
   this is my first script
$export test                       // Make environment variable
$bash                              // Create a subshell
$echo $test    // Display test variable in child shell
   this is my first script
$unset test   // Remove or unset environment variable

$echo $test   // Show no value of test variable in child shell

$
```

When you are using the *unset* command to convert/unset an environment variable, then remember not to place the dollar ($) sign before environment name, as mentioned in the above example. If you are going to reset a particular global environment variable into local variable within a child shell, then it is applicable to only that child shell. The global environment variable is still accessible and available in the parent shell as well in all other subshells, except the shell in which it was reset.

3.8.2.5 *How to Set the PATH Environment Variable*

The PATH environment variable plays a key role to search and execute commands as well as programs of Linux system. The PATH environment variable is also referred to as Command Home Location that comprises a sequence of directory paths of various commands and executable programs, and each one is separated by a colon (:) sign. When a user enters a command on the command line, then the PATH variable informs the shell where to search command for execution. If the shell is unable to locate the entered command, as mentioned in the PATH variable, then it displays an error message, as shown below:

```
$ hello_prog
-bash: hello_prog: command not found
$
```

It is also to be mentioned that if the directory path of the entered command is not mentioned in the PATH variable, then it also gives the same error message as above.

Generally, the above error message associated with *hello_prog* is shown because the directory path of executable program of *hello_prog* is not mentioned in the PATH environment variable. Such condition may happen in many cases, mostly related to user-generated program in Linux system. Therefore, you need to make sure that the PATH environment variable contains the absolute path of all directories, where the executing command or application programs are kept.

As mentioned earlier, the PATH variable comprises various individual directories' paths that are separated by a colon. If the directory path of your executing program is not included in PATH variable, then before executing program, you may simply add the directory path of concerned program into PATH variable string. Let's assume that the above *hello_prog* resides in the /home/rit/scripts directory. You can add this directory to the PATH variable string to execute *hello_prog* by using the following steps:

```
$ echo $PATH

    /usr/local/sbin:/usr/local/bin:/usr/sbin:/usr/bin:/sbin:/bin:/usr/
src/games:/usr/local/rit:/home/abc

$ PATH=$PATH:/home/rit/scripts     // Add to the PATH environment variable

$ echo $PATH
/usr/local/sbin:/usr/local/bin:/usr/sbin:/usr/bin:/sbin:/bin:/usr/src/
games:/usr/local/rit:/home/abc:/home/rit/scripts

$ hello_prog
This is my first shell script program
$
```

After the addition of directory path of *hello_prog* to the PATH environment variable, you may execute *hello_prog* from any part of the file system hierarchy or structure, as shown below:

```
[Linux@Sanvik ~]$ cd /usr/local/rit
[Linux@Sanvik rit]$ hello_prog
This is my first shell script program
[Linux@Sanvik rit]$cd /
[Linux@Sanvik /]$ hello_prog
This is my first shell script program
[Linux@Sanvik /]$
```

3.9 Various Shell Types

To know the name of the current login shell, use the following command:

```
$ echo $SHELL
/bin/bash
```

As per the above output, the current shell is bash. Along with bash, some other shells are also popular on Linux platform, such as *sh*, *bash*, *ksh*, *tcsh*, *csh*, and *zsh*. You can simply switch from one shell to another shell by typing a new shell name on the command line of the current shell. You can see the file */etc/shells* for a complete overview of known shells on Linux system. For example, if you want to work on csh, then simply type the command **csh** to invoke csh, which further leads to tcsh. The conventional shell prompt of csh is the % symbol. But in some Linux distributions, the shell prompt may retain the **$** symbol:

```
$ tcsh
%
% echo $SHELL
/bin/tcsh
```

All Linux users prefer to install and use one or more shells on their system. If you are familiar with the functionality and environment of one shell, then you can easily learn and understand any other shells by just taking the reference from its respective shell's man page, i.e., its help page (e.g., type **man bash** on shell prompt to see the complete information about bash). In the following sections, we are going to briefly describe some popular Linux shells.

3.9.1 Bourne Shell (sh)

The Bourne shell or **sh** is a popular and oldest shell, developed by Steve Bourne in 1979. Initially, it comprised few simple internal commands, but later, it came with rich features and utilities in the seventh edition of Unix and was adopted as a standard shell for other Unix-like operating systems. The key features of the sh have already been discussed in Section 3.5. If you want to switch from the current working shell to the sh, then simply type the *sh* command on shell prompt and press the Enter key to go in on sh, as given below:

```
$sh
$echo $SHELL
/bin/sh
```

3.9.2 Bourne Again Shell (bash)

The bash is another popular Unix shell written by Brian Fox and Chet Ramey as part of the GNU Project, i.e., open-source software. The standard features and compatibility are derived from the sh along with some more useful features from ksh and csh. One of the key features of bash development was that it ultimately adheres to the POSIX shell specification and is possibly the best shell to use. Due to this, bash is more popular among all Linux distributions and is easier to use from command line. As it is known, bash is an improved version of sh; therefore, all commands that work in sh environment also work in bash environment with some more functionality improvements. These include both programming and interactive uses such as command line editing, shell functions, aliases, job control, command history, integer arithmetic, and script initialization. You may run almost all shell scripts (written in sh) without modification in bash [3].

Since it is a GNU open-source software package, bash has been adopted and used as the default shell on most Linux distribution systems such as Red Hat Linux, Fedora, Debian, Ubuntu, and Mint. The bash manual page ($man bash) comprises all information and features of the shell along with the list of all built-in commands [4]. As the GNU Project, developers are working continuously for upgrading bash features and also protect and promote the freedom to use, study, copy, modify, and redistribute bash software packages.

Due to this, bash program incorporates various other features that other shells are not offering. To switch from the current shell to bash, simply type **bash** on shell prompt, as given below:

```
$bash
$echo $SHELL
/bin/bash
```

Some other popular shells available for Linux/Unix include the following:

- **csh or C shell**: The C shell (csh) was initially created by Bill Joy to run on most Berkeley Unix systems, i.e., BSD Unix. When Linux came into existence, this was offered as an alternative shell along with other present shells such as sh and ksh. The csh includes almost all essential commands of the sh but varies in terms of shell programming feature. The csh was created after the sh, and its improved version is widely used in many Linux distributions, known as tcsh, in which most of the commands are taken from csh.

- **tcsh or Tenex C shell**: It is an open-source and improved version of the Berkeley Unix C shell (csh). It comprises all common csh commands along with added features (such as new built-in commands, terminal management, and system variable), enhancing user accessibility and speed. Due to this, it is also referred to as the Turbo C shell. To see the latest package details, manuals, FAQ, and other resources, you may visit its official webpage.*

- **ksh or Korn shell**: It was developed by David Korn at AT&T Bell Labs in 1980 and is a superset of sh and csh along with POSIX shell standard specifications. Initially, it was offered as proprietary software, but later on, it is available as open-source software under AT&T Software Technology (AST) Open Source Software Collection. Due to this, many other open-source shell alternatives were developed, such as bash, pdksh, zsh, and mksh. For writing a program, ksh offers a complete and powerful programming language, including the feature of other scripting languages, such as awk, perl, icon, and tcl. It also supports complex programming characteristics such as associative arrays and floating-point arithmetic. To see the latest package details, manuals, FAQ, and other resources, you may visit its official webpage.†

- **zsh or Z shell**: It is another expanded version of sh with many enhancements by including the features of bash, ksh, and tcsh and adding many other unique functions. The first version of zsh was written by Paul Falstad in 1990 and provided as an open-source Unix shell. The key objective of zsh was to offer an advanced programming platform for shell developers. It provides rich features such as mathematical operations with the support of floating-point numbers, structured command sets, function creation, and automatic expansion on command line. In the first edition of Mac OS X systems, zsh was included as the default shell; however, presently, bash is utilized as the default shell in Mac OS. To see the latest package details, manuals, FAQ, and other resources, you may visit its official webpage.‡

---

* **tcsh**: www.tcsh.org.
† **ksh**: www.kornshell.com.
‡ **zsh**: www.zsh.org.

3.10 Command Execution

As described in the command format in Chapter 2, while entering the command on shell prompt, you may also provide command options and arguments as per the need of task execution. Before proceeding, it is necessary to highlight that many symbols and characters are treated or defined as special characters, also known as metacharacters, by each shell. It means that each shell interprets these special characters in unique ways during the execution of commands if the commands include such special characters. Therefore, special characters are used to perform some particular operations. Such symbols are given in Table 3.2.

The bash has many features that help us to customize the typed command line prompt for a particular task. You can scroll down all your previously executed commands with the keyboard arrow keys. If you fail to remember the name of the command or program that you need to execute to perform a particular task, then you may seek help from shell by using command line completion. You may simply type the first letter or some portion of a command and thereafter press the Tab key. Let us assume that you want to see the list of all commands that start with "l" (small L) character; then, simply type "l" and press the Tab key, and bash will answer by displaying all the available commands starting with the letter l, as given below:

# l <TAB>				
laser	less	locale	look	last
lesskey	lastb	let	locate	lpq
lrp	ln	lockfile	lpr	lndir
log	lp	lprm	login	lptest
logname	ls	local	logout	

TABLE 3.2

Special Characters

Character	Description
\	Escape character. It is used to remove the particular effect of a special character and use as a normal character in your operation, so you must "escape" its special feature with a backslash first.
/	Directory separator. It is used to separate directory names in pathnames, except the first "/" which represents the root directory, for example, /usr/src/linux.
.	Current directory.
..	Parent directory.
~	Show user's home directory.
*	Signifies 0 or more occurrence of a character or set of characters in a filename. For example, xyz*22 can represent the files xyz22, xyzSan22, xyzabcd22, etc.
?	Signify a single character in a filename. For example, file?.txt may represent file1.txt, file2.txt, and file3.txt, but not file33.txt.
[ ]	It is used to represent a range of values, e.g., [0–9] and [A–Z]. For example, file[0–3].txt represents the names file0.txt, file1.txt, file2.txt, and file3.txt, but not file22.txt
|	"Pipe." This is used for communication between two commands by redirecting the output of one command into another command as input, for example, ls -l | more—the output of the *ls –i* command is the input for the *more* command.
>, >>, <	Redirection operators

(Continued)

TABLE 3.2 (*Continued*)

Special Characters

Character	Description
;	It acts as a command separator that permits to execute multiple commands on a single line.
&&	It also acts as a command separator, but with the condition that the second command will execute if the first one executed without errors.
&	Command execution in background.

It is a very useful shell feature in case if you are unable to remember complicated command names. For detailed help of each command, you may use the *info, help,* or *man* command **[4]** as given below:

```
# help <command_name>  or  # info <command_name>  or  # man <command_name>
```

3.10.1 Sequence Commands

Suppose you wish to execute more than one command in bash and don't prefer to wait for the completion of current command before submitting the next one. Basically, there are two methods to address this problem. First, you type all such commands into a file and save it as a script file and execute the file, i.e., create a shell script and execute it. Second, you may just type all the commands in a sequence on command line, where each command is separated with a semicolon (;) character. It is a control operator or metacharacter of the bash. The basic syntax of the sequence command is given below:

```
$command_name1; command_name2; command_name3; ... command_nameN
```

Commands separated by a semicolon (;) character are executed sequentially, and the shell stays for each command execution before starting the next one in turn. The exit status of these sequence commands is depending upon the completion of the last command execution in the given sequence over the prompt. For example, the exit status of the below sequence commands is depending upon the completion of the *ls* command execution:

```
$ cal; date; ls
```

- **Running numerous commands altogether**: If more than one command are independent to each other, then no need to wait for the completion of each command execution before start the next one in sequence. You can execute all the commands at once by putting an ampersand (&) at the end of each command, as shown below:

  ```
  $command_1 & command_2 & command_3 & ...... command_N &
  ```

The basic function of the ampersand (&) metacharacter is to send the command for execution in background. All commands are executed independently of whether the first command is executed successfully or not.

Use "&&"in place of "&" to execute the second command when the first command is executed successfully.

```
For example:

$ ls & pwd    // Run pwd regardless of the exit status of ls.
$ ls && pwd   // pwd will be executed only if ls succeeded.
```

3.10.2 Grouped Commands

In the bash, there are two approaches to combine the list of commands to be executed as a unit. Once commands are grouped, redirection operators may be used to redirect the output of the whole command list.

- **Using parentheses (...):**
  If the command list is placed within the parentheses (...), then a subshell environment is created for each command. Each command from the list is executed in that subshell; hence, variable assignments of each command are not affected by other commands due to the creation of individual subshell feature of each command in this mechanism. The basic syntax is given as follows:

  ```
  (command; command; command;...)
  $ (command; command; command;) >> output_filename
                    // Redirect the output
  ```

- **Using curly brackets {...}:**
  If the list of commands is grouped and placed between curly brackets, then the commands are executed by the current shell. In this mechanism, no individual subshell is generated for each command. The semicolon or newline separates the commands in the list, as shown in the below syntax:

  ```
  {command; command; command;}
  $ (command; command; command;) >> output_filename
  ```

3.10.3 Chained Commands

Command chaining is a way of combining various commands so that each one can execute in sequence based on the mentioned operator that separates them. Such operators decide how the commands are to be executed. A list of frequently used operators and their functions are highlighted below:

- **Semicolon (;):** The subsequent commands will execute despite the exit status of the preceding command.
- **Logical AND (&&):** The command will execute only if the preceding command executes successfully.
- **Logical OR (||):** The subsequent commands will execute only if the preceding command fails.
- **Redirection (>, <, >>):** These are the redirection operators that are used to redirect the output of a command or a group of commands into a file.
- **Pipe (|):** This is used for communication between two commands; i.e., the output of the first command or the previous command acts as the input of the next command in the given list of chain commands.
- **Ampersand (&):** This dispatches the command in background during its execution.

Some other operators are also in use, but they are more conditional in behavior. However, they are very helpful while writing a long and complex set of commands. Due to this, such operators are more frequently found in shell scripts. Some are given below:

- **Concatenation (\\):** It is used to split the given command sets between various lines.
- **Logical NOT (!):** It is used to negate an expression within a command.
- **AND-OR (&& - ||):** It is used as a combination of AND and OR operators.

3.10.4 Condition Commands

The shell executes every command in a sequential manner. This approach is perfectly fine for situations where you want all of the commands to be processed in the predefined order. But you might want to execute a certain portion of commands on the basis of some logics or conditions. There are various structures to define the conditions that are used to control the various command executions in a shell script or over the command line, and they are known as conditional commands. Such commands are described in detail in Chapter 8.

Further, conditional commands permit you to repeat and have a logical flow control over some set of commands. It also allows you to alter the direction of execution flow on the basis of some condition or logic and to execute certain commands over the other set of commands. Bash provides quite a good number of conditional or structured commands that are categorized into two parts: conditional blocks and conditional loops.

- **Conditional blocks (if, if-then-else, test):**
  1. **If-then statement:**

     The most fundamental type of structured command is *if-then statement*, and the format of the same is given below:

     ```
     if command
     then
     commands
     fi
     ```

  2. **If-then-else statement:**

     The if-then-else statement offers another group of commands in conditional block structure as shown below:

     ```
     if command
     then
     commands
     else
     commands
     fi
     ```

  3. **Nested ifs statement:**

     To check for multiple conditions in your script, you may apply another version of *if-else* section, known as *elif*, instead of individual separate *if-then* statements for each condition. In the following structure format, *elif* continues an *else* section with another *if-then* statement:

     ```
     if command1
     then
     commands..
     ```

```
elif command2
then
commands..
elif command3
then
commands..
fi
```

4. **The test command**:

   The **test** command is used to provide a technique to test different conditions, and the syntax of the same is given below:

   ```
   test condition
   ```

The test command evaluates the series of various parameters and values that are provided in the form of condition. The *test* command may calculate three basic categories of conditions [1]:

a. *The test* command: for string comparison
b. The *test* command: for file status
c. The *test* command: for number comparison

a. **The *test* command: for string comparison**:

   The test command is used to perform comparisons between two string values. Table 3.3 lists some string comparison operations.

b. **The *test* command: for file status**:

   The test command is also used to check the status of files and directories of a Linux system. Table 3.4 lists some test command operations.

c. **The *test* command: for number comparison**:

   The test command also permits you to perform comparisons between two numbers as mentioned in Table 3.5.

5. **Compound condition testing**:

   You can use the following Boolean operators to combine more than one test condition and accordingly check them as one whole condition. There are three basic and widely used Boolean operators:

   - [*condition1*] && [*condition2*]: The whole condition is true only if both conditions are met with the true condition.

TABLE 3.3

String Comparison

Comparison	Description
string1= string2	Check if string1 is equal to string2
string1!= string2	Check if string1 is not equal to string2
string1> string2	Check if string1 is greater than string2
string1< string2	Check if string1 is less than string2
-n string1	Check if string1 has a length greater than zero
-z string1	Check if string1 has a length of zero

TABLE 3.4

File Comparison

Comparison	Description
-e filename	Check if the *filename* is available
-d filename	Check if the *filename* is available and it is a directory type
-f filename	Check if the *filename* is available and it is a file type
-s filename	Check if the filename is available and it is not an empty file.
-r filename	Check if the *filename* is available and it is a readable file.
-w filename	Check if the *filename* is available and it is a writable file.
-x filename	Check if the *filename* is available and it is an executable file.
filename_1 –nt filename_2	Check if filename_1 is newer than filename_2
filename1 –ot filename2	Check if filename_1 is older than filename_2

TABLE 3.5

Numeric Comparison

Comparison	Description
num1 –eq num2	Check if num1 is equal to num2
num1 –gt num2	Check if num1 is greater than num2
num1 –lt num2	Check if num1 is less equal to num2
num1 –ge num2	Check if num1 is greater than or equal to num2
num1 –le num2	Check if num1 is less than or equal to num2
num1 –ne num2	Check if num1 is not equal to num2

- [*condition1*] || [*condition2*]: The whole condition is true only if any one of the conditions is a true condition.
- ! [*condition1*]: The logical NOT condition reverses the return value of the condition.

- **Conditional loops (while, until, for):**
  - **while command**: It repeats the execution as long as its test condition is true (exit code is 0).
  - **until command**: It repeats the execution as long as its test condition is false (exit code is not 0).
  - **for ((expression; expression; expression))**: It starts the loop by evaluating the first arithmetic expression; repeats the loop as long as the second expression condition is successful; and at the end of each loop, evaluates the third arithmetic expression (Table 3.6).

3.11 Standard Input/Output Redirection

After typing the command on shell prompt, the shell will execute the command and display the result on the prompt. But on many occasions, you may want to store or save the output in a file instead of just displaying it on the output screen. The bash offers some operators that redirect the output of a command into a file which is stored at any location.

TABLE 3.6

Loop Control Structure

Loop control structure syntax: *while, until,* and *for*	
while command *do* command *done*	Repeat the execution of an action as long as its test command condition is true.
until command *do* command *done*	Repeat the execution as long as its test command condition is false.

for variable *in*
  list-values
do
  command
done
	Repeat the loop till variable operand is repeatedly assigned each argument from the list values.

This is known as output redirection. Redirection may also be used to take the input from a file to an executing command. This is known as input redirection. The metacharacters < and > are used for input and output redirection, respectively [5]. Such characters have particular meaning to the shell, and they are reserved for special conditions.

- **Output redirection:**

  It is one of the most basic and important types of redirection operator which sends or redirects the output of a command to a file. In bash, it is represented by using the ">" sign. The syntax of output redirection is given as follows:

  ```
  $command > outputfile
  For example-
  $cal > xyx // Output of the cal command stored in xyz file
  ```

If the output file, named xyz, already exists, then the redirect operator overwrites the existing file with the current output of the executing command. To avoid such situation or to append the output of the executing command with the existing content of the xyz file, then use the ">>" sign as the operator for redirection, as given in the following syntax:

```
$command >> outputfile  // Append the output of command with file data
```

```
Further, if you want to save or keep the output of several commands into
a single file, type all commands as a single line on command prompt, as
shown below. For example, the output of the cal, pwd, ls, time, and ps
commands will be stored in the file_output.txt file which is kept in the
/usr directory. $ (cal; pwd; ls; time; ps)  > /usr/file_output.txt
```

It is to be mentioned here that if you don't use parentheses ( ) to group all these commands together, then redirection applies only on the output of the last command; i.e., here the last command is ps, then the output of ps will redirect into the *file_output.txt* file (not the output of whole commands). Therefore, in such a case, group all commands in parentheses before the execution of all and their output redirection.

- **Input redirection:**

    The operation of input redirection operator is just the opposite of output redirection. In the input redirection, it takes the content of a file and redirects it to the mentioned command on command line as input. The symbol of input redirection is the "<" operator. The syntax of input redirection is given below:

```
$command < inputfile
```

For example, the hello.txt file takes as input redirection for the *wc* command:

```
$ wc < hello.txt
      2    11    60
$
```

3.12 Pipes

After input and output redirection, another powerful concept of connecting commands to each other during execution is pipe, represented by (|) metacharacter. The communication between commands through pipe offers a way to connect commands to give more detailed and filtered output [6]. As per the following piping syntax of commands, the output of command1 is the input of command2, the output of command2 is the input of command3, and so on.

```
$ command1 | command2 | command3 ........| commandN
```

For example, command line which comprises pipes to connect various commands:

```
$ cat /usr/src/XYZ | sort | more
```

Here, the output of *cat* has been passed directly to the input of *sort*. In the same way, the output of *sort* has been passed directly to the input of the *more* command and the output of the more command is the final output to be displayed on the screen. In this way, pipeline is formed, and the shell, not the commands, sets up this interconnection. As mentioned in the piping syntax, i.e., N, there is no limit on the number of commands in piping, but keep into the consideration of each shell's behavior on character limit of the command line.

3.13 tee Command

The *tee* command is known for its similarity to a plumbing structure which creates a "T" fitting conjunction with pipes and filters. It reads standard input and copies it to both standard output and one or more files, effectively duplicating its output [7]. With this command, you can split the output of a program and save the intermediate output data before the data is altered or processed by another command or program.

The syntax of the tee command is shown below:

$ ls –l | tee file.txt | more

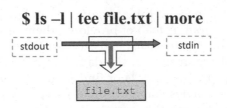

```
tee [ -a ] [ File ... ]

    File ...—a list of files, each of which receives the output.
    -a—append the output rather than overwriting into a file.
For example:

$ls -l | tee /usr/file1.txt file2 file3 | more
```

The output of the ls –l command will store in file1, file2, and file3 and act as standard input for the more command.

For example, if you are running three commands *cmd_1*, *cmd_2*, and *cmd_3* by using pipeline, as shown below:

```
$ cmd_1 | cmd_2 | cmd_3 > file_1
```

After execution, the final output of the pipeline is placed into *file_1*. However, if you are concerned to keep transitional output of each command in a separate file, then you can store it only by using the *tee* command as given below:

```
$ cmd_1 | tee file_cmd_1 | cmd_2 | tee file_cmd_2 | cmd_3 > file_1
```

After execution, the final output will be stored in file_1, but in this way, you can store the output of cmd_1 into the file_cmd_1 file, and subsequently, the output of cmd_2 can be stored in the file_cmd_2 file.

3.14 xargs Command

The **xargs** command may not be very popular with Linux user community, but the fact is that it's an extremely useful command when combined with other commands such as *find* and *grep*. It builds and executes command line from standard input [8]. It converts input from standard input into arguments to a command. The syntax of the xargs command is given as follows:

```
xargs [options] [command] [initial-arguments]
```

Few main options are as follows:

```
-n maxargs : Maximum number of additional arguments to maxargs for every
             request to the mentioned command.

-p: Interactive mode.  Intimation to user for every execution of command
```

xargs executes a *command* along with initial arguments and thereafter adds the remaining arguments from standard input instead of providing them directly. The **xargs** command passes all such arguments to the *command* and executes the *command* multiple times (each argument once) to use all arguments on standard input. The arguments are normally a long list of filenames (e.g., produced by **ls** or **find** or **grep**) that get passed into **xargs** via an **execution pipe (|)**. It separates standard input by spaces or blanks. In the following example, the standard input is piped to *xargs* and *mkdir* command execute for each argument to create four folders named dir_1, dir_2, dir_3, and dir_4:

```
$echo 'dir_1 dir_2 dir_3 dir_4' | xargs mkdir
// Create four separate directories in the current working directory

$find *.txt | xargs cat
```

In the second given example, the *find* command searches the entire current directory and prints all files having the .txt extension. This output acts as a standard input to the *xargs* command, and each file is passed as an argument to the cat command in order to display the contents of each file respectively.

3.15 Backslash (\) and Quotes

As mentioned in Table 3.2, the following characters have special meaning in bash and are also known as metacharacters:

```
\     '     "     `     <     >     [     ]     ?     |     ;     #     $     ^     &
*     (     )     !     {     }     /
```

But sometimes you need to use metacharacters as normal characters on the command and don't want the shell to interpret them as metacharacters. There are three ways to avoid shell interpretation of metacharacters:

- Backslash (\)
- Single quotes(' ')
- Double quotes (" ")
  - **Escape the metacharacter with a backslash (\):** The backslash is used as the escaping character on command line that disables or escapes the special meaning of the subsequent metacharacters. For example, to remove the special meaning of wildcard, we need to provide a \ (backslash) before the wildcard. In the pattern *, the backslash (\) tells the shell that the asterisk must be treated and matched as a normal character instead of being interpreted as a metacharacter. Therefore, when two or more special characters appear together, you need to put a backslash just before each special character; for example, if you want to remove the special meaning of the * character of ***$***, then you must escape each * as follows: **\$*.

```
$ rm ant\*        /// Remove the ant* filename; it does not remove
anything attached with ant such as ant1, ant2, antXY, and anta.

$ echo \$450     // Remove the special meaning of $ and print $450

$ mv day\&nightfilename whole_dayfilename    // Disable the special
meaning of &
```

- **Removing the space with \\:** To remove the space between two words or after the word or before the word, \\ (backslash space) is used. For example, for removing the effect of space in the file named "my home.txt":

```
$ cat my\ home.txt    // Ignore the space character and treat my
home.txt as one filename and print its content.
```

Removing the \\ itself: Similarly to remove "\\" itself, we have to use \\\\ (double backslash, i.e., two times backslash).

```
        $ echo \\
          \                // print only this \.

        $ echo The newline character is \\n
        The newline character is \n
```

But sometimes the use of escape is very difficult, for example:

```
        $ echo san > | < and % is a good boy
```

In the above example, we need to use four "\\" in front of each of these four metacharacters to disable their special meaning. For such conditions, quotes are used to easily remove the effect of metacharacters.

- **Use single quotes or apostrophes (' ') around a string:** Single quotes treat all characters as normal characters except the backslash (\\). It means single quotes disable the special meaning of metacharacters inside the quotes. But the parameter expansion, arithmetic expansion, and command substitution are not carried out within single quotes; however, it is possible with double quotes, for example:

```
    $ echo '$USER $((5+4)) $(cal)'
    $USER $((5+4)) $(cal)    // Output with single quotes
```

- **Use double quotes or quotation marks (" "):** Double quotes disable the special meaning of metacharacters inside the quotes and protect all characters except the backslash (\\), dollar sign ($), and back tick (`). It means that word splitting, pathname expansion, tilde expansion, and brace expansion are curbed, but parameter expansion, arithmetic expansion, and command substitution still take place. Using double quotes, we can handle filenames comprising spaces. For example, call a file with the name *one two words.txt*. If you use this file on the command line, the shell will split the filename into three words which would be considered as three separate arguments rather than the one anticipated single argument.

```
$ ls -l one two words.txt
ls: cannot access one: No such file or directory
ls: cannot access two: No such file or directory
ls: cannot access words.txt: No such file or directory
```

But, with the use of double quotes, we can prevent the word splitting and get the anticipated result as follows:

```
$ ls -l "one two words.txt"
-rw-rw-r-- 2 sanrit sanrit 14 2018-05-20 15:13 one two words.txt
```

As mentioned, the parameter expansion, arithmetic expansion, and command substitution may be carried out within double quotes, for example:

```
$ echo "$USER $((5+4)) $(cal)"

sanrit 9 May 2018
Su Mo Tu We Th Fr Sa
       1  2  3  4  5
 6  7  8  9 10 11 12
13 14 15 16 17 18 19
20 21 22 23 24 25 26
27 28 29 30 31
```

3.16 Building Shell Commands

Linux is a completely user-oriented operating system. It is not necessary that you must be a good programmer to write your command to perform a particular task. After having familiarity with various commands and their features, you can make your own commands by using them. It is hoped that now you are comfortable with the process of command execution on CLI, i.e., type command on shell prompt, execute, and view its result. Suppose you need to execute multiple commands to perform a task; then, you need to type and execute all commands one by one on command line. This is not a very efficient way to execute a set of commands again and again. Instead, you may write and save all such commands into a text file and make it an executable file. This executable file is now treated as your new command. In a simple way, this command (newly created executable file) is purely the combination of multiple commands that you use frequently. Look at the following example. If you want to run four commands together to know some basic information, then traditionally, type each command on command line as given below:

```
$cal            // Display the current month calendar
$date           // The current date with time
$uname-a        // Complete system information
$whoami         // The name of the user who is using the system
```

Now you can make a new file, namely *info_file,* and type all the above commands into it and save. You may use any text editor for this purpose, such as vi/vim editor, as given below:

```
cat
date
uname-a
whoami
```

Info_file

After creating the file, namely *info_file,* then you add the execute permission to it by using the chmod command, as given below:

chmod +x info_file

Now, you are able to execute the file *info_file* on your shell that gives the same result.

sh info_file

Now, *info_file* is your new command and gives the combined result of four commands, as explained. This way, you can make your own commands which is certainly a hassle-free way to execute a set of commands in a simple one go, which reduce the chance of error. You may perform many more such tasks by using shell script files. The subsequent section gives a conceptual overview of shell scripts.

3.17 Shell Scripts

In bash, in addition to the execution of commands on command line, a programming feature may also be used to execute commands, known as shell scripting. In a simple way, a shell script is a file comprising a series of Linux commands to perform a specific task. As discussed earlier, each Linux shell is a powerful CLI along with a scripting language interpreter to the system. The shell reads each script file and takes out the commands sequentially as mentioned in the file and executes them accordingly. Therefore, whatever can be done on the command line can also be done through scripts, and vice versa.

Mostly, each shell offers many programming tools that may be utilized to create shell programs or scripts. Similar to the other programming languages, you may also describe variables in a script file and assign values to them. You may also provide a way where a user interactively enters a value to each variable during the execution of shell script. The shell also provides loop, conditional operators, and control structures that help you to take right decisions and command looping during the command's execution. You can also build arithmetic or comparison expressions to perform various operations. The following key steps are required to effectively make and run a shell script:

- **Create a script:** A shell script is a regular text file that requires a text editor to write and create it. The *vi/vim* or other text editors may be used to create shell scripts.
- **Provide execute permission to script:** After creation, you need to provide the execute permission to each shell script by using the *chmod* command.

- **Store in the PATH environment variable**: You store the script file in such a folder, so that the shell can find it easily for execution. The shell automatically searches certain directories to locate executable files when no explicit pathname is mentioned or provided. Generally, the */bin* directory is one of the directories that the system automatically searches. But all the directories listed under the PATH environment variable are searched automatically by the shell for execution. So, for ease, you will always place your shell script files in such directories or places that are listed in the PATH variable.

In the bash, the shell script file has *.sh* extension. If you are comfortable with the programming concept, then you might find that shell programming is simple to learn and write. Chapter 8 provides more details about the shell script concept and programming features.

3.18 Summary

A shell is a program and is also referred to as command language interpreter that acts as an interface between the user and Linux kernel. The shell program starts when a user logs in into the system and closes when a user logs out. The login prompt sign indicates the type of user: The # sign indicates a superuser login and the $ sign denotes a normal user login. In csh, the normal user prompt is indicated by the % sign. A shell is less interactive as compared to the GUI, but several users believe that the shell environment is much more powerful than the GUI. There are some basic operations performed by shell in its interpretive cycle of command execution, namely accepting the command, interpreting, executing, and giving the output of the command, and thereafter waiting for another command on command line for execution in the same way.

Global variables, also known as environment variables, are available and visible in all shells including all subshells. To display the global environment variables, use the **env** or **printenv** command on shell prompt. The **echo** command is used to display the value of a variable and others on output screen. The environment variables are accessible and visible in all active shells and subshells. By using the *export* command, users may convert any local variable into a global variable. The PATH environment variable is very important in the aspect of executing the command. If the path of the entered command is mentioned in the PATH variable, then the command will execute; otherwise, the system will display the error message *command not found*. Presently, a good number of shells are available; some popular shells are *sh, bash, ksh, tcsh, csh,* and *zsh*. The SHELL environment variable keeps the information of the installed shell in the system. Users can also execute more than one command by command grouping using parentheses (…).

Conditional commands permit you to repeat and have logical flow control over some set of commands. Standard output redirection helps to store the output of a command into a file. In the same way, the input redirection takes the input from a file for executing a command. The external *tee* command duplicates its standard input to both standard output and one or more files. The *tee* command helps to store the intermediate outputs into the files when you are executing a series of sequence commands. The *xargs* command converts the input from standard input into arguments to a command. The traditional way to execute a set of commands is to enter each command on command line and execute them sequentially. But thorough shell scripts, you can run the same set of commands interactively by writing all the commands in a file and executing the file. Such files are known as shell scripts. So, you need a text editor to write and create such shell script file. The subsequent chapter explains the features and usability of *vi/vim* editor along with covering the basic features of *emacs* and *gnome* editors.

3.19 Review Exercises

1. Discuss the role of the shell in Linux and why the shell is known as a command interpretive language?
2. What are the different types of commonly used shells on a typical Linux system?
3. Describe shell variable and shell interpretive cycle with an example? Why doesn't the shell perform type checking of variable?
4. Which command is used to print the environment variable? Highlight the key difference between global and local variables.
5. Write a command syntax to set multiple values of an array as environment variable and to display an entire array value.
6. How does the shell allow us to use any variable without declaring it? What is the value of an undeclared variable in shell?
7. Describe the key difference between the "printenv" and "env" commands.
8. Which file is used to save shell commands? How to unset a local variable from a global environment variable and what will be the effect after this conversion?
9. How can you add the path of a new program at the location "/home/sanrit/program" to the system PATH environment variable?
10. Write the series of commands to create the local variable VAR and assign it "this_is_variable" and make it an environment variable, and after that, remove the VAR environment variable.
11. Write a chain command in which you can create a directory called "new_directory" and check in the newly created directory.
12. If you execute the below command:

    ```
    # cat  file_1.txt
    Cat: Command not found      // error message
    ```

    Explain the various causes of this error message and possible solutions to remove it.
13. Explain the benefit of group commands and chained commands.
14. Highlight the key advantages of the C shell over the Bourne shell.
15. List down various metacharacters supported by bash with examples.
16. Which of the following is incorrect: var=32
    a. If [$var >= 32]
    b. If ($var >= 32)

17. What is the output of the following:
    a.   echo hello\n
    b.   echo "hello\n"
    c.   echo -e "hello\n"

18. Why is the tee command known as the plumbing command? List down some key applications.

19. Draw the tree structure of the following command:

```
ls  /  | tee file1.txt | tee file2.txt | more
```

20. How can you customize the shell prompt? Discuss. Which command helps to identify the current shell and how you customize it?

21. Write a command to check that a file exists and the existing file is a directory or executable.

22. How can you send the output of the "cal" command to a file named "caloutput"? Suppose the output file "caloutput" already exists. How can you overwrite or append output data with the existing file with the current output?

23. Explain the execution output of the following command line:

```
#   ls cat ps who >> file.txt
```

24. Define the purpose of input redirection with an example.

25. There are two files named file1.txt and file2.txt as given below:

```
Linux
Yourself
```
file1.txt

```
Linux Yourself
The Shell
```
file2.txt

Write the output of the following command operated on file1 and file2:

```
ls file* | tee output.txt | wc -l
wc -l file1.txt| tee file2.txt
cat file2.txt
```

26. Write the commands for the following:
    i.   Multiline output from the ls command into a single line using xargs
    ii. To generate a compact list of all Linux user accounts on the system using xargs.

27. What is the output of the following program [hint: ${#arr[*]} represents the length of the array]:

    ```
    arr=( text1 text2 text3)

    a)   i=0
         While [ $i -lt ${#arr[*]} ]
         do
         echo -n $i ->
         echo ${arr[$i]}
         i=$( expr $i + 1 )
         done
    b)   for i in ${!arr[*]}
         do
         echo -n $i ->
         echo ${arr[$i]}
         done
    ```

28. What is the difference between using if after an if block and using elif after an if block?

29. Execute the below command:

    ```
    $echo 'math result book' | xargs touch
    $find /bin/san *.cpp | xargs rm
    ```

30. How to print the following string in the same manner as it is provided with the echo command:

    ```
    $ echo san > | < and % is a **good $ student.
    ```

References

1. Richard Blum and Christine Bresnahan. 2015. *Bible- Linux Command Line & Shell Scripting.* John Wiley & Sons.
2. Richard Petersen. 2008. *Linux: The Complete Reference.* The McGraw-Hill, 1.
3. Carl Albing, JP Vossen, and Cameron Newham. 2007. *Bash Cookbook.* O'Reilly Media.
4. Michael Kerrisk, Linux man pages in the World Wide Web. http://man7.org/linux/man-pages/.
5. Matthias Kalle Dalheimer, Terry Dawson, Lar Kaufman, and Matt Welsh. 2003. *Running Linux.* O'Reilly Media.
6. Michael Kerrisk. 2010. *The Linux Programming Interface.* No Starch Press, Inc.
7. William E. Shotts, Jr. 2012. *The Linux Command line: A Complete Introduction.* No Starch Press, Inc.
8. J. Purcell. 1997. *Linux Complete: Command Reference.* Red Hat Software, Inc., 2.

4

vi Editor

Learning the Linux command line is becoming more popular day by day among all users. The shell itself is highly interactive by creating shell scripts to avoid repetitions. Unix/ Linux has several editors for creating your shell scripts, programs in C, C++, or other programming languages, and text files. Editors are like Notepad where you simply create files that contain data, source, or sentence. The *vi* and *vim* (improved version of *vi*) editors are perhaps the most popular pervasive text editors available in Linux systems. *vim* is involved in nearly every Linux distribution. Several other editors are also available, such as emacs, gedit, elvis, nvi, and vile. Broadly, there are two types of editors: line editors and screen editors. Line editors, such as *ed* and *ex*, show a line of the file on the output screen, but screen editors, such as *vi* and *emacs*, display a section of the file on the output screen. When you start learning and understanding the various commands and features of the vi editor, then you appreciate that vi is well designed and extremely efficient. Due to this feature, a system administrator or coder uses only limited keystrokes to instruct *vi* to do complex tasks of editing whether on text file or code documents.

4.1 Introduction

vi is the most suitable and widely used screen-oriented text editor, originally developed for Unix system. vi is the short form of *visual editor* and pronounced as "vee-eye." It was envisioned to allow editing with a moving cursor on a video terminal. Earlier visual editors, line editors, were used that operated on a single line of text at a time. Bill Joy created the original code of vi and released it with the BSD system. The original vi editor was not incorporated by most Linux distributions. They somewhat switched to an enhanced version of vi, referred to as vim (vi improved), written by Bram Moolenaar and first released in 1991.

This chapter describes the vi commands, various operation modes, how to insert and delete text, operation on regular expressions, command combinations, coding in the vi editor, how to get help, and other relevant information for using vi. The vi editor provides a rich feature to support programmers for creating and compiling the code [1].

4.1.1 Invoke vi

If your system does not have or installed vim, then refer to the installation help file of vim to install the vim editor. If you are running vim on Unix or Linux, then execute the following command to enter the vim editor:

```
[sanrit@Linux ~]$ vim
```

A screen similar to that shown in Figure 4.1 would be displayed.

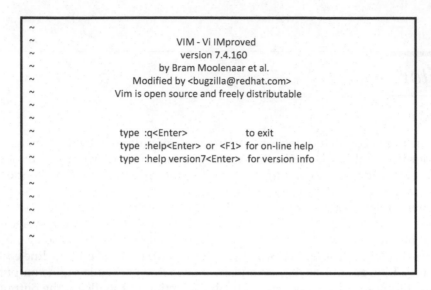

FIGURE 4.1
Home screen template of the vim editor.

Now you can perform various operations in the vi editor, such as file creation, adding data, editing data, and saving data in the respective file. After completing your tasks in vi, then learning how to exit is also important. To exit from vi, you enter the following command (first press the Esc key followed by colon character and then type q):

```
:q                    // Exit from vim
```

NOTE: To quit and exit without saving, then press the Esc key and then type the below-mentioned command to exit from vim. The user login shell prompt will be displayed again.

```
:q

[sanrit@Linux ~]$          // Login shell prompt display
```

When you type colon (:) followed by pressing the Esc key, the vim cursor will move to the lowest line of the editor screen. The command q! informs the *vim* editor to quit without saving the content of the current work.

If you want to start the vim editor, then simply type vim on terminal and press the Enter key. This will start the vim editor without opening any file.

```
[sanrit@Linux ~]$ vim              <Enter>
```

In case you would like to start the vim editor for editing a file, referred to as *file.txt*, then you would enter the following command:

```
[sanrit@Linux ~]$ vim file.txt
```

Because *file.txt* is a new file, you get a blank window. Figure 4.2 shows how your screen will display a blank file in the vim editor. The tilde (~) lines indicate that there is no content in the file. In other words, when vim runs with a new file, it displays tilde lines; otherwise, it will show the file content, and then by using the various modes of the vi editor, you can perform the various editing operations on the file. At the bottom of a screen, a message

```
█
~
~
~
~
~
~
~
~
"file.txt" [New File]
```

FIGURE 4.2
The vim editor opening a blank file (file.txt).

line indicates the name of the file as *"file.txt,"* and *[New File]* means this is a new file. If you open an existing file, then the message line indicates the name of the file along with the content of that file by including the total number of lines and characters. Such message line information is momentary and will disappear when you start typing the characters.

There are various ways to open a file in the *vim* editor by using the following command options (*vi* or *vim*) in shell prompt:

- *vi file.txt*: Start the vim editor with the given filename, *file.txt*
- *vi file1 file2*: Start the vim editor with the given files sequentially, *file1 file2*
- *view file*: Start vim on the given file in read-only mode
- *vi -R file*: Start vim on the given file in read-only mode
- *vi -r file*: Start vim to recover the given filename with recent edits after a crash
- *vi + file*: Start vim and open the given filename at the last line
- *vi +n file*: Start vim and open the given file directly at line number n

By default, *file.txt* will be stored in the current working directory of the user; otherwise, you give the full absolute path of the directory in Linux directory where you want to store the file. Similarly, you can open any file from any directory in a Linux system for editing, for which you can simply enter the full absolute pathname of that file. For example, if you wish to see the file content, namely *details file* which is kept in the /sanrit directory, then type the full absolute path of the *details file* as given below:

```
$ vim /home/sanrit/details
```

4.2 Modes in vi Editor

The vim editor is a modal editor that is different from other editors. It means that the vim editor works differently, dependent on which mode you are presently working. Generally, it operates in two modes: insert mode and command mode. In insert mode, whatever you type is stored as text in your file, whereas in command mode, every keystroke signifies a vi command. Invoke vi with the name of existing file or with a new filename or without filename, it begins with command mode where every pressed key

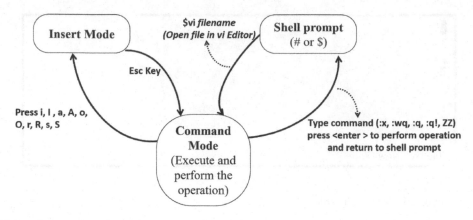

FIGURE 4.3
Various modes of operations in the vim editor.

act as a command [2,3]. In this chapter, except text editing, all the operations of the vim editor are performed in command mode including ex mode or last line mode. All the modes have their own features, and the way to switch between the modes is clearly described in Figure 4.3.

- **Command mode:** This mode permits you to perform various operations on files, such as executing commands, saving files, quitting files without saving, moving the cursor, cutting, yanking and pasting words or lines, and finding and replacing text or pattern. This mode takes all that you type as a command. When you press a key, then it will not be shown on the screen, but it may perhaps perform a function such as moving the cursor or searching a text. You cannot use the command mode to enter or replace text. In command mode, the last line is used for executing vim commands and displaying editor-generated messages. In command mode, it accepts all the commands of the vim editor and further shifts to the **ex mode** or **last line mode** (the third mode inside the command mode) by typing a *: (colon)*, which shown in the last line of the vim editor. For example, if you want to quit the vim editor, then enter the following command:

```
:q[Enter]      // Must be in command mode and then type colon (:)
               followed by the command and press the Enter key to
               execute the typed command
$              // Exit from the editor and return to shell prompt
```

- **Insert mode:** This mode enables you to enter text in your file. It means that whatever you type becomes input text in your file. To go into the insert mode, press the I or i key. After press <i> character, you will find, in the lower left side of the screen, with display INSERT that means you are in insert mode, as shown in Figure 4.4.

    After entering the insert mode, you enter the text into your file. Thereafter, if you want to perform any operation on file such as search, delete, replace, save as, or save, then you switch from the insert mode to the command mode by pressing the Esc key. Suppose you forget your present mode of working. Then, simply press the Esc key twice to know it; it will also take you back from the current mode to the command mode.

FIGURE 4.4
Notation of the insert mode in the vim editor.

4.3 Start, Edit, and Close Files

The vim/vi editor is widely to edit any text file. Basically, there are two ways to start the vim editor: first, without any specific file; and second, with the files.

```
$ vi                    // Start vim without any file
```

In the first way, invoke the vi editor without file as shown, this will create an empty file, in which you typed text as file content and save with new filename.

```
$ vi fruits.txt         // Start vim with file, namely fruits.txt
```

In the second way, it opens the vi editor with mentioning the filename *fruits.txt*. If the typed file exists, then the vi editor starts with the content of the file. If the typed file does not exist, then the vi editor creates a new file named *"fruits.txt"* in the current working directory and opens the editor. The open editors have an empty screen with tilde (~) characters displayed on the left side of the screen which indicate no content or text in the file, not even blank lines. The bottom last line of the editor's screen is known as prompt line or status line that displays the name and status of the file. The vi commands are case-sensitive in Unix/Linux; i.e., characters "a" and "A" are treated as different characters in the vim/vi environment. Hence, you should be careful while typing the name of the file or performing the file editing work in the input mode of the vi editor.

New files may also be created in other directories rather than the current working directory. For example, to create, to store, and thereafter to open a file named *"fruits.txt"* in the directory /home/sanrit/, then simply type the following:

```
$ vi /home/sanrit/fruits.txt     // Create and store in the
sanrit directory, and then open in the vi editor
```

The same method is used to open an existing file which is not residing in the current working directory: Simply type vi followed by the absolute path of the existing files. For example, to open an existing file named *"mango.txt"* from the directory */home /sanrit/local/ mango.txt*, then type the following:

```
$ vi /home /sanrit/local/mango.txt
```

4.3.1 Editing

After opening a file in the vi editor, then switch to the insert mode to perform the editing operation on the file. Once the file editing operation is over, you may exit from the inert mode by pressing the Esc key and enter the command mode; thereafter, you can perform various operations on your file as desired by using various vim/vi commands, as explained in the following section.

4.3.2 Saving Your Work and Quitting

At any time, you can stop the editing work on a file and quit the vi editor and return to the shell prompt. To save the typed text of a file or perform any other operation, you must switch from the insert mode to the command mode by pressing the Esc key and then type a : *(colon)* to enter the ex mode or last line mode (the third mode inside the command mode). After this, the colon (:) character should appear at the bottom of the vi screen, as given below:

```
:
```

During the editing operation in vi or any other editor, the content of the original file is not altered as such, but the copy of editing content is placed in a buffer (a temporary procedure of storage). The original file content will update only when you save your work from buffer to disk. If you want to save the modified file and continue the editing, type **:w** in command mode.

```
:w <Enter>
```

The file content will be written to the hard drive (on original file), with the confirmation note at the bottom of the vi editor screen, as given below:

```
"file1.txt" [New] 5L, 62C written
```

We perform the following frequently used operations with the buffer:

- Save and quit (type ZZ in caps without a :).
- Save and quit (:x and :wq).
- Discard all changes and quit (:q and :q!).

After quitting, vim returns control to the shell and displays shell prompt. Suppose that during your editing, you wish to ignore all the edits after the last saved session. Then, you can return to the previous saved version of the file by executing the following command:

```
:e! <enter>
```

When we refer to saving a file, it means saving the content of buffer on disk space. Some key commands that are used to save the file content and exit from vi are shown in Table 4.1.

4.3.3 Open and Recover File from a Crash (:recover and -r)

Due to some unavoidable causes, such as power anomaly and inappropriate shutdown of the system, you could not save the changes in a file. Then, there is no need to be panic, as

TABLE 4.1

List of Basic Opening and Closing vi Commands

Command	Description
:w	Save file and continue in editing mode
:w!	Write (save) file, overriding protection
:wq	Save file and exit from vi
:x	Save file and exit from vi
ZZ	Save file and exit from vi
:n1,n2w newfile1	Write from line n1 through line n2 as *newfile1*; for example, if n1=10 and n2=20, then write from line 10 to line 20 as *newfile1*
:n1,n2w>> file1	Write from line n1 through line n2 and thereafter append to file1
:10,25w>> file1	Write from line number 10 through line number 25 and append to file1
:w %.new	Write the existing buffer named *file* as *file.new* (*similar to* Save As in MS Word)
:.w xyz.txt	Write the current line to file *xyz.txt*
:$w xyz.txt	Write the last line to file *xyz.txt*
:q	Quit the file when no changes are made to the file
:q!	Quit the file without saving the changes, i.e., overriding protection
Q	Quit vi and invoke the execution mode
:e file2	Start editing *file2* without leaving vi
:n	Edit the next file without leaving vi
:e!	Open the last saved edited version of the current file (similar to *Revert in MS Word*)
:e #	Return to start editing the most recently edited file

vi stores most of its buffer information in a separated hidden swap file. In such a situation, this swap file remains on the disk and vi will recover the content that was being edited when the system crashed, by using the following command:

```
$ vi -r filename   // Recover file
```

Similarly, you can type **:recover** in the last line of the vi editor command mode to recover the file.

4.4 Various vi Commands

The subsequent sections explain you the basic commands of the vi editor that you must know to edit a file and perform various operations of the vi editor. Most editing comprises few elementary operations, such as text insertion, text deletion, and moving text during the cutting and pasting operations [4]. vi performs all these operations in its own exclusive way. Now you can try out some vi commands that make you convenient to work with the vi environment.

TABLE 4.2

List of **Cursor Moving Commands**

Command	Description
j *or* [down arrow key]	Cursor moves one line down from the current position *(like the Return key function)*
k or [up arrow key]	Cursor moves up one line from the current position
h *or* [left arrow key]	Cursor moves to the left one character from the current position *(like* the Backspace *key function)*
l *or* [right arrow key]	Cursor moves to the right one character from the current position *(like* the Spacebar *key function)*
0 (zero)	Cursor moves to the beginning of the current line
$	Cursor moves to the end of the current line
w	Cursor moves forward one word
b	Cursor moves back one word
:0<Return> *or* 1G	Cursor moves to the first line of the file
:n<Return> *or* nG	Cursor moves to line number n
:$<Return> *or* G	Cursor moves to the last line or up one line from the current position
G	Cursor moves to the end of the file

4.4.1 Moving the Cursor

After writing some text contents in the vi editor, you are able to move the blinking cursor across the displayed text on the vi editor, by using four basic commands. These commands allow you to move any preferred location to modify or edit the text, including insertion and deletion operations. Table 4.2 lists some basic cursor positioning commands.

The commands mentioned in Table 4.2 can be used individually or by adding an integer value. This value must be mentioned before the command that denotes the number of moves of the cursor in characters or words or lines. For example, type the below command in command mode:

```
4j          // Cursor moves down four lines from the current position.

5h          // Cursor moves to the left up to the fifth character from the
               current position.

6w          // Cursor moves forward to the beginning of the sixth word from
               the current position
```

Straightaway, you can move to any desired line by using the G command preceded by the line number. For example:

```
7G          // Cursor moves to the seventh line from the first line of text.
```

4.4.2 Inserting or Appending Text

You can edit any file in the insert mode of the vi editor. There are various ways to enter the insert mode of the vi editor. It is already described that by typing the i (insert) command, you can put vim in insert mode. But you can also follow other ways that allow you to enter the insert mode and add or delete text, such as the following:

- Insert (i and I)
- Append (a and A)
- Replace (r, R, s, and S)
- Open a line (o and O)

The simplest way to insert the new text within the existing text is that you just move the cursor to the place where you want to enter the text and then type the i (insert) command to put vim in input mode, and then enter the new text; thereafter, press the Esc key to return vim into the command mode.

Alternatively, you can also use the append command (a, A) to enter the text. The A command is more suitable because it places the cursor at the end of the line before entering into insert mode to add text. You can also use the open line command (o, O) to enter the insert mode to add text. The o command inserts a blank line just below the cursor's current line before entering the insert mode. The O command works in the same way "o" works, except it opens a blank line above the current line of the cursor. When you finish the editing work, press the Esc key to return to the command mode. Table 4.3 shows various commands to enter the insert mode of the vi editor.

Just keep in mind that after you finish the text entry using any of the mentioned commands (except r), you must return to the command mode by pressing the Esc key.

4.4.3 Replacing Text

Along with the text insertion and deletion operations, you may also perform the replacement operation to change the existing text with a new one. To perform the replacement operation, vi offers various replacement commands that are able to replace several lines as well as a single character of the text (Table 4.4).

4.4.4 Undoing Mistakes

If you make a mistake and delete a character, line, or word, then simply use the *u (undo)* command to reverse the mistake. You can use this command many times to reverse your

TABLE 4.3

List of Editing Commands

Inserting-cum-Editing Commands	
Insert Command	**Description**
i	Insert text before the cursor place
I	Insert text at the beginning of the line
a	Append text after the cursor place
A	Append text at the end of the current line
o	Open a newline for inserting text just below the cursor place
O	Open a newline for inserting text just above the cursor place

TABLE 4.4

List of Replacing Commands

Command	Description
r	It is applied to replace a single character, and the editor remains in command mode. For example, press rj, which replaces the current cursor location with the character "j."
R	Replace characters, starting from the present cursor location, until the Esc key is pressed.
s	Replace the current text and enter the insert mode.
S	Replace the entire line irrespective of the cursor position; the existing line completely vanishes.
cw	Replace and change the word from the current position of the cursor.
c motion	Change text from the current cursor location to the location mentioned by *motion*.
cNw	Change N words, starting with the character under cursor, until the Esc key is pressed. For example, c5w changes five words from the current cursor location.
cG	Delete the line from the cursor position to the end of the line and enter the insert mode.
cc	Change or replace the entire current line.
Ncc *or* cNc	Change (replace) the next N lines, beginning with the current cursor line, until the Esc key is pressed.

actions; however, vim can undo only the most recent change made by you. In the same way, if you undo a command and now you don't want to undo, then simply execute the *:redo* command. The *:redo* command redoes a change like CtrlR does. You can also execute the redo command many times in a row like the undo command. Table 4.5 highlights various commands to reverse or repeat your actions.

4.4.5 Screen Navigation

Table 4.6 lists the commands that allow you to navigate the vim editor screen (or window) to move up or down.

4.4.6 Deleting Text

The vi editor provides a range of options to delete text during file editing. By using the "x" command, you can delete a specific character by placing the cursor over it. The

TABLE 4.5

List of Undo and Repeat Commands

Command	Description
.	Repeat the last edit command
u	Undo the last edit command
U	Undo all recent alterations to the current line
:redo	Do the reverse of the undo command
J	Join two lines

TABLE 4.6

List of Screen Navigation Commands

Command	Description
Ctrl+f	Move ahead one screen
Ctrl+b	Shift backward one screen
Ctrl+d	Shift down (forward) one half–screen
Ctrl+u	Move up (back) one half–screen
M	Shift to the middle line on screen
H	Shift the cursor at the top line of the current screen (nH: move to n lines below the top line)
L	Shift the cursor at the bottom line of the current screen (nL: move to n lines above the last line)

command "x" may be started with a number, specifying the total number of characters that are to be deleted from the current location of cursor. You can delete a word by putting the cursor at the first letter of word that you want to delete, and then type the command dw (delete word). Table 4.7 lists some deletion commands with their description.

TABLE 4.7

List of Delete Commands

Command	Description
x	Delete the present cursor position character
nx	Delete n characters starting from the present character
X or (Shift-x)	Delete the character before the cursor location
nX	Delete the previous n characters to the left of the current position
dw	Delete from the current cursor location word
dd	Delete the current line
Ndd	Delete N lines from the current line (e.g., 3dd will delete three lines from the current line)
d motion	Delete the text between the cursor and the target of motion
d$	Delete from the cursor current position to the end of the current line
d0	Delete from the current position to the beginning of the line
Shift-d or D	Delete everything from the cursor to the end of the line
dG	Delete from the current line to the end of the file
dnG	Delete from the current line to the nth line of the file
u	Undo the last line deletion
Shift-u	Undo all changes in the current line
ndw	Delete the next n words starting from the current word
ndb	Delete the previous n words starting from the current word
ndd	Delete n lines from the cursor current line
:n,md	Delete all lines from the nth line to the mth line; for example, the :3,6d command will delete the third, fourth, fifth, and sixth lines of the file.
"np	Retrieve the last nth deletion

4.4.7 Cutting, Pasting, and Copying Text

The vi editor provides various commands for the cut, paste, and copy operations. The "d" command not only deletes text, but also performs the cut operation. When we use the "d" command for deletion, the deleted text is copied and stored in buffer. The text can be recalled from buffer and pasted with the "p" command at the desired location in the file. To paste the text after the cursor position, use the "**p**" command. Similarly, to paste the text before the cursor position, use the "**P**" command.

In vim/vi editor, the copy operation is referred to as yank (y). Therefore, the "copy and paste" operation is referred to as "yank and put"; similarly, "cut and paste" is referred to as "delete and put." The "y" command is used to "yank" (copy) text; in the same way, the d command is used to cut or delete text. To copy one line of text, use the **yy** or **Y** command. Note that Y does the same thing as yy. You may also yank (copy) several lines of text by using the **nyy** command, where n is the number of lines, counting from the existing cursor place. The cut–paste operation is comparable to moving text from one location to another. You can also move text with the "**m**" command, which moves one line of text from the existing cursor position to a new position. Table 4.8 lists some commands for the copy, paste, and move operations.

TABLE 4.8

List of Some Commands for the Copy, Paste, and Move Operations

Command	Description
yy or Y	Copy the entire line of the current cursor.
yw	Yank one word at the cursor position in forward direction; for example, 4yw is used to copy four words forward including the cursor position word.
Nyw or yNw	Copy N words from the current cursor position; for example, 4yw or y4w is used to copy four words forward from the current position.
y *motion*	Copy text from the current cursor location to the location mentioned by *motion*.
yG	Copy text from the cursor location to the end of the file.
yNG or NyG	Copy text from the current line to the nth line from the beginning of the file. y3G or 3yG: from the current line to the third line of the file y1G or 1yG: from the current line to the beginning line of the file.
nyy	Copy n lines starting from the current line position. For example, 10yy means the 10th line will be yanked counting down from the current cursor position from the given line. vi will give the message 10 lines yanked.
y$	Copy text from the current cursor position to the end of the current line.
y0	Copy text from cursor position to the starting of the current line.
:m n	Move the current line and paste it after line number n; for example, :m 2 will move the current line where the cursor is pointed after line no. 2.
:a,bmn	Move (cutting) lines from a to b and paste them after line number n; for example, :2,4m5 will move the second, third, and fourth lines and paste them after line no. 5.
:a,btn	Copy lines from a to b and paste them after the nth line from the current cursor position; for example, :2,4t5 will copy the second, third, and fourth lines and paste them after the fifth line.
p	Place the copied (yanked) text after the cursor position.
P	Place the yanked text before the cursor position.

4.5 Global Replacement

The previous section discussed how to replace certain text or line around the cursor location. But sometimes the user may feel to replace a word or string of characters all over the file by using a single command. The global replacement commands help in such a situation and make replacements throughout a file for a provided pattern. Before replacing any pattern globally (whole file), there are various ways to search a pattern from file. The ex editor (ex mode of the vi/vim editor) or line editor provides a more powerful way to make such search and changes within a file for a given pattern or string.

- **Ex mode**

    It is the execution mode or line editor mode of the vi/vim editor. In general, vi is known as visual mode to open a file and perform various operations. But you can also open a file directly in line editor mode or ex mode where you can perform various operations on a file without opening it. The ex mode gives you the editing commands with better flexibility and choice; for example, you can move text among files, transfer text from one file to another, and edit chunks of text which is larger than the available space of a single screen. Therefore, with the help of global replacement commands, you can make various changes throughout a file based on the input pattern or string.

    How to enter the ex mode

    You can open a file in the vi/vim editor by typing vi/vim followed by the name of the file on shell prompt, as given below:

    ```
    $ vi hello.txt    // Open hello.txt in vi visual mode for editing
                         and other operations.
    ```

    But if you want to perform various vi command operations on a file except editing, then you can open a file in the ex mode or last line execution mode of the vi editor. If you open the hello.txt file in ex mode, then a message is displayed including the total number of lines and characters in the file along with the colon command, as given in the below box. It is worth mentioning here that all ex mode commands must be preceded by a colon. Further, it is also important to mention that the vi editor includes all ex mode commands along with other commands that are not preceded by a colon.

    ```
    $ ex hello.txt
    "hello.txt" 12 lines, 356 characters
    :
    (You will not see any lines unless you use the ex command to see
    the lines of the hello.txt file)
    ```

    One of the most basic commands is p for printing the output on the screen. For example, if you type 1p at the prompt, it will print only the first line of the file on the screen:

    ```
    :1p      // Type 1p or only 1; print the first line of the file
    :N       // Type N or Np; print the Nth line of the file
    :3,6p    // Print from the third line to the sixth line of the file;
                p is optional
    :n1,n2   // Print from the n1 line to the n2 line of the file; p is
                optional
    :=       // Print the total number of lines of the file.
    ```

```
:.=      // Print the line number of the current line of the file.
:/pattern/=    // Print the line number of the first line that
                  matches the pattern.
:1,12#       // Display the line numbers from line 1 to line 12; use
                the # sign for momentarily displaying the line
                numbers for a set of lines
:50;+6 p     // Print the 50ᵗʰ line plus six lines below it; i.e., it
                will print line numbers 50, 51, 52, 53, 54, 55, and
                56.
:%t$         // Copy all lines and place them at the end of the file
:.,$d        // Delete lines from the current line to the end of file
:%d          // Delete entire lines from the file.
:.,.+15d     // Delete lines from the current line to the next 15
lines.
```

Searching for a string or pattern (using / and ?)

In the vi editor, pattern searching may be performed in both forward and backward directions. It can be repeated many times. It is initiated from the command mode by pressing a / followed by search pattern and pressing the Enter key. For example, if you are looking for the string *hello*, enter the string after the /:

```
/hello [Enter]   // From cursor position, start search in forward
                    direction and show or display the first
                    occurrence
                    of the hello pattern.
n                // Repeat the search in the same forward direction of
                    original search and show the next instance of the
                    pattern after each pressed "n."
?hello [Enter] // Start search in backward direction from cursor
position and show or display the first occurrence of the hello
pattern.
N                // Repeat the search in the same backward direction of
                    original search and show the next instance of the
                    pattern after each pressed "N."
:s/pattern       // In vi editor command mode, type a colon (:), "s,"
forward slash (/), and search string pattern. It will highlight the
first match for "pattern" in the file. Further, to see additional
occurrences of the same string "pattern" in the file, type n. It
will highlight the next match, if it exists.
```

Global searching for a string or pattern (using g and g!)

The previous section described how to use */pattern* and *:s/pattern* to search each occurrence of input *pattern* from files in the vi editor. However, the ex mode of the vi editor offers a global command, **g**, that searches for a *pattern* for all occurrences in one go and displays all lines comprising that pattern, if found in the file. The command **:g!** does the reverse of **:g**. By using **:g!**, you can search each line from a file that does not hold the input pattern and display the same on the screen. Some examples of global command operations, on all lines, are given below:

```
:g/pattern     // Search and show all lines comprising pattern

:g!/pattern   // Search and show all lines that do not comprise
pattern
```

```
:g/pattern/p   /          / Search and show all lines in the file
                          comprising pattern.

:g/pattern/nu                 // Search and show all lines in the file
                              that comprise pattern along with the
                              line number of each found line.

:20,100g/pattern/p      // Search and show any lines between the line
                        numbers 20 and 100 in the file that contain
                        pattern
```

Substitution—global search and replace [:s (substitute) and :g (global)]

The global search and replacement operation (referred to as substitution in vi) is another powerful feature of the vi editor which is achieved with the ex mode's s (substitute) command.

In a global replacement, the ex mode actually utilizes two ex commands—**:g** (global) and **:s** (substitute). The syntax of the search and replace operation is given as follows:

```
:address/Old_pattern/New_pattern/flags
```

Syntax Element	Meaning
:	Starting of an ex command
Address	This identifies the characteristic and range of lines for the operation. For example: % means all lines of file (first line to last line) 1,5 means from the first line to the fifth line 2,$ means from the second line to the last line in the file .,$ means from the current line to the last line in the file **Note:** If the range of lines is not mentioned, then the operation is performed only on the current line
Old_pattern	Existing pattern of characters in the file
New_pattern	Pattern to be exchanged with the old pattern
/Old_pattern/New_pattern/	Old_pattern replaced with the new pattern
Flag	g—replace all occurrences of pattern in the file (If g is not mentioned, then only the first search instance is performed and the string is replaced with a new string). c—confirm replacement

- **:s (substitute):** The syntax of the substitute command is as follows:

```
:s/old/new/
```

```
For example:    :s/orange/apple/
(After the execution of the command, substitution happens with the first
occurrence of orange pattern with apple pattern in the current line. The
/ (slash) is the delimiter that differentiates the various parts of the
command. (The last slash is optional in command line.)
```

- **:g (global):** The substitute command with g has the syntax *:s/old/new/g*

```
For example:    :s/orange/apple/g
(This change happens with all occurrences of orange pattern with apple
pattern in the current line. The g option stands for global, which means
all occurrences of the given pattern in the current line of a file.
If there is no mention of g, then the substitution will be carried out
for the first occurrence in the current line.)
```

TABLE 4.9

List of Some Examples of Search and Replace

Command	Description
:40,80s/many/more/g	Replace each occurrence of *many* with *more* from line 40 to line 80
:1,$s/xyz/ABC/g OR :% s/xyz/ABC/g	Replace each occurrence of *xyz* with *ABC* from the whole file. You can also use % rather than 1,$ which denotes all lines in a file.
:3,20s/xyz//g	The target pattern is non-compulsory. If you leave it as blank, then you will delete all instances of the source pattern *(xyz)* from line 3 to line 20.
:$s/int/long/g	Replace all occurrences of *int* with *long* in the last line of the file.
:.s/print/printf/	Replace only the first occurrence of *print* with *printf* in the current line; the metacharacter "." indicates the current line.
:g/xyz/cmd	Run cmd on all lines of the file that comprise the pattern *xyz*.
:g/str1/s/str2/str3/	Find the lines comprising *str1* throughout the file, and replace *str2* with *str3* on those lines.
:v/xyz/cmd	Run cmd on all lines of the file that do not match the pattern *xyz*.

By adding the line address before the :s command, you can spread the search from a single line to many lines. Table 4.9 shows some examples.

Interactive substitution

In some instances, you may want to selectively replace a string in file, for which you must specify a substitution command with user confirmation before replacing the pattern; for this, you can add the *c (confirmatory)* parameter as flag at the end of the command. For example:

```
:%s/long/line/gc      (Replace the pattern long with line with user
confirmation on every instance of occurrence of long throughout the file)
:5,$s/mess/mass/gc     (Replace the pattern mess with mass with user
confirmation on every instance of occurrence of mess form the fifth line
to the last line of the file)
```

Before performing each substitution operation, vi/vim halts the operation and asks confirmation from the user of the substitution operation *:%s/long/line/gc* with the following message:

```
replace with line (y/n/a/q/l/^E/^Y)?
```

The function of each character mentioned in the above confirmation message is described in Table 4.10.

TABLE 4.10

Description of Various Confirmation Characters

Confirmation Character	Description
y	Accept and perform substitution for the current instance of occurrence
n	Skip the substitution for the current instance of occurrence
a	Do the substitution for the present and all subsequent instances of occurrence.
q	Quit (q) the substitution operation.
l	Do the current substitution and then quit.
Ctrl-E, Ctrl-Y	Scroll down and scroll up to view the substitution operation.

4.6 Command Combinations

The previous sections have already introduced the basic vi editing commands, such as inset, cut, copy, paste, delete, undo, and move. The vi editor also provides some more advanced ways of editing operation by using command combination. Basically, there are two ways to combine the commands:

- In the first way, commands are combined by using number and movement of text. You have already seen some examples in the previous sections of this chapter, such as the **dw** command to delete a word, which combines the delete (**d**) command with word (**w**) text object. In the same manner, if you want to delete four words, you can use **4dw** or **d4w**; similarly, you can use **3dd** to delete three consecutive lines from the current line.

  The universal syntax/format of command combination is given below:

  ```
  (number) (command) (text object) or (command) (number) (text object)
  ```

  In this format, number is the optional numeric argument. Text object is a text movement command. You can try out some frequently used commands such as c (change), y (yank/copy), and d (delete) in combination with the text object command, as given in Table 4.11.

TABLE 4.11

Text Object Command Combined with Number

Text Object Command	Description		Examples
h	Move left one space	5h	Move left five spaces
j	Move down one line	3j	Move down three lines
k	Move up one line	6k	Move up six lines
l	Move right one space	4l	Move right four spaces
w	Move the cursor in forward direction one word	5w	Move the cursor in forward direction five words
b	Move the cursor in backward direction one word	4b	Move the cursor in backward direction four words
e	Move the cursor to the last character of the current word	2e	Move the cursor to the second character of the current word
H	Move from the current location to the top of the screen	4H	Move to the fourth line from the top of the screen
L	Move the cursor from the current location to the bottom of the screen	3L	Move the cursor to the third line from the bottom of the screen
+	Move the cursor to the first character of the next line	4+	Move the cursor to the first character of the fourth next line
-	Move the cursor to the first character of the preceding line	3-	Move the cursor to the first character of three preceding lines
G	Move the cursor from the current location to the last line in the file	8G	Move the cursor to eight lines from the current position
$	Move the cursor to the end of a line		

(Continued)

TABLE 4.11 (*Continued*)

Text Object Command Combined with Number

Text Object Command	Description		Examples
(	Move the cursor to the beginning of the current sentence	4(	Move the cursor to the beginning of four previous lines
)	Move the cursor to the beginning of the next sentence	4)	Move the cursor to the beginning of the next four lines
{	Move the cursor to the beginning of the current paragraph	2{	Move the cursor to the beginning of the second previous paragraph
}	Move the cursor to the beginning of the next paragraph	3}	Move the cursor to the beginning of the third next paragraph
$	Move the cursor to the end of the current line	4$	Move the cursor to the end of the next four lines

You may also try out some text object commands with change (c), copy (y), and delete (d) to see the effect of command combination. Table 4.12 lists some more options of command combinations.

- **Command combinations by using vertical bar (|)**

  Further, you can also use vertical bar (|) to combine multiple commands on the same ex mode prompt. This vertical bar is also known as a command separator which separates commands from each other. For example:

```
:2,7d | s/suu/sun/
```
    Delete lines from line number 2 to line number 7 and thereafter
    make a substitution of sun for suu in the current line, i.e., in
    line number 8.

```
:3,6 m 9 | :/SUN/d | g/XYZ/nu
```
        Move lines from 3 to 6 and place after line 9, and then delete
        the line containing SUN pattern. After this, display all lines
        containing pattern XYZ with line number for each line found

```
:%s/html/htm/gc  | %s/mp3/mp4/gc  | %s/GIF/jpge/gc
```
        The % symbol indicates that the operation is performed on the
        entire file and three substitution operations are done on the
        entire file with confirmation: html for htm, mp3 for mp4, and
        GIF for jpge, respectively.

TABLE 4.12

Text Object Command Combined with Number

Delete (d)	Copy or Yank (y)	Change (c)	Perform the Delete or Copy or Change Operation as per the Given Description
dL	yL	cL	The bottom of the screen from the current position
dH	yH	cH	The top of the screen from the current position
d-	y-	c-	The previous line from the current position
d4+	y4+	c4+	The next four lines from the current position
d4|	y4|	c4|	Move to the fourth column of the current line
d6G	y6G	c6G	Move to the sixth line of the file

4.7 vi Programming

Linux is an ideal environment for the development of various applications with the help of numerous programming languages, such as C, C++, and Java, with JDK, Python, Ruby, and perl others, you just name any programming language, Linux has it development environment. It provides a complete set of programming language development packages including their standard libraries, programming tools, compilers, and debuggers that you require to write and execute the code. You can use the vi editor (also vim, emacs, gedit, and other editors) for writing the code; thereafter, you can compile or debug the code with the respective programming language complier or debugger tool. The most commonly used compiler on Linux is *gcc*. The GCC (GNU Compiler Collection) comprises with the development tools including front end for C, C++, Objective-C, Ada, Go, and Fortran programming languages. We are going to introduce some examples and brief the basic steps of writing, compiling, and executing the code of various standard programming languages such as C, C++, Java, Python, and Ruby on a Linux system. We believe that you are familiar with the programming concept of the following languages:

- **C language:**
  The GNU C compiler, gcc, is one of the most versatile and advanced compilers for C and C++ languages.

Step 1: Write the C code in vi.

```
$ vi hello.c
```

Write a C program and save the file with .c extension, like hello.c

Step 2: Compile the program.

```
 $ gcc -o hello hello.c
```

Call the GNU C compiler to compile the file hello.c and create a separate output file, named *hello*, by using the option (-o) of the hello.c file. The executable file *hello* will be created only after successful compilation of your code.

Step 3: Run the program.

```
$./hello     // Execute hello.c program
```

Or

```
$ /a.out     // Execute hello.c program, if it is the last compiled file
in shell prompt.
```

NOTE: *a.out* is a common system output file, which keeps the executable code of the last compiled file in shell prompt. For example, if you compile three individual files of the same language or different languages, such as add.c, hello.c, and multiply.cpp, then the a.out file will keep the executable code of the last complied file, i.e., multiply.cpp.

- **C++ language**:
  The GNU C compiler, gcc, is also for C++ language, but you can also use the g++ compiler.

Step 1: Write the C++ code in vi.

```
$ vi hello.cpp
```

    Write a C++ program and save the file with .cpp extension, like hello.cpp

Step 2: Compile the program.

```
$ g++ -o hello1 hello.cpp
```

    Call the g++ compiler to compile the file hello.cpp and create a separate output file, named *hello1*, by using the option (-o) of the hello.cpp file. The executable file *hello1* will be created only after successful compilation of your code.

Step 3: Run the program.

```
$./hello1     // Execute hello.cpp program
```

Or

```
$ /a.out     // Execute hello.cpp program, if it is the last compiled file
in shell prompt.
```

- **Java language**:
  The javac compiler is used to compile the java programming code.

Step 1: Write the Java code in vi.

```
$ vi HelloWorld.java
```

Write a Java program and save the file with .java extension, like
HelloWorld.java

Step 2: Compile the program.

```
$ javac HelloWorld.java
```

Call the javac compiler to compile the file HelloWorld.java and then
HelloWorld.class, a bytecode file will be generated to be used for
execution. Class name is the same as the filename, HelloWorld

Step 3: Run the program.

```
$ java  HelloWorld
```

Execute the generated class file but don't put the .class extension,
just the class name

- **Python language**:
    Python is an interpreted programming language. It uses the *python* interpreter
    for python2 version and *python3* interpreter for python3 version, respectively.

Step 1: Write the Python code in vi.

```
$ vi hellocity.py
```

Write a Python program and save the file with .py extension, like
hellocity.py, and also make sure that the first line of your code must
have the following line: #!/usr/bin/python.

Step 2: Compile the program.

```
$chmod +x <filename>.py       i.e., $chmod+x hellocity.py
```

Make code file executable by using the chmod command

```
python   hellocity.py
```

```
python <filename>.py             // For Python 2.x version
python3 <filename>.py    // For Python 3.x version
```

Step 3: Run the program.

```
$ ./hellocity.py
```

```
// For execution, use ./<filename>.py, and Python is an interpreted
language, which is why the compilation and execution steps are slightly
different from C/C++.
```

- **Ruby language**

  Ruby is an interpreted, object-oriented programming language. It uses the *ruby* interpreter to interpret the Ruby programming code.

Step 1: Write the Ruby code in vi.

```
$ vi myworld.rb
```

```
    Open text editor and write a Ruby program and save the file with .rd
    extension, like myworld.rb, and also make sure that the first line of
    your code must have the following line: #!/usr/bin/ruby
```

Step 2: Compile the program.

```
$chmod +x <filename>.rb.        i.e., $chmod+x myworld.rb
```

```
Make code file executable by using the chmod command
```

```
ruby myworld.rb
```

Step 3: Run the program.

```
  $ ./myworld.rb
```

```
// For execution, use ./<filename>.rb
```

4.8 vim (vi Improved) and nvi (New vi)

The standard editors on Unix system are **vi** and **ex**. Initially, the **vi editor** was created on an older line editor, known as the **ex** editor. Thereafter, numerous improved versions of the vi editor were announced, including **nvi, vim, vile,** and **elvis,** but still vi is widely operational and popular. vim is one of the most popular enhanced versions of vi. On some Linux distributions, the vi command refers to vim in a vi-compatible mode. vim stands for "vi improved" and is an improved version of the vi editor. It was written by Bram Moolenaar in 1991. vim incorporated numerous new features in addition to the existing vi feature that ease the editing of code on various new programming languages. Such features are considered to be the essential part of modern text editors.

nvi stands for "new vi," and it was written by Keith Bostic at Berkeley Software Distribution (BSD). In the initial days of 1990, during the development of 4.4BSD, the BSD developer needed a new version of vi that could be easily and freely available and supplied with BSD version of Unix, which is why BSD developers redeveloped the traditional vi editor, known as nvi, and provided it as part of 4.4BSD. Later, it is the default vi editor of all popular BSD systems, such as NetBSD, OpenBSD, and FreeBSD. It also fulfills the POSIX Command Language and Utilities Standard [5].

On many Linux distributions, vim is installed as a default version of the vi editor, i.e., becomes synonymous with vi and runs when you invoke vi. In Linux systems, the installation library directory of vim is */usr/bin/vim*. vim is also known as a ubiquitous text editor that offers many extra features for efficiently creating and performing all kinds of text editing including ease of use, graphical terminal support, color, syntax highlighting, and formatting and composing an email along with extended customization. Moreover, the visual mode and graphical interfaces are the key facilities provided by vim that allow you to use mouse to highlight text. You can find the complete details of vim on its official webpage [1], and the latest stable version of vim is 8.2 at the time of writing this chapter. Table 4.13 provides some vim editing commands with their description.

TABLE 4.13

vim Editing Commands

Command	Description
vim	Normal execution of vim; everything is default.
vim filename	Create a file, named *filename*, and then open it.
view	Start in read-only mode.
gvim, gview	This is the GUI version of vim that starts a new window. You can also open a new window by using the "-g" argument.
evim, eview	Execute the GUI version of vim in "easy mode," like almost a normal text editor.
x	Remove the existing cursor position character.
dw	Remove the existing cursor position word.
dd	Remove the existing cursor position line.
d$	Delete the last position of the line from the existing cursor position.
u	Undo the preceding edit command.
a	Append the data just after the existing cursor position.
A	Append the data to the end of the line of the existing cursor.
v	Select text in visual mode, i.e., one character at a time.
V	Select text in visual mode, i.e., one line at a time.
r char	Replace a single character at the existing cursor position with another character mentioned as *char*.
R text	Overwrite the current cursor position data with another text mentioned as *text*, until you press the Escape key.

4.9 GNOME Editor: gedit

GNOME is one of the popular desktop environments for Linux users. If you are using a GNOME desktop on a Linux system, then the **gedit** editor comes as the default text editor, which is a part of the GNOME desktop environment comprising few advanced features. Most GNOME desktop environments comprise gedit in the *Accessories menu panel* item of the operating system. If you are unable to locate gedit through the menu panel, you may initiate it from command line prompt in a GUI terminal emulator and open one or more files under different tabs, as given below:

```
$ gedit addition.sh hello_prog.c
```

Once you begin the gedit editor comprising several files, then it uploads all the files into individual buffers and displays each file under individual tabbed window within the main editor window. The main window editor of gedit is shown in Figure 4.5.

4.9.1 Key Features of gedit

In addition to the highlighted editor window, the gedit editor operates both toolbar and menu bar features that permit you to establish various characteristics and configure settings. The toolbar offers fast access to menu bar APIs. The following menu bar APIs are accessible:

- **File:** Creating, saving, and printing files.
- **Edit:** Text editing or manipulation of text in the given file.
- **View:** Configuring the text highlight mode and editor features to display window
- **Tool:** List of installed tools in gedit.
- **Search:** Finding and replacing text in the given file.
- **Help:** Providing a complete help manual
- **Documents:** Managing various open files.

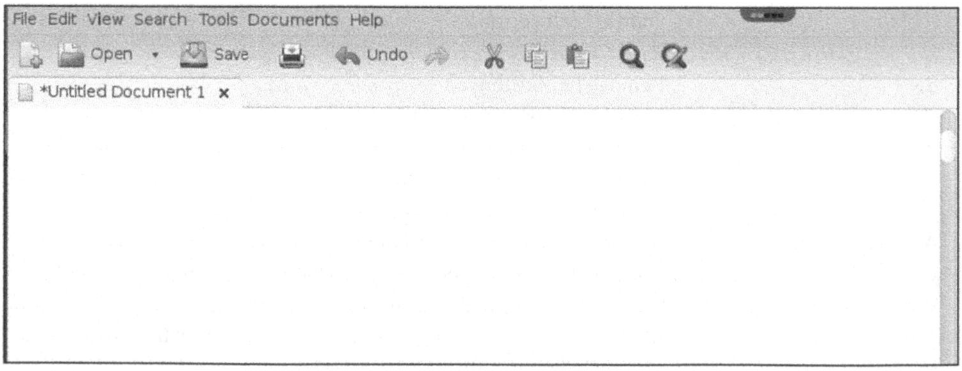

FIGURE 4.5
The gedit editor home panel.

Further, gedit is a free and open-source software package, accessible under the GNU General Public License. You may visit the official GNOME gedit webpage* for further details. It is also available for Mac OS X and Microsoft Windows with the following key features:

- Complete support for internationalized text Unicode Standard (UTF-8)
- Support of several programming languages, including C, C++, Java, HTML, XML, Python, and perl.
- Providing undo/redo utility
- Providing the facility of editing files from remote locations
- Print and print preview support
- Support of clipboard services such as cut/copy/paste.
- Searching and replacing with the support of regular expressions
- Jumping to a specific line
- Text wrapping/line numbers/right margin with customization of fonts and colors
- Highlighting the current line/bracket matching/backup files
- Availability of online user manual
- Providing a flexible plugin system that may allow to dynamically add new advanced features and extend the functionality into the gedit editor.
- One major limitation is that multimedia content is not allowed to be inserted.

4.10 Emacs Editor and Commands

Emacs is one of the famous and powerful text editors utilized by Linux/Unix systems. It is the oldest editor among the various available editors, but occupies the second place in popularity after vi. The development of the first version of emacs was started in the mid-1970s at the MIT AI Lab and still active in 2019. Initially, the emacs editor was used as a console editor, considerably like vi, but later, a graphical environment was added to it. Presently, the emacs editor provides both modes: console mode and graphical mode. In graphical mode, emacs uses a graphical window, and in text terminal or console mode, it occupies the full terminal screen. Usually, these terminal screen or graphical window, reside in emacs, is referred with term frame in which you can perform all emacs operations. Normally, emacs begins with only one frame, but later, you may add many extra frames as per the need.

It is worth mentioning here that emacs is not merely an editor; rather, it provides a fully built-in programming language environment that can perform significantly with a richer editing feature than merely the basic insertion and deletion operations of text. It may regulate sub-processes, automatically indent programs, open various files, etc. The major part of the editor is written in the Lisp programming language, except only the most basic and low-level sections of emacs are written in the C language. Therefore, emacs is a complete programming language and offers a deep extensible and customizable capability that allows users and developers to easily modify the behavior of emacs commands

---

* GNOME gedit webpage: https://wiki.gnome.org/Apps/Gedit.

and entirely create new commands and applications for the editor. New commands and applications are merely programs written in the Lisp language that are operated by emacs full-featured Lisp interpreter. The Lisp interpreter is written in the C language. You can also manage, read, and send email, access the Internet, read news, etc. Emacs is a free and open-source project and one of the oldest editors that is still under advancement.

Basically, two separate types of emacs editors are available, namely GNU emacs and XEmacs, which came from the same emacs legacy and have mostly the same features. The most wide-spread and highly ported version of emacs is GNU emacs [6]. GNU emacs was written and developed by Richard Stallman as part of the GNU Project. Many websites and documents provide the emacs overview, its history, and related matters, but you may visit the official GNU emacs editor webpage* for complete details. The subsequent section briefly describes various GNU emacs console commands and how to start and quit the emacs editor [7].

4.10.1 Starting and Quitting emacs

- **Starting emacs**: At your shell prompt, type <emacs> followed by the filename, to create a new file or open an existing file. The following command is used to start emacs:

  ```
  #emacs <name_of_file>
  ```

  If the mentioned file is available in the specified place of system, then it displays the file content in emacs frame. If the mentioned file is not available, a blank frame appears, and a file is created as soon as information is typed on the frame. By default, emacs routinely saves your work at regular intervals. The auto-saved files are designated or named as #name_of_file #, to differentiate them from regularly saved files.

- **Quitting emacs**: Whenever you feel that your work is finished, you can quit emacs by the following key combination:

  ```
  <Control>-x <Control>-c
  ```

  If your buffer workspace contains some unsaved changes, then emacs will ask: Save file *filename*? (y or n). If you answer y, your unsaved changes will be saved and then quit.

- **Exploring the emacs commands (in console mode)**

  Emacs commands use two types of keys: **Ctrl key** and **Meta key**. The Meta key is generally the **Alt key** or the **Escape key** that is denoted by M-. The Ctrl key is denoted by C- and pressed along with another character that defines the operation.

  If you are going to use the Escape key as Meta key, then this key is pressed and released and thereafter the next key is pressed. On the other hand, if you are using the Alt key as the Meta key, then just press it similarly as the Ctrl or Shift key. It means that you should press both the keys simultaneously with the other key(s) that follows. For example, C-f means that you hold the Ctrl key and type/press "f." Table 4.14 provides the most common emacs commands with their description. For a complete list of such commands, visit the GNU emacs editor webpage and GNU emacs manual documentation file [6].

---

* GNU emacs editor webpage: https://www.gnu.org/software/emacs.

TABLE 4.14

List of Some Emacs Command Functions

Command	Command Name	Description
Cursor-Movement Commands		
C-f	forward-char	Move cursor in forward direction one character (right).
C-b	backward-char	Move cursor in backward direction one character (left).
C-p	previous line	Move cursor to the previous line (up).
C-e	end-of-line	Move cursor to the end of the line.
C-n	next line	Move cursor to the next line (down).
M-f	forward-word	Move cursor one word in forward direction (right).
M-b	backward-word	Move cursor one word in backward direction (left).
C-a	beginning-of-line	Move cursor to the beginning of the line.
C-v	scroll-down	Move cursor down one screen.
Esc-v	scroll-up	Move cursor up one screen.
Deletion Commands		
C-k	kill-line	Delete from the cursor location to the end of line.
C-w	kill-region	Delete the current region.
C-d	delete-char	Delete the character under cursor location.
C-y	yank	Restore the text that you have deleted.
Del	backward-delete-char	Delete the previous character from the cursor location.
C-d	delete-char	Delete the character under cursor location.
File Commands		
C-x C-s	File save	Save file and continue editing (not quit emacs).
C-x C-c	Save and quit	Quit emacs by giving an option to save or not save the unsaved changes.
C-x C-w	Write	Write file, choice to change the filename.
C-x i	Insert	Insert a file.
C-x C-f	Editing	Edit another file.
C-x u	Undo	Reverse the last operation.
Search and Replace		
C-s	Search forward	Type the text after the command and then search; pushing C-s again searches for the next occurrence.
C-r	Search backward	Type the text after the command and then search; pushing C-r again searches for the next occurrence.
C-g	Abort the current search	Cancel the current search or go back to the earlier found search item if search was done again.
Esc-x	Replace string	Replace all occurrences of first string with second string.
Esc-%	Replace string	Replace the first string with the second string, but query for each incidence (n for the next, and y for replacement).
Emacs Help		
C-h t	Tutorial Help	Provide help with interactive tutorial
C-h i	Help with Information	Provide help in documentation form

```
C = Control key = Ctrl key
M = Meta key = Alt key or Esc key
```

4.11 Summary

When you are thinking about creating a shell script or any text file, you require some kind of text editor. Presently, numerous text editors are available for Linux/Unix distributions. These text editors help to create a shell script and write application programs in the text file of various languages. These text editors are similar to the Notepad application of Windows operating system. The most popular editor in Linux environment is vi, which has also been incorporated into the various Linux distributions as the vim editor. The vim editor is an improved vi editor. Apart from the vi editor, several other editors are also available, such as emacs, gedit, elvis, and nano. The vi editor works in three modes: insert mode, command mode, and self-prompt mode. You can use the vi editor by using various commands that you can insert in the insert mode. vi provides strong features of cut, paste, and copy operations. You can also replace certain text or line from a particular location of file content. Along with this, you can also perform searching and substitution of a given pattern. This chapter also elaborates on how to write a program in C, C++, and Java with the help of vi editor and compile it as well. Other popular editors, namely nvi, gedit, and emacs, have also made a visible impact and existence to the Linux world. The gedit editor is a basic text editor that offers some advanced characteristics such as highlighting the code syntax and line numbering along with the accessory's menus. The gedit editor provides a GUI that starts with multiple files and displays each individual file under a tabbed window inside the main editor window. The vi editor provides many advanced features of text searching and replacement.

4.12 Review Exercises

1. Write down the basic mechanism by which an editor works.
2. Highlight the key parameters of line editors and screen editors with an example.
3. How can a local user install the vi editor on a Linux machine?
4. Discuss the various modes of the vi editor and how to open a file in the vi editor.
5. In which mode the vi editor opens when it is invoked from shell? Write down a few commands to enter the insert mode from the command mode.
6. Both the *:q[Enter]* and *:q![Enter]* commands are used to exit the vi editor. What is the difference between them?
7. If the vi editor displays <"myfile.txt" [New] 5L 32C written> at the bottom of the editor screen, what is the meaning of this statement?
8. What is the difference between <vi -r filename> and <vi -R filename>?
9. What happens when you press the Esc key twice in the vi editor?
10. Suggest the commands that allow you to edit the text even when you are in command mode.

11. How does the vi editor restore the file data that was lost due to some power anomaly or inappropriate shutdown? Explain with an example.

12. How to perform various vi commands on a file without opening it?

13. Suppose at present you are working in the vi editor and editing the file "myfile.txt" and you are in the insert mode. Write down the commands to do the following:

    a. To save this file and come back to the insert mode

    b. To remove all your recent edits and return to the insert mode

    c. To recover your crashed file if during editing your system shuts down due to power failure

    d. To copy the content of the line from line number 2 to line number 5 and paste it to the same file starting from line 7 to line 9.

    e. Suppose that there are occurrences of the word "orange" in the current line:

        i. Replace its first occurrence in the line with "red."

        ii. Replace all the occurrences in the line with "red."

        iii. Replace all the occurrences in the file with "red."

14. What do "%"and "g" specify in the following command:

    ```
    < :%s/search_string/replacement/g >
    ```

15. In the command mentioned at Q. 14, do necessary modifications   to take user confirmation on each replacement.

16. Justify why Linux is used in the development of various applications in various programming languages and also brief the importance of the a.out file.

17. Suppose you are writing a program in C and you have named the variable in an unprofessional manner like you use the "*a*" variable for storing an *average* value and "*s*" for storing *sum.* What will you do to replace all these shortcut names to their respective descriptive names in one step?

18. Write down the compilation procedure of Python and Java language codes in Linux.

19. If you want to make a text editor like vi, which programming language will you prefer for its development and why?

20. Write a C program that will work like a replacement command of the vi editor.

21. Compare the vi editor with the following:

    a. Genome editor (gedit).

    b. emacs.

    c. vim editor.

22. Some quick questions:

    a. How can you insert a line above the fourth line and below the second line of a file?

    b. Which command helps to clean your screen?

    c. How to write from line 14 through line 28 and append to myfile2.txt?

    d.   Suppose you are in shell now. Write a command to open two files named "file1. txt" and "file2.txt."

    e.   How to start editing the file *"myfile2.txt"* without leaving the current one?

23. Write the vi command statement to search the error characters in printing function and replace it with printf. The C program code file is given below:

```
#include<stdio.h>

void main()
{
        ---
        ---

            ---
  prootf("hello");
        ---
        prnitf("world");
}
```

References

1. Vim – the ubiquitous text editor in the World Wide Web. http://vim.org.
2. Sumitabha Das. 2012. *Your UNIX/Linux: The Ultimate Guide*. The McGraw-Hill.
3. Richard Petersen. 2008. *Linux: The Complete Reference*. The McGraw-Hill.
4. Steve Oualline. 2007. *The Vim Tutorial and Reference*. Newriders.
5. Arnold Robbins, Elbert Hannah, and Linda Lamb, 2008. *Learning the vi and Vim Editors*. O'Reilly Media, Inc.
6. Richard Stallman. 2019. *GNU Emacs Manual*. Free Software Foundation.
7. Jeremy D. Zawodny. 2001. *Emacs Beginner's HOWTO*. Available under GNU General Public License.

5

Regular Expressions and Filters

We have seen that the vi editor works with text file with rich features and operations. However, this chapter is going to discuss and elaborate the various text manipulation tools offered by Unix/Linux system. Text manipulation tools are becoming more popular day by day among all Linux users. Broadly, the text manipulation tools are referred to as filter commands that use both standard input and standard output for various operations. Filters would be very helpful in text manipulation by including the regular expression features that may easily perform some tasks automatically on text such as inserting, modifying, arranging, and eliminating.

Regular expressions are backed by several command line tools and programming languages to simplify the solution of text manipulation problems. Regular expressions somewhat vary among available tools and programming languages. Therefore, we keep following the POSIX standard for regular expressions in our discussion. Hence, before discussing the regular filters, the subsequent sections highlight the concept of regular expressions and how to produce the same. This chapter describes simple filters and regular expression tools, but Chapters 6 and 7 describe in detail the features of advanced filters sed and awk, respectively.

5.1 Regular Expressions

A regular expression, also referred to as *regex* or *regexp*, is a defined way of explaining complex patterns in text that are used by Linux utilities to filter text. A regular expression is represented by a set of characters that is used to search and match patterns in a string. These patterns consist of two types of characters: normal text characters, referred to as literals; and metacharacters, which have special meaning. For example, in regexp ".txt," "." is a metacharacter that matches every character, except a newline and "txt" are plain text characters that are literals. Hence, this regexp matches "a.txt," "xyz.txt," "aab.txt," or many more combinations for output patterns, and all such patterns must have the .txt extension.

Some of the most powerful Linux utilities, for example, grep, sed, and awk programs, use regular expressions for text searching or data matching. If the given data matches the input pattern, then it will agree for processing; otherwise, it will terminate the processing. Further, regular expressions are also used in several other tools for text processing, such as vi, tr, rename, emacs, egrep, gawk, Python, tcl, and perl languages. A biggest concern is to understand the implementation of regular expressions using various character sets. Every character of a regular expression acts as a fundamental tool that understands regular expression patterns and utilizes those patterns to match text [1]. Broadly, the Linux community has divided regular expressions into two categories by following the POSIX standard:

- The POSIX Basic Regular Expression (BRE)
- The POSIX Extended Regular Expression (ERE)

5.1.1 What's the Variance between BRE and ERE?

It is commonly helpful to study regular expressions as their own language, where literal text acts as words and expressions. The metacharacters are defined as the syntax of the language. With BRE, the characters ., ^, $, [ ], and * are recognized as metacharacters and all other characters are referred to as literals, as mentioned in Table 5.1. But in the case of ERE, the following metacharacters are added: ( ), { }, ?, +, and |. These characters along with their associated functions are given in Table 5.3. The characters ( ) and { } are treated as metacharacters in BRE if they are escaped with a backslash, whereas with ERE, preceding any metacharacter with a backslash causes it to be treated as a literal, as given in Table 5.4. By default, the GNU version of *grep* program supports BREs, but with the -E option, it supports EREs. Further, the sed program supports only the BRE set, but the gawk program recognizes the ERE patterns. In the following sections, we are going to cover grep, egrep, and other filter tools to show the features of both categories of regular expressions.

5.1.2 Meaning of Various Characters and Metacharacters in Regular Expressions

Regular expression is an influential language for matching patterns of partial words, whole words, or even set of several words. It is interpreted by the command, not by the shell. We used quote characters to avoid the shell interpretation of metacharacters in its own way. You may construct a simple regular expression for matching text characters in a data stream by using metacharacters and literals, and even a complex expression can also be created. This section demonstrates the meaning of various metacharacters and literals in regular expressions.

- **BRE metacharacters**:
  The BRE character set is used by grep, sed, and awk filters. Some key metacharacters are mentioned in Table 5.1.
  - **BRE Special Character Classes**:
    Linux permits the utilization of POSIX "special character classes" that are mentioned within the square brackets comprising a set of characters. They are entered in an enclosed format, such as [:and:]. For example, [:alnum:] matches a single alphanumeric character regardless of the case. Some BRE special character classes are mentioned in Table 5.2.
- **ERE metacharacters**:
  The ERE character set includes a few more symbols in addition to BRE metacharacters that are used by some Linux applications and utilities. ERE metacharacters are used by the grep, egrep, and awk filters. The sed tool does not support the ERE metacharacters. Some key ERE metacharacters are mentioned in Tables 5.3 and 5.4.

5.2 grep Family

Unix/Linux has provided a special family of global search commands for regular expressions to handle search needs. The main tool of this family is grep (**g**lobal **r**egular **e**xpression **print**), which is used to refer to the name of the grep family. The other commands that belong to the grep family are egrep (extended grep), fgrep (fixed grep), pgrep (process grep), rgrep (recursive grep), etc.

TABLE 5.1

List of Some Key BRE Metacharacters

Metacharacter	Function		Example with Description
^	Beginning of line anchor	"^home"	Match all lines beginning with home
$	End of line anchor	"home$"	Match all lines ending with home. ^home$: home as the only word in line ^$: lines containing nothing (blank lines)
.	Match any character at one place of the character	"h..e"	Match lines comprising an h, followed by any two characters and ending with an e. If there are three dots, then match any three characters between h and e.
*	Match zero or more occurrences of the characters	"*home"	Display all lines with zero or more characters, of just immediate previous characters of the given pattern "home," for example, 0abchome, ahome, 12aahome, and xyzhome. g*: nothing or g, gg, ggg, etc. hh*: h, hh, hhh, hhhh, etc. .*: zero or any amount of characters
[ ]	Match one character in the set	"[Hh]ome"	Match lines containing home or Home.
[^]	Match one character not in the given set	"[^B–G]ome"	Match and display lines that do not comprise B through G followed by *ome*.
[0–9a–g]	Match one character	[0–9a–g]XY	The first character is a part of the list [0123456789abcdefg] followed by XY, for example, 0XY, 4XY, fXY, and aXY.
[a–e]	Match one character in the given range	[a–e]XY	Possible output: aXY, bXY, cXY, dXY, and eXY
[aeiou]	Match words with a vowel character		Any word that comprises a vowel character, for example, apple, elephant, orange, and India
[^aeiou]	Use ^ to specify the negation of the given set		Match any word that does not comprise vowels. [^a-zA–Z]: any nonalphabetic character
\<X	Match characters starting with X	"\<home"	Match all lines comprising a word that starts with *home*, for example, *homealone*
X\>	Match characters ending with X	"home\>"	Match all lines comprising a word that ends up with *home*, for example, *newhome*
\<X\>	Match the same mentioned character X		"\<vik": Match words like vikas or vik. "vik\>": Match words like ritvik or vik. "\<vik\>": Match the word vik.
\(..\)	Tag the matched characters	"\(home\)ing"	Tag the marked portion or sub-pattern, covered between \(home\) into a particular holding space or registers. The text matched by sub-pattern can be remembered later by the escape sequences \1to\9, because up to nine sub-patterns may be stored on a single line. Here, the pattern home is stored in register 1, and to be remembered later as \1.
X\{m\}	Match the preceding element "X" if it occurs exactly m times. For example, b\{4\} matches only "bbbb."		
X\{m,\}	Match the preceding element "X" if it occurs at least m or more times. For example, b\{3,\} matches "bbb," "bbbb," "bbbbb," "bbbbbb," and so on.		
X\{,n\}	Match the preceding element "X" if it occurs not more than n times.		
X\{m,n\}	Match the preceding element "X" if it occurs at least m times, but no more than n times.		

TABLE 5.2

List of BRE Special Character Classes

POSIX	Description
[:alnum:]	Alphanumeric characters irrespective of the case. [a–z, A–Z, 0–9]
[:alpha:]	Alphabetic characters [a–z, A–Z]. All cases including upper and lower
[:blank:]	Space and tab characters [, \t]
[:cntrl:]	Control characters [ASCII characters from 0 to 31 and 127]
[:digit:]	Numeral digits [0–9]
[:graph:]	Visible characters (anything except spaces and control characters) [ASCII characters from 33 through 126]
[:lower:]	Lowercase alphabetic characters [a–z]
[:print:]	Visible characters and spaces (anything except control characters) [ASCII characters from 33 through 126+space]
[:punct:]	Punctuation (and symbols) [!" \ # $ % & ' ( ) *+, \ - . /:; <=>? @ \ [\\\] ^ _ ` {	} ~]
[:space:]	All whitespace characters, comprising a space, newline (\n), tab (\t), vertical tab (\v), carriage return (CR) (\v), and form feed (\f).
[:upper:]	Alphabetic characters [A–Z]—uppercase only.
[:word:]	Word characters (letters, numbers, and underscores) [A–Z, a–z, 0–9]
[:xdigit:]	Hexadecimal number system digits [A–F, a–f, 0–9]

TABLE 5.3

List of ERE Metacharacters

Metacharacters	Description
X+	Match one or more occurrences of character X
X?	Match zero or one occurrence of character X
Exp_1	Exp_2	Match Exp_1 or Exp_2; for example, PDF	JPEG matches PDF or JPEG
(x1 | x2)x3	Match x1x3 or x2x3; for example, (loc | wal)ker matches locker or walker
( )	Apply the pattern match mentioned in the enclosed brackets.
X{m}	Match the previous element "X" if it occurs exactly m times
X{m,}	Match the previous element "X" if it occurs at least m or more times
X{,n}	Match the previous element "X" if it occurs not more than n times
X{m,n}	Match the previous element "X" if it occurs at least m times, but no more than n times

TABLE 5.4

Comparative Syntax of Some Metacharacters in BRE and ERE

Basic Regular Expression (BRE)	Extended Regular Expression (ERE)	Description
*	*	Zero or more occurrences
\?	?	Match zero or one occurrence
\+	+	Match one or more occurrences
\{m,n\}	{m,n}	Match at least m times, but not more than n times
|			Follow the logical OR operator in matching
\(regex\)	(regex)	Grouping of specified matches

But "grep," "egrep," and "fgrep" are three main commands of this family that make Linux users select one or more as per the need. The grep command is used to search globally in a file for the given regular expression and display all lines that comprise the search pattern [2]. The fgrep and egrep commands are basically modifications of grep. The egrep command is referred to as extended grep, supporting more regular expression metacharacters, i.e., ERE. The fgrep command, named as fixed grep, is also called fast grep and considers all characters as literals. It means that in fgrep, the regular expression metacharacters do not have special meaning; they simply match themselves.

- **grep: search through pattern**:
  grep is a tool that was created from the Unix world around the 1970s. grep searches one or more inputs for a pattern via files and folders and subsequently checks which lines in those files match the given regular expression (regex). grep displays the selected pattern, maybe line numbers or filenames, where the pattern matches the regular expression. By default, only matching lines will be displayed on screen by ignoring the nonmatching lines. When multiple files are specified in searching, then grep displays the name of the file as prefix to the output lines. Entirely, it is a very useful tool to locate information stored anyplace on your system. Almost all Linux distributions, BSD, and Unix have this command as part of GNU open-source tools. For further details, you can visit its official website.*

The grep command uses the following syntax by including regular expression (regex) as pattern:

```
grep <options> <pattern> filename(s)
OR
grep <options> <RegEx> filename(s)
```

The regular expressions (regex) are comprised of normal text characters and special characters referred to as literals and metacharacters, respectively. You can also use escape sequence that allows you to use metacharacters as literals. There are two basic ways to use grep: first, with the above-mentioned grep command syntax, and second, with the cat command, as follows:

```
grep regex filename       // grep searches the selected regex from the
mentioned filename.
cat filename | grep regex // grep searches the selected regex from the
output of the cat command.
```

Table 5.5 lists the commonly used POSIX options with the grep command. Linux supports all these options in its variants.

5.2.1 grep Associated with Exit Status

The grep command is extremely advantageous in making shell programs/scripts. It is due to the return of an exit status which specifies whether the search pattern has been found or not. Generally, the grep returns 0 as exit status if the given pattern is found and selected, 1 as exit status if the pattern is not found and located, and 2 as exit status if any other error occurs such as the input file being not located. On the other side, sed and awk don't use the exit status to specify the search operation status, i.e., success or failure. They report failure only if there is a syntax error in a command. For example:

---

* GNU grep: http://www.gnu.org/software/grep.

TABLE 5.5

List of POSIX grep Command Options

grep Command Options	Description
-c	Display the count of occurrences
-i	Ignore case for matching
-n	Display the line numbers along with each line
-l	Display only filename list
-h	Display only matched lines, but not filenames
-v	Don't display lines matching expression
-E	Read regex or pattern as an extended regular expression; this will make the behavior of grep as egrep
-e expression	Identify an *expression* used during the search of the input and display all matches that comprise the mentioned *expression*. It is highly useful when several -e options are used to identify various *expressions*, or when an *expression* begins with a dash ("-") Specify expression as option
- f filename	Get patterns from *filename*, one per line
-x	Match whole line only (fgrep only)
-F	Match multiple fixed strings (fgrep type)

```
$ grep "sanrit" /home/userlist.txt
$ echo $?   // In the Bourne shell (sh) and Korn shell (ksh) environments
 1          // sanrit is not located in the mentioned file.
```

In the above example, grep searches for the "sanrit" pattern in the given */home/userlist.txt* file, and if the searching operation becomes successful, then grep returns 0 as exit status. If the "sanrit" pattern is not located in the *userlist.txt* file, then grep returns 1 as exit status. Suppose the *userlist.txt* file is not located/available in the home directory. Then, grep returns 2 as exit status.

- **Some Examples of the *grep* Command with Regular Expressions**
  - grep hello file1.txt
    This will print all lines containing the regular expression "hello" from file1.txt.
  - grep ABC e*
    This will print all lines comprising the pattern or regular expression ABC in all files whose name begins with the character "e." The shell explores e* which applies to all filenames that begin with an "e".
  - grep '^n' file1
    This will display all lines that begin with an n character. The caret (^) represents the beginning of line announcer.
  - grep hello welcome file1
    As per the defined syntax, the first argument is pattern and the remaining are treated as arguments that are referred to as filenames. So, grep will search for *hello* from files *welcome* and *file1*. If you want to search for *hello welcome* pattern, then type the pattern as shown in the below example:

    ```
    grep 'hello welcome' file1
    ```

    This will print all lines containing the pattern *hello welcome* from file1. You can use either single or double quotes to make the search pattern of more than one word with a space.

- grep '8$' file1
    This will display all lines finishing with 8. The dollar sign ($) represents the end of line announcer.

- grep '^[me]' file1
    This will display all lines starting with either character "m" or "e." The ^ sign represents the starting of the line, and either of the characters mentioned in the bracket will be matched.

- grep '[A-Z]\{7\}' file1
    This will print all the lines that comprise at least seven successive uppercase letters.

- **The grep Command with Pipes**
    The grep command may also be used with pipe. Here, grep gets the input from pipe instead of from the given files. For example:

```
$ls -l
-rw-rw-r--          sanrit 8      Aug 9 11:01    file3.txt
-rw-rw-r—1          sanrit 24     Aug 7 21:36    file3.txt
drwxrwxr-x          sanrit 6      Aug 11 15:16   file3.txt
drwxrwxr-x          sanrit 6      Aug 11 15:14   datafile1
drwxrwxr-x          sanrit 6      Aug 3 11:55    hello1

$ ls -l | grep '^d'
drwxrwxr-x          sanrit 6      Aug 11 15:16   file3.txt
drwxrwxr-x          sanrit 6      Aug 11 15:14   datafile1
drwxrwxr-x          sanrit 6      Aug 3 11:55    hello1
```

The output of the ls -l command is the input for grep "^d," and it will print all lines that begin with d, i.e., print all directories.

- cat filename | grep 'regexp1' | grep 'regexp2' | grep 'regexp3'
    You can use more than one pipe to execute to create a more refined way of searching.

- grep 'regexp' filename1 > filename2
    This will store the output of the search in another file, *filename2*.

- **Extended grep (egrep or grep -E):**
    The functionalities of the grep and egrep commands are similar. The key advantage of egrep is that it uses additional metacharacters, such as ?, +, { }, |, and ( ), over the BRE metacharacters to generate a more complex and controlling search pattern or string. These additional metacharacters are not supported by the grep command, but with the -E option, you can perform the same operations with the grep command also (in case it is not supported by your system, then use egrep) [3]. In BRE, the characters ^, $, ., [ ], and * are identified as metacharacters and the rest are taken as simple texts (literals). The details of additional metacharacters are given in Table 5.3. A brief description of additional metacharacters of egrep is provided in the following section.

    - **? (Match zero or one occurrence of the preceding character):**
        Any character preceding the question mark (?) may or may not look into the target string. For example:

```
egrep 'beautifu?l' filename
```

In the above example, the search is successful for both "beautifl" and "beautiful" because it will treat the presence or absence of the character "u" in the same way.

- **+ (Match one or more repetitions of the preceding character):**
  The plus sign (+) will look at the preceding character and allow an infinite amount of repetitions when it goes for matching strings. For example:

```
egrep 'hello1+' filename
```

This command will match "hello1," "hello11," "hello1111," and "hello111111," but will not match "hello":
  {n, m}

- **{ } (Match a character or element for the defined number of times):**
  The curly brackets { } are used to evaluate the number of times a character should be repeated before a match occurs. The metacharacters { and } are used to define the minimum and maximum numbers of compulsory matches. For example:

```
egrep 'hello{3}' file1
```

This will match any string that contains exactly "hellooo."

```
egrep ' hello{3,}' file1
```

This will match at least three repetitions, such as hellooo, hellooooo, and helloooooo, and there is no upper limit of matching of character "o."

```
egrep ' hello {, 3}' file1
```

This will match at most three repetitions, such as hello, helloo, and hellooo.

```
egrep ' hello{3,7}' file1
```

This will match between three and seven repetitions of character "o," such as hellooo, helloooo, hellooooo, helloooooo, and hellooooooo.

- **| (Match either element):**
  Here, the pipe (|) signifies the "or" operation. This is a very important metacharacter that permits the user to join various patterns into a singular expression. For example, if a user needs to find either of two employee names in the given file, then the following command statement is required:

```
egrep 'Employeename_1|Employeename_2' filename
```

It will search and match all lines comprising either "Employeename_1" or "Employeename_2."

- **( ) (Grouping characters):**
  The text strings can be grouped together using parentheses. For example, if you are searching for lines that comprise either "patant" or "patent," then you must type the following command:

```
egrep 'pat(a|e)nt' filename
egrep 'qualit(y|ies)' filename // Display lines containing
   either string
"quality" or "qualities."
```

- **Fixed strings (fgrep or grep -F)**

  fgrep is known as fixed string, but it is also referred to as "fast grep" due to its better performance as compared to grep and egrep. fgrep is very useful when you are required to search for a string that comprises good numbers of regular expression metacharacters. In such cases, fgrep permits you to specify matching patterns as a list of fixed-string values that are separated by newline characters. Therefore, fgrep searches for fixed-character strings in a file or files. "Fixed character" means the string is interpreted literally and there is no need to escape metacharacters mentioned in regular expression. It means the fgrep command never considers any metacharacter, and assumes that it does not exist, so you don't require to escape each metacharacter with a backslash to remove its special interpretation by shell. All characters represent themselves in entirety. For example, a plus (+) sign is merely a plus sign, and an asterisk (*) sign is merely an asterisk sign. It is helpful for searching for certain static content in an accurate approach. The syntax of the fgrep command is given as follows:

  ```
  fgrep <option> <string_pattern> <filename>
  ```

  Executing the fgrep command is similar to executing the grep command with the -F option. The fgrep options are mentioned in Table 5.6.

5.3 Other Regular Filters (with Examples)

5.3.1 cat: Concatenate Files and Display the File Contents

The file management is one of the key operations of an operating system. You may require to perform some elementary operations on files, such as creating a file and listing and displaying its content on the screen. The Unix/Linux system provides the cat (which stands for "concatenate") command, which is one of the most frequently used commands to perform various file operations. The *cat* command allows to create single or several files, concatenate (combines) various files into one file, display the content of file, and redirect the output into files.

The basic syntax of the cat commands is as follows:

```
cat [options] [filenames]
```

Some examples of the cat commands are given below:

- **To display the file content:**

  ```
  #cat /etc/passwd  // Display the content of file passwd
  ```

- **To display the content of multiple files:**

  ```
  #cat hello hello1 // Display the content of files hello and
     hello1
  ```

TABLE 5.6

List of fgrep Options

Option	Particulars
-b	Show the block number of the newly found string pattern. The byte number displays the byte offset from where the line starts.
-c	Display only the count of the lines that comprise the pattern.
-e, -string	Used either in the case when the string pattern contains a hyphen in the beginning or in searching for more than one pattern. A newline character may be used to specify multiple strings; you may also use more than one -e option to specify multiple strings, for example: fgrep -e string_pattern1 -e string_pattern2 -e string_pattern3 filename.
-f filename	Outputs of search results are stored into a *newfile* rather than printing directly to the terminal. fgrep -f newfile string_pattern filename
-h	When a search is performed in more than one file, the -h option stops fgrep from showing the names of files before the matched output.
-i	Ignore capital letters contained in the string pattern when matching the pattern.
-l	Display files during matching of the string without displaying the respective lines.
-n	Display the line number before the line that matches the given string pattern.
-v	Match all the lines that do not comprise the specified string pattern.
-x	Display the lines that match the input string pattern in whole. But due to the default behavior of fgrep, generally it does not require to be specified.

- **To create a file with content:**

  ```
  #cat > test1
  ```

  The execution of the above command will be as follows: First, it will create a new file with the name *test1*; thereafter, it will open the file to type/enter the content in it, and after finishing content writing, press Ctrl+D to save and exit. You can see the stored content of the file on the screen by using the cat command. If the *test1* file already exists, then the typed content is overwritten on the *test1* file; to avoid this overwriting, use the append operator.

- **Output redirection with cat:**

  ```
  #cat file_1 > file_2
  ```

  The output of file_1 content is stored into another file, named file_2, rather than displayed on the output screen. But you must be careful that if file_2 exists, then the contents of file_2 will be overwritten by the contents of file_1.

- **To display line numbers in file:**

  ```
  #cat -n file_1
  ```

  It displays the file content along with line numbers to all output lines.

- **To append the content in a file by using the redirection operator:**
  If the file exists and you want to add some content in that file, then use the ">>" (double greater than) symbol in place of the ">" (single greater than) symbol.

  ```
  # cat file_1 >> file_2  // Add the output (i.e., file content
       of file_1) to the content of file_2, no overwriting of
       content
  ```

- **Adding the contents of multiple files into one file**:

  ```
  #cat file_1 file_2 file_3 > file_4
  ```

  The content of *file_1, file_2, and file_3* is over written in *file_4*.

  ```
  #cat file_1 file_2 file_3 >> file_4
  ```

  The contents of *file_1, file_2, and file_3* are added to the content of *file_4*.

The commonly used cat command options are listed in Table 5.7.

5.3.2 The *comm* Command: to Compare Two Sorted Files Line by Line

The *comm* command is used to compares two sorted text files. Let us assume file1 and file2, line by line. The basic syntax of the comm command is given as follows:

```
comm [-123] file1 file2
```

With no options, the comm command gives three-columnar output, in which column 1 comprises lines unique to file1, column 2 comprises lines unique to file2, and column 3 comprises lines common to both file1 and file2. The following options are used to hide a particular column as per the need:

Option	Particulars
–1	To hide column 1 that comprises lines unique to FILE1
–2	To hide column 2 that comprises lines unique to FILE2
–3	To hide column 3 that comprises lines that appear in both files

For example:

```
#comm -12 file1 file2
```

This will display column 3 that comprises the lines common to both files.

```
# comm -23 file 1 file 2
```

This will display column 1 that comprises the lines unique to file1.

TABLE 5.7

List of Commonly Used cat Command Options

Option	Notation Details	Description
-b,	--number-nonblank	Number Display all nonblank lines with number, beginning with 1.
-n,	--number	Display all lines with number, beginning with 1.
-s	--squeeze-blank	Squeeze blank spaces, i.e., join all multiple adjacent blank lines with a single blank line.
-T	--show-tabs	Show tab characters as ^I.
-u	-- Ignored	Ignored; for Unix compatibility.
–E,	--show-ends	Display line ends, i.e., place a "$"sign at the end of each line.

5.3.3 The *cut* Command: Remove Sections from Each Line of Files

The *cut* command is applied to extract some parts from each line of input files and write the result to standard output. This operation can be performed to cut parts of a line by byte position, character, and delimiter. Fundamentally, the cut command is used to slice a line and get the text. It may take multiple files as arguments or input through standard input. The basic syntax of the cut command is given as follows:

```
cut -option file_names
```

When invoking cut, use the -b, -c, or -f option, but only one of them.

The cut command uses byte (-b), character (-c), and field (-f) options, but only one of them at a time. Each option is associated with one or more numbers or ranges (two numbers separated by a dash), referred to as LIST. Every LIST is finished up with one or many numerical ranges that are separated by commas. The cut command reads the input as defined in the below LIST range:

- X: The Xth position of byte or character or field, starting from 1.
- X-: From the Xth position of byte or character or field to the end of line.
- X-Y: From the Xth to the Yth position, both including byte or character or field.
- -Y: From the first place to the Yth position in the line of byte or character or field.

If no FILE is mentioned or only a – is given, then the cut command reads standard input. The first byte or character, or field, is numbered with 1. Some key cut selection options are listed in Table 5.8.

5.3.4 The *expand* Command: to Convert Tabs into Spaces

The expand command is applied to change tabs into spaces of input files. Basically, this command converts tabs into an equivalent amount of whitespaces. By default, the *expand* command converts all tabs into spaces but keeps backspace characters in its output. The default equivalent number of spaces to each tab is 8. When no file is passed or "-" is passed, then standard input is read.

TABLE 5.8

List of Commonly Used cut Command Options

Option	Notation Details	Description
-b	--bytes byte-list	Print only the bytes from each line as mentioned in byte-LIST.
-c	--characters character-list	Print only the characters from each line as mentioned in character-LIST.
-f	--fields field-list	Select and display one or more fields from each line as mentioned in field-LIST.
-d	--delimiter delim_char	Make use of character *delim_char* as an alternative of a tab for the field delimiter. By default, a single tab character is used to separate fields.
-n	--no split	No division of multibyte characters.
-s	--only-delimited	Don't print lines that do not comprise delimiters.
-	--complement	Complement the set of selected bytes, characters, or fields

On the other side, the *unexpand* command performs the opposite operation of the *expand* command. It converts space characters (blanks) into tabs in each file. The basic syntax of the expand command is given as follows:

```
expand [options] [files]
```

Some key options of the expand command are given in Table 5.9.

5.3.5 The *compress* Command: to Compress Data

The *compress* command is one of the compression tools available in Linux. This command is used to reduce the size of named files using adaptive LZW algorithm. The compression performed by this tool is faster and consumes less memory as compared to other compression tools such as *gzip* and *bzip2*. If no file is mentioned, then the standard input is read and compression will be performed. The compressed file extension that is created by this utility is .Z. For example:

```
#compress file.txt  // Compress and rename the file as file.txt.z
```

On the other hand, the *uncompress* or *zcat* command will restore the compressed files into their original form if they have been compressed using the *compress* tool. The syntax of the compress command is given as follows:

```
compress [-cfv] [-b bits] [file...]
compress -c [-b bits] [file]
```

The key options associated with the compress command are listed in Table 5.10.

5.3.6 The *fold* Command: to Break Each Line of Input Text to Fit in Specified Line Width

The fold command is used to break each line of text at a specified width. In the case of long lines, it breaks lines into various smaller lines by adding a newline at specified width of the given files, or standard input. By default, it breaks lines at a maximum width of 80 columns. It also supports specifying the column width and breaking by numbers of bytes. The syntax of the fold command is given as follows:

```
fold [-option] [list of files]
```

The key options associated with the fold command are listed in Table 5.11.

TABLE 5.9

List of Commonly Used expand Command Options

Option	Notation Details	Description
-i,	--initial	It does not allow to convert tabs after nonblanks.
-t,	--tabs=N	By default, expand converts tabs into the corresponding number of spaces. But this option sets tab numeric (N) characters apart, instead of using the default characters apart, i.e., 8.
-t,	--tabs=LIST	This uses comma-separated LIST of specific tab positions.

TABLE 5.10

List of Commonly Used compress Command Options

Option	Notation Details	Description
-b bits	--Size	The code size is limited to bits, which must be in the range 9...16. The default is 16.
-c	--Standard Output mode	To perform the compress or uncompress operation and direct the output to the standard output. With this option, no files are modified, and it is very helpful only when it is utilized with output redirection operator (>) and directs the output to a different file, for example: # compress -c userlists.txt>newfile.Z With the -c option, the output file is not created after compression as *userlists.txt*.Z; rather, the output is transferred to *newfile.Z*.
-f	--Force mode	It forces to overwrite the output file without prompting for user confirmation, even if it does not truly reduce the size of the file, or if the equivalent compressed file already exists.
-v	--Verbose mode	Print the percentage reduction (how much compression is done) of each file. This option is overlooked if the -c option is also applied.
-r	--Recursive mode	To compress all files and subdirectories of the input directory in a recursive manner. It will be very useful if it is jointly applied with the -v option.

TABLE 5.11

List of fold Command Options

Option	Notation details	Description
-b,	--bytes	To count bytes as width rather than columns. It counts all tabs, backspaces, and carriage returns as one column, similar to other characters, for example: #fold -b16 hello.txt // It breaks each line in a 16 bytes width.
- n, -wn,	--width=*n*	Use WIDTH=n columns instead of 80, for example: # fold -w 15 hello.txt It breaks the line after 15 characters.
-s,	--spaces	Break at spaces to avoid word split. if the line contains any blanks, the line is broken after the last blank that falls within the maximum line width. If there are no blanks, the line is broken at the maximum line width. for example # fold -w 20 -s hello.txt. • -w tells the width of the text, where 80 is set as maximum by default. • -s tells to break at spaces, and not in words.
-c,	--characters	Count characters rather than columns.

5.3.7 The *head* Command: to Display the Starting Part of File

Sometimes you want to see some few lines from the beginning of the file. For this purpose, the *head* command is very useful. The *head* command is applied to print the first few lines from the top of one or more files. If more than one file is mentioned, then it will print a header comprising the file's name followed by the output for each file. By default, it shows the initial 10 lines of a file from the top. With no FILE, or when FILE is -, it will read the standard input. The syntax of the *head* command is given as follows:

```
head [options] [files]
```

Some key options of the head command are described in Table 5.12.

TABLE 5.12

List of Commonly Used head Command Options

Option	Notation Details	Description
-c N	--bytes N	Display the initial N bytes of each file, where N is a nonzero integer. It may also followed by k or m to print the initial N kilobytes or megabytes, respectively.
		b 512-byte blocks.
		k 1-kilobyte blocks.
		m 1-megabyte blocks.
-n Num	--lines Num	As per the POSIX mandate, the -Num option is used to identify a line count and display the initial N lines from the top; by default, Num is 10.
		# head -n 8 hello1.txt
		Print the first eight lines from hello1.txt.
-q	--quiet	Don't display filename as header.
-v	--verbose	Always display filename as header.

5.3.8 The *more* Command: File Checking Filter for Control Viewing

The more command is a filter for paging through text one screenful at a time. The more command was originally written by Daniel Halbert in 1978. The cat, more, and less commands are used to display the contents of a file on the screen. But the cat command displays the entire text of a file to the screen at once. If the file is large and can't fit on a screen, then the entire file text goes quickly. It displays only the ending part of the file texts that may be accommodated on the screen.

The *more* and *less* commands are used to overcome this limitation by displaying one screen of text at a time and then stop. After this, to view the remaining text of the file on the screen, press the Enter or Spacebar key. Thereafter, press the same key until you will see the whole file on the screen.

The more command shows a label displaying that you are still in the more command execution along with the percentage value in the lower left corner, reflecting how far through the text file you are; for example, -- *more* -- (60%) means that you have viewed 60% of the file content and 40% is still to view. When --*more*--(100%)--, it means that you have viewed whole file and reached at the end of a file..

The syntax of the more command is given as follows:

```
more [options] [filename]
```

For best utilization of the more command, you may use it with other commands by using pipe, for example:

```
# command1 | more
# ls -l /etc | more // Display the content of the /etc directory on one
                       screenful; then, press the Enter key for each
                       subsequent line.
```

The options associated with the more command are described in Table 5.13.

5.3.9 The *less* Command: to Scroll and View Text

The *less* command is utilized to view the file content to the standard output (one screen at a time). It was written by Mark Nudelman in 1985 and is a program like the *more* command,

TABLE 5.13

List of more Command Options

Option	Descriptions
-number	Set the number of lines that is to be displayed on the screen. The number must be an integer.
-d	Display instruction message to the user with "[Press space to continue, 'q' to quit]" and display "[Press 'h' for instructions]" if a wrong key is pressed, instead of ringing a bell.
-l	Do not stop after any line is found which comprises a ^L (Ctrl-L).
-f	Count logical lines instead of screen lines.
-p	Do not scroll. Instead, clear the entire display or screen and then show the text.
-c	Do not scroll. Instead, paint each screen from the top and clear the remainder of each line as it is displayed.
-s	Compress several blank lines into one blank line.
-u	Do not show underlines.
+/ string	It specifies a *string* that is to be searched in every input file before each file is shown.
+*number*	Start displaying the text of each file at line number *number*.

but it has many additional features. The *less* command allows both forward and backward scrolling of the displayed text as compared to the *more* program.

The command syntax is given as follows:

```
less [options] [file_name]
```

When the less command starts displaying the file content, and if the file content is larger than the screen size (i.e., not accommodatable in one screen), then the user may scroll up and down, and press the Q key to exit from the less command. Table 5.14 lists some keyboard commands that are used by the less command.

5.3.10 The *nl* Command: to Number the Lines in a File

The *nl* command is used for numbering the lines of the given files. It copies each mentioned file to the standard output, with line numbers added to the lines [4]. The line number is

TABLE 5.14

List of the less and more Command Options

Command	Description
Page Up or b	Move one page in backward direction
Page Down or Spacebar key	Move one page in forward direction
G	Go to the end of the file.
1G or g	Go to the starting of the file.
N	Repeat the previous search in backward direction.
/pattern	Search forward to the subsequent occurrence of *pattern*.
?pattern	Search for a *pattern* that will take you to the preceding occurrence.
&pattern	Display only matching lines that match the *pattern*, rather than displaying all lines.
n	Repeat the previous search.
j	Navigate forward by one line.
k	Navigate backward by one line.
q	Quit.

TABLE 5.15

List of Markup Notations of Special Strings

Markup Notation	Description
\:\:\:	Start of logical-page header
\:\:	Start of logical-page body
\:	Start of logical-page footer

TABLE 5.16

List of *nl* Command Options

Option	Description
-b STYLE	Use *STYLE* to set text body line numbering
-d CC	Use *CC* to separate logical pages
-f STYLE	Use *STYLE* to number footer lines
-h STYLE	Use *STYLE* to number header lines
-i NUMBER	Increment line number at each line by the given *NUMBER*
-l NUMBER	Join of *NUMBER* empty lines counted as one
-n FORMAT	Set line numbering format to *FORMAT*
-p	Do not reset page numbering at the starting of each logical page
-s STRING	Add *STRING* to the end of each line number to create a separator
-v NUMBER	Set the initial line number of each logical page to *NUMBER*. The default is 1
-w NUMBER	Set the width of the line number to *NUMBER*. The default is 6

reset to 1 at the top of each logical page. The *nl* command endorses the concept of logical pages when numbering. This permits *nl* to rearrange (make a fresh start) the numerical sequence when numbering, and it is feasible to set the starting number to a certain value. A logical page is divided into a header, body, and footer. The *nl* command is used for assigning numbers to each part of text, including special header and footer options which are usually omitted from the line numbering. The header, text body, and footer are enclosed by the special string's markup notations, which are given in Table 5.15.

The syntax of the nl command is given as follows:

```
nl [OPTIONS] [FILE_Names]
```

Table 5.16 lists some key options that are associated with the *nl* command.

The CC, STYLE, and FORMAT are described below with Table 5.17 and 5.18:

- The *CC* is two delimiter characters for separating logical pages. A missing second character implies a colon (:), for a backslash (\), two backslashes (\\)
- *STYLE* is one of:

TABLE 5.17

List of STYLE Options

a	Number all lines
t	Number only nonempty lines
n	Number no lines
pBRE	Number only lines that comprise a match for the basic regular expression

- *FORMAT* is one of:

TABLE 5.18

List of FORMAT Options

ln	Left justified, no leading zeros
rn	Right justified, no leading zeros
rz	Right justified, leading zeros

5.3.11 Perl: Practical Extraction and Report Language

Perl is a general-purpose, interpreted, and dynamic programming language and was initially developed by Larry Wall in 1987. Perl is used to scan random text files, extract information from such text files, and give output based on that information. Linux offers many powerful tools for system management and text processing such as sh, sed, awk, and perl. Chapters 6 and 7 provide complete details of the sed and awk tools.

Perl stands for "Practical Extraction and Report Language." Generally, it is used in place of shell scripts for many administration tasks. Perl is a language that is capable of automatically editing multiple files with a wide range of text-processing capability. Over time, it improved its capability to scan random text files, extract information for those text files, and produce reports based the desired information so quickly. It's also a preferable language for many system management tasks and Web development. You can use any editor to create a perl script or program file and store the file with the .pl extension, for example, hello.pl to identify that *hello.pl* is a perl programming file. This section just gives an overview of perl, but explaining it is outside the scope of this book. You may visit the *perl home page** for details or use the *perldoc -h* command for help and its detailed documentation.

5.3.12 The *pr* Command: Formatting Text Files for Printing

The *pr* command is used to format the text files before printing to make them look better. It converts a text file into a paginated and optionally multicolumn version, with headers, footers, and page fills.

Generally, while printing the text, you may prefer to separate the pages of output with some lines of whitespace to provide a top and bottom margin for each page. Further, these whitespace lines may be used to insert a header and footer on each page. The header may comprise date and time, filename, and a page number.

The basic syntax of the pr command is given as follows:

```
pr [options] [file]
```

Table 5.19 lists some basic options associated with the pr command for paginating or columnating file(s) before printing.

5.3.13 The *split* Command: to Break a File into Parts

The split command is used to break a file into various parts in Linux/Unix systems. It creates one or more spilt files as the output of the given input file. The default size of each split file is 1000 lines. The output filename is optional, and the default output filename prefix is

---

* Perl home page: www.perl.org.

TABLE 5.19

List of pr Command Options

Option	Description
-l *num*	Set the number of lines per page to *num*. The default number of lines per page is 66. There is no header and footer for page if the page length is less than 10 lines.
-w *num*	Set page width character columns to *num*. The default value of page width is 72.
-n	Lead each line with a line number.
-h *str*	Replace the filename with the string *str* in the header.
-d	Double-space output.
-m	Parallelly print all files, one in each column.
-t	Eliminate headers, footers, and margins totally.
-num	Produce output in *num* columns.
+num	Start printing from page *num*.
-r	No warning message if the input file could not be opened.

x. If the output filename is mentioned with the spilt command, then the spilt output files are named with the given output filename followed by a group of letters, generally. So concatenating the output files in sorted order by filename produces the original input file. The basic syntax of the *split* command is given as follows:

```
split [option] [inputfile] [outputfile]
```

(Here, outputfile is optional.)

Let us assume that the input file is defined as *File_1* and the output file is defined as *File_2*. Then by default, the split command breaks the input file *File_1* into *File_2*PREFIX, where PREFIX ranges from aa to zz. If the output file *File_2* is not mentioned, then "x" will act as the output filename. In accordance with this, the split output files are named as xaa, xab, xac, xad, ... xzz.

In this simple example of split, let us assume *myfile* is comprised of 3,000 lines. Then:

```
#split myfile      // By default, it will break myfile and give three
  output files of 1000 lines each, namely xaa, xab, and xac.

#split myfile mypart // It will break myfile and give three output files
  of 1000 lines each, namely mypartaa, mypartab, and mypartac.
```

Some key options associated with the split command are listed in Table 5.20.

5.3.14 The *strings* Command: to Print the Strings of Printable Characters in Files

The strings command displays printable character sequences from each given input file. With no options, the strings command shows all printable characters that are at least four characters long (or the number given with the options below) and are followed by an unprintable character from the given input files. The strings command is mainly beneficial for determining the contents of non-text files.

The basic syntax of the strings command is given as follows:

```
strings [options] file_name(s)
```

TABLE 5.20

List of split Command Options

Option	Description
-l num	Put *num* of lines into each split file as output.
-a *N*	Use suffixes of length *N*. By default, *N* is 2.
-b size	To fix the size of split files in output. The size of each split file is set to *size* bytes. It is a nonzero integer. Some other units may also be specified to define size as follows:
	b 512 bytes;
	k 1 kilobyte;
	m 1 megabyte; and so on
	For example:
	# split -b{bytes} {file_name} // In bytes
	# split -b nK {file_name} // In kilobytes
	# split -b nM {file_name} // In megabytes
	# split -b nG {file_name} // In gigabytes
	Here, n is the numeric value.
-C size	Set to the max size of lines in each output file to *SIZE* bytes.
-d	Use numeric suffixes instead of alphabetic suffixes in spilt output files, for example, x00, x01, x02, and x03 instead of xaa, aab, xac, and so on.

Possibly, the most commonly used option with the strings command is *-n*, followed by an integer. This integer indicates that the strings command displays all printable characters that have at least the length of the mentioned integer.

For example, the given command statement would display all strings from the files named *file_1* and *file_2* which comprise at least two characters:

```
#strings -n 2 file_1 file_2
```

Some key options of the strings command are listed in Table 5.21.

5.3.15 The *tail* Command: to Display the Ending Part of a File

The *tail* command performs the operation just opposite to the *head* command. The *tail* command is used to print the first few lines from the bottom of one or more files. If more than one file is mentioned, then it will print a header comprising the file's name followed by the output for each file. By default, it displays the last 10 lines of a file from the bottom. With no FILE, or when FILE is -, then it will read the standard input. The syntax of the tail command is given as follows:

TABLE 5.21

List of strings Command Options

Option	Description
-a,	Scan entire files.
f,	Display filename prior to each string.
-v,	Display the version number of strings on the standard output and exit.
-n *num*	Display all printable characters that are at least *num* characters long, instead of the default 4.
-t radix	Display the offset within the file before each string. The single character argument indicates the *radix* of the offset—o for octal, x for hexadecimal, and d for decimal.
-T bfdname	Indicate an object code layout other than the system's default layout.

```
tail [options] [files]
```

Two types of option formats are accepted by the tail command. In the latest version of the tail command, the numbers are recognized as arguments to the option letters, but in the old version of the tail command, the plus (+) or minus (−) sign is used with optional number, lead by any option letter. Let us assume that if a number (N) begins with plus (+) sign, then the tail command starts printing with the Nth item from the start of each file, rather than from the end. For example, -c +*num* will display all the data after skipping *num* bytes from the starting of the stated file, and -c -*num* will display the last *num* bytes from the file. Some key options of the tail command are listed in Table 5.22.

5.3.16 The *tac* Command: to Concatenate and Print Files in Reverse

The *tac* command is the reverse of the cat command. The function of the *tac* command is also opposite to the function of the *cat* command. It displays text files to standard output with lines in opposite order; for example, the end line is displayed first, then the second end line, and so on, of the given input files. The syntax of the *tac* command is given as follows:

```
tac [OPTION] [FILEs]
```

Some key options of the tac command are listed in Table 5.23, and Table 5.24 shows an example to highlight the basic difference between the cat and tac commands.

TABLE 5.22

List of tail Command Options

Option	Notation Details	Description
-c *N*	--bytes *N*	Display the last *N* bytes of each file, where N is a nonzero integer. It may also be followed by k or m to print the last *N* kilobytes or megabytes, respectively. b 512-byte blocks. k 1-kilobyte blocks. m 1-megabyte blocks. For example: #tail -c 512 hello1 OR #tail -c -512 hello1 It prints the last 512 bytes from the hello1 file. #tail -c +512 hello1 It prints everything after skipping 512 bytes from the starting of the hello1 file.
-n *Num*	--lines *Num*	As per the POSIX mandate, the -*Num* option is used to identify a line count and display the last *Num* lines from the bottom; by default, *Num* is 10. # tail -n 15 hello1.txt OR #tail -n -15 hello1 It prints the last 15 lines from the bottom of the hello1.txt file. # tail -n +15 hello1 It starts printing from the 15th line number to the end of the hello1 file.
-f	--follow	This option is very useful for system administration to monitor the progress of the log files as the system runs in a real-time manner. It continuously prints extra lines on console, written or added by another process to the file after it is opened, i.e., follow the file content until the user breaks the command.
-q	--quiet	Don't display filename as header.
-v	--verbose	Always display filename as header.

TABLE 5.23

List of tac Command Options

Option	Notation Details	Description
B	--before	Attach the separator to the beginning of the record that it precedes in the file.
-r	--regex	The separator is a regular expression.
-s string	--string separator	Use *string* as the separator instead of newline.

TABLE 5.24

Difference between the cat and tac Commands While Opening a File

# cat hello.txt	# tac hello.txt
Hi,	and age?
Who are you?	What's your name?
What's your name?	Who are you?
and age?	Hi,
#	#
#	#

5.3.17 The *tee* Command: to Duplicate Standard Input

The *tee* command is an external command. It duplicates its input and saves one copy in one or more specified files and sends the other copy to the standard output. The *tee* command acts as a "T" junction in a piping of commands. For more details, see Section 3.13.

The basic syntax of the *tee* command is given as follows:

```
tee [options] files
```

The most widely used option with the *tee* command is "-a." By using this option, the tee command does not overwrite into the mentioned file; rather, it appends the output into the mentioned files.

```
-a  : Don't overwrite, but append the existing content of specified files
```

For example:
Let's assume you are executing three commands, namely X1, X2, and X3, and store the final output in *textfile_1*:

```
$ X1 | X2 | X3 > textfile_1
```

The above sequence of commands stores the final output in *textfile_1*. However, you may be interested to keep the output of the X2 command, an intermediate result, in another file, *newfile_2*, in the intermediate result of X2. The tee command helps in such a condition, as given below:

```
$ X1 | X2 | tee newfile_2 | X3 > textfile_1
```

In the above statement, the final output that is stored in textfile_1 will be similar to the previous example, but here, the intermediate results of X2 will also be stored in another file, namely *newfile_2*, by making one more copy of the output of X2 using the tee command.

5.3.18 The *tr* Command: to Translate Characters

The tr command, which stands for translation, is a command filter that is used for translating or deleting characters in a line. It helps in a variety of conversions of characters including uppercase to lowercase and one alphabet to another, in squeezing of repeating characters, in deleting specific characters, and in many other operations. It can also perform character-based search and replace operations. It may also be used with pipes (|) to support more complex translation.

The basic syntax of the *tr* command is given as follows:

```
tr [options] [SET1] [SET2] standard input
```

The tr command accepts two sets of characters, referred to as [SET1] and [SET2], generally of the same length. A SET is basically a string of characters, comprising the backslash-escaped characters. It replaces or converts the characters of the first set [SET1] with equivalent characters from the second set [SET2]. These two character set fields may be defined in the following three ways:

- A character range, for example, A-Z.
- A specified list, for example, ABCDEFGHIJKLMNOPQRSTUVWXYZ.
- POSIX character classes, for example, [:lower:].

It is worth mentioning here that the *tr* command gets input only from the standard input; it does not accept input file as argument. The input must be redirected from a file or a pipe. For example:

```
$ echo "hello and welcome" | tr a-z A-Z
  HELLO AND WELCOME
```

Some options of the tr command are listed in Table 5.25.

5.3.19 The *uniq* Command: to Report or Omit Repeated Lines

This *uniq* command helps us to remove duplicate lines from the given sorted input file. It works for sorted file only because it eliminates duplicate lines that adjoin each other.

It sends the output to the specified file. But if the output file is not mentioned, then it sends to the standard output. In the absence of the input file, it takes input from the standard input. It is often used as a filter. This command is also known as a tool for omitting repeated lines from a sorted file. The basic syntax of the uniq command is given as follows:

TABLE 5.25

List of tr Command Options

Option	Description	Particulars
-c	--complement	Do character complement of SET1, i.e., replace with all characters that are not in SET1.
-d	--delete	Remove/delete all occurrences of input characters that are mentioned by SET1; don't translate.
-s	--squeeze-repeats	Replace instances of repeated characters with a single character.
-t	--truncate-set1	Truncate SET1 to the length of SET2 prior to the processing of standard input.

TABLE 5.26

List of uniq Command Options

Option	Notation Details	Particulars
-c	-- count	Print the list of repeated lines, led by a number which indicates the number of repetitions of a particular line.
-d	-- repeated	Print only repeated lines.
-w *num*	--check-chars=*num*	Compare only the initial *num* characters in a line.
-u	--unique	Show only unique lines.
-s *num*	--skip-chars= *num*	It ignores (doesn't compare) the first *num* characters of each line while filtering uniqueness.
-i	-- ignore case	Overlook alteration cases during duplicate line checking.
-f *num*	--skip-fields= *num*	Permit to ignore the initial *num* fields of a line. Here, fields are defined as a set of characters that are separated by tabs or spaces.

```
uniq [OPTION] [INPUT [OUTPUT]]
```

In the above syntax, INPUT refers to the input file and OUTPUT refers to the output file. The various options associated with the uniq command are given in Table 5.26.

5.3.20 The *sort* Command: to Sort Lines of a Text File

The *sort* command is used to sort the data in ascending or descending order. The sort command arranges the contents of standard input, or one or more files mentioned on the command line, and directs the results to standard output. By default, the sort program arranges lines by using the ASCII value as reference, like the first whitespace, then numerals, then uppercase letters, and then lowercase letters. The GNU *sort* command has no limits on input line length or restrictions on bytes allowed within lines.

The basic syntax of the *sort* command is given as follows:

```
sort [options] [files]
```

The sort command can take several files as input on the command line. Further, it is also feasible to combine several files into one sorted file. For example, if you have three files and prefer to merge and convert them into a single sorted file, then you may use the below command syntax:

```
sort file_1 file_2 file_3 > Combined_sorted_list
```

The sort command performs various fascinating operations on input files with the help of its numerous options. Some key options are mentioned in Table 5.27.

5.3.21 The *wc* Command: to Count Lines, Words, and Characters

This command is very useful to print the total numbers of bytes, characters, words, and lines of each input file, or standard input if no input files are mentioned. The *wc* command prints the output as counts, in the sequence of *lines, words, bytes*. By default, it displays all three counts. The basic syntax of the wc command is given as follows:

```
wc [options] [files]
```

Table 5.28 lists the various options that may perform certain operations and give output accordingly.

TABLE 5.27

List of sort Command Options

Option	Notation Details	Description
-b	--ignore-leading-blanks	Ignore leading spaces or blanks in lines and sorting based on the first non-whitespace character on the line.
-c	--check	No sorting, but it checks whether files are already sorted, and report if not sorted
-C	--check=silent	No sorting, like the -c option, but it does not report if data is not sorted.
-d	--dictionary-order	Sort as per the dictionary order; doesn't consider special characters
-f	--ignore-case	Overlook uppercase/lowercase
-g	--general-numeric-sort	Sorting in numeric order.
-i	--ignore-nonprinting	Overlook nonprinting characters that are outside ASCII range between 040 and 176.
-M	--month-sort	Sort by calendar month order, and use three-character month names, such as JAN, FEB, and MAR. In sorting, it treats JAN< FEB.
-m	--merge	Combine or merge two sorted input files.
-n	--numeric-sort	Sort using arithmetic value order.
-o file	--output =*file*	Redirect output into a specified file.
-r	--reverse	Reverse the sorting order, i.e., descending order by default.
-nr	-- reverse -numeric sort	Perform sorting with numeric data in reverse order, similar to the combination of two options: –n and –r.
-S size	--buffer-size=*SIZE*	Specify the size of the required memory to be used.
-T *dir*	--temporary-directory=*dir*	Specify the pathname of directory location to be used to store temporary working files, instead of $TMPDIR or /tmp.
-t *c*	--field-separator=*c*	Separate fields with *c* rather than nonblank-to-blank changeover
-u	--unique	Display only matching lines of the input file, and show the first incidence of two identical lines.
-z	--zero-terminated	End all lines with a NULL character, not with a newline.

TABLE 5.28

List of wc Command Options

Option	Notation Details	Description
-c	--bytes	Display byte count only.
-l	--lines	Display line count only.
-L	--max-line-length	Display the length of the longest line only.
-m	--chars	Display character count only.
-w	--word	Display word count only.

5.3.22 The *zcat* Command: to Display Contents of Compressed Files

The zcat command helps to view or display the content of a compressed file without decompressing it. This command enlarges a compressed file to standard output, so that you can look its contents. The zcat command is a part of the gzip, gzcat, and gunzip commands. All these commands help to compress or expand files. The gzip -d or gunzip or zcat command is used to uncompress/restore the compressed files into their original form. The function of the zcat command is similar to gunzip -c, and both share the same options as mentioned for gzip/gunzip. The basic syntax of the gzip, gunzip, and zcat commands is given as follows:

TABLE 5.29

List of Various zcat Command Options

Option	Notation Details	Description
-#	--fast,--best	To control the compression speed by specifying "#"as a digit. For example, -1 or--fast indicates the fastest compression technique, i.e., less compression, and -9 or --best shows the slowest compression technique, i.e., most compression. The default value of compression level is -6.
-c	--stdout	Send output to standard output, and do not change input files.
-d	--decompress	Decompress, similar to the *gunzip* command.
-f	--force	Force compression or decompression.
-l	--list	Display detailed information of each compressed file, with subsequent fields, such as the size of the compressed file, the size of the uncompressed file, compression ratio, and the name of the uncompressed file. The uncompressed size of the files that are not in gzip format is mentioned as -1.
-h	–help	Print help screen and quit.
-n	–no-name	By default, do not save the initial filename and timestamp during compressing.
-N	–name	By default, save the initial filename and timestamp during compressing.
-q	–quiet	Suppress all warnings.
-v	–verbose	Print the name and percentage decrease for each file compressed or decompressed.

```
gzip [options] [files]
gunzip [options] [files]
zcat [options] [files]
```

Table 5.29 lists the various options associated with the *zcat* command.

5.4 Summary

If a user prefers to manipulate data files in shell script, then they must be familiar with regular expressions. The regular expression concept is supported by many programming languages, Linux utilities, and command line tools. A regular expression is represented by a set of characters that is used to search and match pattern in a string. These patterns are comprised of two types of characters: normal text characters, referred to as literals, and metacharacters. There are two most popular regular expressions defined by POSIX: POSIX BRE and POSIX ERE. The Unix/Linux system also offers a special global search command for regular expressions referred to as the grep (global regular expression print) family. The POSIX ERE is generally used in programming languages that manipulate regular expressions for text filtering. Similarly, the gawk program also uses BRE to manage its regular expression pattern, which will be discussed in Chapter 7. A filter is a program or command that accepts textual data and then transforms it in a particular way and gives output. Some basic regular filter commands are also described, such as comm, cut, expand, head, tail, more, and less. Some more advanced filters, such as sed and awk, and their features are described in detail in Chapters 6 and 7.

5.5 Review Exercises

There are three files, namely *selection_sort.c, Name1.txt,* and *Name2.txt,* which are given below:

```
#include<stdio.h>

void selection_sort(int* arr,int size){
for(int i=0;i<size-1;i++){
int min=i;
for(int j=1;j<size;j++){
if (arr[min] > arr[j]){
min=j;


}
}
int temp=arr[i];
arr[i]=arr[min];
arr[min]=temp;
}
}
```

selection_sort.c.txt

Name	Contact No.	Email	Roll No.
Apaar	5555555555	appar555@gmail.com	CS001
Ayush	3333333333	ayush333@gmail.com	CS002
Deepak	111111111	deepk111@gmail.com	CS003
Hemant	2222222222	hemat222@gmail.com	CS004
Pranjal	6666666666	pranjal666@yahoo.com	CS005
Lucky	8888888888	lucky888@gmail.com	CS006
Pranjal	777777777	prjanal7771@rediffmail.com	CS007

Name1.txt

```
Apaar
Hemant
Lucky
Pranjal
Arun
Rishi
```

Name2.txt

Consider the above files for solving the following questions, as applicable:

Q1: Write the **grep** command to print all lines comprising "include" of the file *"selection_sort.c."*

Q2: What will be the output for the shell command **grep** **"{\|}"** **queue.c** if *"selection_sort.c"* is present in your current working directory?

Q3: Solve Question no. 2 using the **egrep** command and write the function of the following command statement:

- grep –v 'Sanrit' file1
- grep –i 'Meenakshi' file2
- egrep '[A–Z]...[0–9]' file1
- egrep '^X[0–9]?' file2

Q4: Write the **fgrep** command to print the lines containing comments along with line number from *"selection_sort.c."*

Q5: Write the **grep** command to print the lines containing scanf and printf in *"selection_sort.c."*

Q6. Write the **fgrep** command to calculate the number of lines that don't have comments.

Q7. Write the **cat** command to create file "q.txt" containing lines from *"selection_sort.c"* along with line number.

Q8. Write the **comm** command to print names present in "name1.txt" but not in "name2.txt."

Q9. Write the **cut** command to print the first five characters from each line of "name1.txt" and queue.c.

Q10. Write the **expand** command to convert the Tab key function to six spaces.

Q11. Write the **compress** command to compress a file and also print the percentage reduction in the size of a file.

Q12. Write the **fold** command to wrap a line to a width of 40, and line should only be wrapped at spaces in "queue.c."

Q13. Write the **head** command to print the first three lines of the files "name1.txt" and "name2.txt."

Q14. Write the **less** command to view "queue.c" with line number.

Q15. Write the **more** command to view code for main function from "queue.c."

Q16. Write the **nl** command to print *"selection_sort.c"* with line number.

Q17. What is the command to execute any file in the perl language?

Q18. What will be the output of **$ pr -3-h "questions" file.txt** if *file.txt* contains the below text?

```
hello buddy
how are you?
are you good?
```

file.txt

Q19. What does the command **$ split -l 300 file.txt new** mean?

Q20. Explain the meaning of the command **$ strings -n 2 test** where test is a binary file.

Q21. How to print the last 30 lines of a file if the file location is LOC and filename is file.txt?

Q22. What is the use of the **tac** command? Give an example with the help of an input file and then the output file.

Q23. What gets stored in file2.txt if the output of the command **$ wc file.txt | tee file2.txt** is as follows: 5 5 32 file?

Q24. Write the command to delete "e" from the sentence "Hello Linux" using the **tr** command.

Q25. How to print only the unique lines in file.txt if the file is already sorted?

Q26. Write a command to arrange file.txt having names of your class in reverse alphabetical order and save the output to a new file named student.txt.

Q27. If a user wants to compare two files, namely file1.txt and file2.txt, by calculating the number of lines, word count, byte count, and character count, which command should the user use in Linux environment?

Q28. To view the content of a zipped file, which command will you use without uncompressing it?

Q29. Give a real-life example where you encounter the use of regex.

Q30. Make a regex that can suitably define the file *"selection_sort.c."*

Q31. What are the strings that a computer checks when the following regular expression is searched: *ab+c*.

Q32. What do the following regular expressions match? (i) a.*b, (ii) ..*, and (iii) ^}$

Q33. Write a brief description of the following character classes:
(i) /s, (ii) /S, (iii) /d, (iv) /D, (v) /w, (vi) /W, and (vii) /b

Q34. What sequence of characters is going to be matched with the following regular expressions on the given string: "99999 3333 7867867557 2ye3yr73rfueg7g 8y7gyg333 yugygg 23 333"

   i. ^/d{3}

   ii. /d{3}$

Q35. Briefly explain the use of the following in a written regular expression:

   1. +, *, { }

   2. .

   3. ?

Q36. The list of names of your classmates is as follows:

   1. Rajeev Kumar

   2. Samuel Shamrock

   3. Anrid Kaul

   4. Brijesh Sank

   5. Rajvir Sharma

Find the following using regex:

   a. Name of your classmate with surname Kumar.

   b. Name of your classmate starting with R.

   c. Name of your classmate whose first name starts with Λ and surname starts with K.

Q37. Write the command to find the list of students whose roll number has an odd number from the file "Name1.txt."

Q38. Justify the statement "the tee command is also called T-junction command" with a suitable example.

References

1. Richard Blum and Christine Bresnahan. 2015. *Bible- Linux Command Line & Shell Scripting.* John Wiley & Sons.
2. John Bambenek and Agnieszka Klus. 2009. *grep Pocket Reference.* O'Reilly Media, Inc.
3. Sumitabha Das. 2012. *Your UNIX/Linux: The Ultimate Guide.* The McGraw-Hill.
4. William E. Shotts, Jr. 2012. *The Linux Command Line: A Complete Introduction.* No Starch Press, Inc.

6

Advanced Filters: sed

Chapter 4 elaborated on how to do editing of text files by using various text editors available with Unix/Linux system. You can easily manipulate text of a file with the help of many simple commands of these editors. In the case of the vim text editor, basically you can use keyboard commands to insert, replace, or delete text in the file. But sometimes you need to repeat a series of keystrokes for refining a desired task. In such a situation, the Linux system provides some interesting tools to avoid or minimize the repetition of keystroke for data-editing purpose. Such tools are called command line editors in the Linux community, for example, sed and awk. These tools apply some set of rules to generate *regular expression* or *regex* for text manipulation. Both sed and awk identify as pattern-matching languages. Regular expressions are supported by numerous command line tools and programming languages to streamline the solution of text manipulation problems. This chapter focuses on the stream editor (*sed*) tool, and the subsequent chapter covers the *awk* filter tool.

6.1 Pattern-Matching Programming Language

Several Linux text-processing functions provide you the way to search for, and in some cases change, text patterns instead of fixed strings. These functions comprise editing tools such as ed, vi, ex, sed, and gawk programming language along with the grep and egrep commands. The text pattern, often called regular expression, comprises literals and metacharacters [1]. Pattern matching is an act of checking the given sequence of characters in the whole source file. Here, the pattern itself is a character or a set of characters which should meet in the file. The work of pattern matching may be done manually or with the help of some programming languages/tools. Some prominent programming languages or tools that help in pattern matching are awk, perl, and sed. Here, we take two pattern-matching programming languages, i.e., sed and awk, and describe their features along with advantages and limitations. sed is a stream-oriented editor that can be used for automatic editing action in one or more files together, whereas awk takes text file like any other database to manipulate it. That is why, among the available text-editing programs, awk is one of the powerful pattern-matching languages. These two tools, sed and awk, are widely used by developers, system administrators, and users, i.e., everyone who loves to work with text files. This chapter describes the various features and functions of sed, and the subsequent chapter describes those of the awk programming language.

6.2 sed Overview

sed, which stands for stream editor, is a non-interactive and most prominent text-processing tool. It was developed by Lee E. McMahon of Bell Labs during 1973–1974 and runs on various Unix/Linux distributions. Initially, McMahon wrote a general-purpose line-oriented editor, but after many additions with various features, it converted and become widely known as sed. The function of stream-oriented editor is opposite to that of a normal interactive text editor. In the case of interactive text editors, for example, vim, you can interactively make use of keyboard for insertion, deletion, or replacement of texts in the data file.

But stream editors, for example, sed, can manipulate data in a stream of data, based on the set of rules or commands that you may provide over command line or that are kept in the file, before the processing of data by the editor. Basically, sed takes inputs from files as well as pipes, but it may also accept input from keyboard.

In sed, it takes one line of data from the input file and matches/executes it with the provided sed commands, and after processing, it does the required changes in input line as per the command rules; thereafter, the editor redirects the output on standard output. After executing all editor commands on the entire input line data and giving output, it takes the next line of data from the input file and applies the editor commands to match/execute on it, and thereafter does the required changes and gives output. These steps are performed by the sed editor to process all lines of the input file before terminating the execution. So basically, sed is an ideal tool for performing a series of edits to one or more files.

The sed editor may be used as an editing filter that allows to make changes quickly in file data as compared to an interactive editor. In the same way as shell scripts, you can write a sed script or program where you can mention the required sed commands in a specified way to perform a particular task or requirement. In the vim editor, mostly actions are performed manually, such as inserting new text, replacing text, and deleting lines. But in sed, you can mention all the required commands or instructions at one place, i.e., in a file, known as sed script. So simply, execute sed script to perform all editing action on input file in one go. There are numerous ways to use the sed editor. Some are as follows:

- Substitution of text
- Printing of selective text files
- Editing a part of text files
- Automating editing activities on one or more files
- Easing similar editing tasks on numerous files
- Non-interactive editing of text files
- Creating/writing a sed script program

It was one of the initial editing/text-processing tools to support and recognize regular expressions and is still popular for text processing, especially with the substitute command feature. But awk and perl are also referred to as common alternative tools for plain-text string manipulation and stream editing.

TABLE 6.1

sed Options

Option	Description
-e script	Specify multiple script commands that apply for processing on input.
-f file	Add sed *script-file* where the set of script commands are stored that to be apply for processing on input.
-n	Disable automatic printing for each command; sed prints the output only when the "*p*" command is used.

6.3 Basic Syntax and Addressing of sed

Basically, sed uses instructions to perform editing on text. Every instruction comprises two parts: The first is pattern or address, and the second part is procedure or action [2]. The pattern part basically involves regular expressions for selecting lines, and the procedure part indicates one or more actions that are to be applied on the selected lines.

The basic **command line syntax** of **sed instruction** is given as follows:

```
sed options script-file file(s)
OR
sed options 'address action' file(s)
```

There are two forms of syntax for invoking sed:

```
sed [-n] [-e] 'command' file(s)
sed [-n] -f script-file file(s)
```

The first form of syntax permits to mention an editing command on the command line, surrounded by single quotes. The second form of syntax permits to mention a script-file, a file comprising sed commands. You can use both approaches altogether one and more times. If no file(s) is mentioned, sed reads from standard input. The option parameters of the sed command permit you to change or amend the behavior of sed instruction. Some options are listed in Table 6.1.

The basic syntax of the **sed command** is given as follows:

```
[address] cmd [arguments]
```

In the above sed command syntax format, *cmd* is a single letter or character symbol of the sed command. You may see all such commands and other details at its official page.* The mentioned [address] can be a regular expression, or a single line number or a set of lines.

The [address] is an optional line address. If [address] is mentioned, then the command *cmd* will be executed on match lines. Some single-letter *cmd*, sed commands are listed in Table 6.2. Some prominent commands are s (substitute), d (delete), a (append), and i (insert).

---

* GNU sed page: www.gnu.org/software/sed/.

TABLE 6.2

List of Some Prominent One-Character sed Commands

One-Character Command	Description
a	Append
c	Replace (change)
d	Delete
e	Execute
i	Insert
P	Print
r	Read
s	Substitute
w	Write
x	Exchange

TABLE 6.3

Syntax of Some Pattern Addresses

Address	Description
/pattern/	Select any line which matches the *pattern*.
\;pattern;	Allow to use semicolon or any other character as the delimiter in place of slash (/). It is basically useful when the *pattern* itself comprises a lot of slash characters. For example: sed -n '\; /home/sanvik/documents/;p'
N	Line number N.
$	Last input line.
No address given	The given command applies on every input line.
One address	Apply to any line that matches the address, because few commands take only one address: a, i, r, q, and =.
Two addresses separated by comma	Apply from the first matching line and then all succeeding lines up to a line matching the second address. Both lines are included.
An address followed by !	The command is applied to all lines that do not match the address.
Caret (^)	Match the starting of the line.
Dollar sign ($)	Match the end of the line.
Asterisk (*)	Match zero or more occurrences of the preceding mentioned character or regular expression.

In pattern matching, the symbol "\n" may be used to match any newline in the pattern space. When a regular expression is provided as an address, the command touches only the lines that are matching that pattern. Therefore, all regular expressions should be surrounded by a slash (/) character. Some of the pattern address formats are listed in Table 6.3 [2] and are further elaborated with some examples in Section 6.5.

- **Grouping Commands**: In sed, brace characters { } are applied to cover one address inside another address or group various commands to operate on the same address. Therefore, you may cover various addresses if you desire to mention a range of lines, and, from within that range, specify another address and so on. The basic syntax of grouping commands is given as follows:

```
[/pattern/[,/pattern/]]{
command1
```

```
command2
}
```

In the above given syntax, the opening curly brace "{" must be placed at the end of the same line command/pattern address, but the closing curly brace "}" must be placed on a newline or its own line. Further, it has also to be taken care that there are no spaces after the braces. You can put several commands to a similar range of lines by enfolding the editing commands within the braces notation. Every command can have its own address, and various levels of grouping are also allowed.

6.4 Writing sed Scripts

If you want to execute many sed commands on an input file, then simply put all commands into a file and execute the file. Therefore, a sed script or program comprises one or more sed commands or actions, passed with the -e, -f, --expression, and --file options. You can run a sed script file on an input file with the following command line syntax:

```
#sed -f sedscriptfile.sed Input_file
```

The above syntax does not create any alteration in *Input_file*. It displays all lines to standard output (basically on the screen) including both modified lines and unchanged lines. If you wish to store this output in a new file, then you can save it by using redirection operator, as given below:

```
#sed -f sedscriptfile.sed < Input_old_file > Output_new_file
```

The redirection symbol ">" redirects the output from standard output to the mentioned file, i.e., *Output_new_file*. The main purpose of storing this output to a new file is to verify the performed operation as well as to use this file for other operations. You can use any other editor, for example, vi, to write such script file which is used as an interpreter script by sed. For example, write a simple sed script to change a lowercase vowel into uppercase of an input file, as given below:

```
#!/bin/bash
s/a/A/g
s/e/E/g
s/i/I/g
s/o/O/g
s/u/U/g
```

Put the above script lines into a file, called *VowelConvert.sed* (save all sed script files with .*sed* extension), and make it executable (chmod +x *VowelConvert.sed*) and run it as given below:

```
$sed-f VowelConvert.sed inputfile > output_file
```

In another example, delete all lines from the 20th position to the 45th position from the given input file. So, here 20,45 defines an address range of lines, d is the delete command.

Then, the sed command deletes the line numbers from 20 to 45 of the given input file, namely *inputfile.txt*, and stores the new edited copy of the input file into another file, called *outputfile.txt*, as given below:

```
sed '20,45d' inputfile.txt > outputfile.txt
```

Such script file has a fixed procedure or action that makes execution in a controlled manner. There are three elementary philosophies of how a sed script file works:

- All sed commands written in script file are applied sequentially to every line of input file. By default, all commands are applied to all lines (globally) unless the mentioned line addressing limit is mentioned or specified.
- The original input file for sed script is unaffected.
- It sends the modified copy of the input file to a new output file, by using an output redirection operator.

Basically, there are four categories of sed script applications:

- In the first category, sed script performs several edits to the same file.
- In the second category, sed script performs a set of search and replacement edits across a set of files.
- In the third category, sed script is applied to obtain relevant information from a file. Here, sed functions are similar in nature with grep.
- In the fourth category, sed script performs edits in a pipeline that are never written back into a file.

Although numerous sed commands are available to perform various operations from simple to complex level, the coming sections are going to explain some basic as well as advanced sed commands.

6.5 Basic sed Commands

Before proceeding to explain the various commands in the coming subsections, there are some key points that must be relooked, related to sed command syntax. A line address is optional with any sed command. It may be a pattern that is described as a regular expression surrounded by slashes, a line number, or a line-addressing symbol. But mostly sed commands can accept two comma-separated addresses that indicate a range of lines. For such commands, it may be defined as follows:

```
[ address] command
```

The basic uses of sed commands are editing one or more files automatically, streamlining the tedious edits to multiple files, and writing a sed script or program. But some commands take only a single line address that may not be utilized to a range of lines. The syntax for such command is as follows:

```
[ line-address] command
```

As mentioned earlier, the sed editor never modifies the data of input file during the operation. It just sends a copy of modified text of input file to standard output or redirects into another file by using the output redirection operator (>). After applying the sed command over input file, you may check it that after operation, the content of input text file is still the same as original.

You can execute multiple sed commands in the command line by using the -e option. The commands must be separated with a semicolon (;) and there should be no space between the end of each command and the semicolon, as given below:

```
$ sed -e ` command1; command2; commmand3' input_file
```

For example:
```
        $ cat vehicle.txt
        The yellow speedy car runs very fast.
        The yellow speedy car is very expensive.
        The yellow speedy car runs very fast.
        The yellow speedy car is very expensive
        $

        $ sed -e 's/yellow/red/; s/fast/slow/' vehicle.txt

        The red speedy car runs very slow.
        The red speedy car is very expensive.
        The red speedy car runs very slow.
        The red speedy car is very expensive.
        $
```

As discussed earlier, if you have several sed commands and need to execute on a given input, then it would be simpler to just write all sed commands into a single file, referred to as sed script, and use the -f option to include this script file with sed, as in the examples given below. Let's assume that you have created a sed script file, namely *vehicle_script.sed*, which comprises three sed substitution commands (*see the subsequent section*). To get the content of this script file, type the below command format:

```
$ cat vehicle_script.sed
   s/yellow/red/                        // sed substitution command
   s/fast/slow/                         // sed substitution command
   s/expensive/economy/                 // sed substitution command

$ sed -f vehicle_script.sed vehicle.txt
The red speedy car runs very slow.
The red speedy car is very economy.
The red speedy car runs very slow.
The red speedy car is very economy.
$
```

6.5.1 Substitute Patterns

The "s" command (referred to as the substitute command) is possibly one of the highly valuable and frequently used commands in sed, having lots of different options. The basic concept of the "s" command is simple: It tries to match the *pattern* to the lines, and if the match is successful, then that portion of the matched pattern is replaced with *replace_pattern*. The substitution flag is one of the key parts of the "s" command that decides the substitution operation occurrence. With no flag, it performs the substitution operation only on the first occurrence of matching pattern in each line. To perform the substitution on various occurrences of the text, use the substitution flag. The pattern is supplied as a regular expression [3]. The syntax of the "s" command is given as follows:

[address] `s/pattern/replace_pattern/flags`

The "s" (substitute) command can be supported by zero or more of the following flags:

1. **n**: The substitution should be done only for the nth occurrence of the *pattern*.
2. **g**: Replace with *replace_pattern* to all matches to the *pattern*; i.e., substitution is performed in all occurrences (globally) of the matched *pattern*, not just the first occurrence as by default.
3. **p**: Print the content of new pattern space if substitution was performed.
4. **w file**: Write or redirect the outputs to the named *file* if substitution was performed.

The substitute command is utilized to the lines that are matching the given address condition. If no address is mentioned, then it is applied to all lines that match the *pattern*, which is given as a regular expression. If the address is provided in regular expression form and no *pattern* is mentioned, then the substitute command performs the same matches as provided in the address. The flags can be used in combination; for example, gp performs the substitution globally on the text lines and prints newlines.

Here, some substitution examples are shown to discuss how the substitution is performed. You can control and decide the number of occurrences of the matching pattern which replace with new text as *replace_pattern*. For example, there are three *"car"* patterns in the first line of the *vehicle1.txt* file, and you would like to replace the second occurrence of the *"car"* pattern in the first line with new pattern as *"taxi"*; then, the following substitute command would be executed for this change:

```
$ cat vehicle1.txt
The car is yellow, speedy car and four vehicle car.
The yellow speedy car is very expensive.
The yellow speedy car runs very fast.
The yellow speedy car is very big.
$

$ sed 's/car/taxi/' vehicle1.txt
The taxi is yellow, speedy car and four vehicle car.
The yellow speedy car is very expensive.
The yellow speedy car runs very fast.
The yellow speedy car is very big
$
```

By default, replace only the first occurrence of the "car" pattern with "taxi"

```
$ sed 's/car/taxi/2' vehicle1.txt
The car is yellow, speedy taxi and four vehicle car.
The yellow speedy car is very expensive.
The yellow speedy car runs very fast.
The yellow speedy car is very big.
```

As a result of mentioning 2 in the place of substitution flag, the sed editor only replaces the second occurrence of the *"car"* pattern in each line. If you wish to replace every occurrence of the *"car"* pattern with *"taxi"* in the whole *vehicle1.txt* file, then use the g (global) substitution flag, as given below:

```
$ sed 's/car/taxi/g' vehicle1.txt
The taxi is yellow, speedy taxi and four vehicle taxi.
The yellow speedy taxi is very expensive.
The yellow speedy taxi runs very fast.
The yellow speedy taxi is very big.
```

The "p" substitution flag enables you to print a line that comprises the matching pattern. You may also combine the "p" flag with the "n" numeric flag to print specific lines where the replacement was performed.

```
$ sed 's/fast/slow/p' vehicle1.txt
The yellow speedy car runs very slow.
```

It displays only the matching pattern line. The substitution flag (w) produces the same output, but it also stores the output in the specified file, as given in the below example:

```
$ sed 's/fast/slow/w  test1.txt' vehicle1.txt
The car is yellow, speedy car and four vehicle car.
The yellow speedy car is very expensive.
The yellow speedy car runs very slow.
The yellow speedy car is very big.

$ cat test1.txt
The yellow speedy car runs very slow.
```

The sed editor displays its standard output on the screen, but it will store only the lines that include the matching pattern, in the specified output file, namely test1.txt.

6.5.2 Replacement Characters

The replacement metacharacters are comprised of ampersand (&), backslash (\), and a number (i.e., \n).The number n indicates the substring component position. The replacement metacharacter has a special meaning. For example, the backslash (\) has a characteristic to ignore or escape the other metacharacters' function in a string. In the case of the pattern substitution syntax, the forward slash (/) is used as a string delimiter, but sometimes we need to use forward slash (/) as simple character; then, we must escape the special meaning of forward slash (/) with the use of backslash (\), as "\ /." Hence in this case, the sed editor would not interpret forward slash (/) as a special character, but to take it as literal.

Brief details of all replacement metacharacters are given below:

\n: Match and replace the nth occurrence of the pattern on each addressed line. Here, n is a single-digit number.

&: Represent the extent of the pattern match by the regular expression.

\: It is applied to ignore the special meaning of ampersand (&) and backslash (\). It may also be utilized to escape the newline and create a multiline replacement string. It is worth mentioning here that two backslashes are necessary to output a single backslash, because backslash itself is a replacement metacharacter.

For example, to perform the substitution of K shell path (/bin/ksh) with C shell path (/bin/csh) in the given input file, i.e., /etc/passwd file, type the following sed command syntax:

```
$ sed 's/\/bin\/ksh/\/bin\/csh/' /etc/passwd
```

Hence, in the above command syntax, the sed editor does not interpret any metacharacter if that metacharacter is just placed after backslash (\), and to note that no space is allowed after the backslash. The backslash is also applied to include a newline in a replacement string. This method works well when we need to implement the forward slash for a single command. But, if we have multiple metacharacters in the command syntax, as in the above substitution example, then it will look more confusing, which leads to make mistakes, and become more complex to execute. To solve this problem, the sed editor allows you to use other characters for the string delimiter in the substitute command, as given below:

```
$ sed 's!/bin/ksh!/bin/csh!' /etc/passwd
```

In this example, the exclamation character (!) is utilized for the string delimiter, making the pathnames very much simpler or easier to read and understand.

6.5.3 Append, Insert, and Change

Similar to the other editors (e.g., the vi/vim editor), the sed editor also provides various editing commands such as append (a), insert (i), and change (c) [4,5]. The syntax and brief details of these editing commands are given below:

```
The append command (a) adds a newline after the specified (or current)
line and then places the supplied text

      append [ line-address]a\
      text

The insert command (i) adds a newline before the specified (or current)
line and then places the supplied text.

      insert [ line-address]i\
      text
```

The change command (c) replaces the specified contents of the pattern space with the supplied text

```
change [ address(es)]c\
text
```

According to the above command descriptions, the *append* command inserts the supplied text after the current line, the *insert* command puts the supplied text before the current line, and the *change* command replaces the text contents with the given text. All the mentioned commands need a backslash (\) to ignore the first end of the line. The text must start from the next line. For entering numerous text lines, every consecutive line must finish with a backslash, except the last line. For example:

```
$ cat vehicle2.txt
The yellow speedy car runs very fast.
The White speedy car is very expensive.
The Red speedy car runs very fast.
The Pink speedy car is very expensive.
$
```

For appending the text:

```
$ sed  '4a\
> This is a car showroom'  vehicle2.txt
```

This command adds a newline after the fourth line number in file "vehicle2.txt," as shown below:

```
The yellow speedy car runs very fast.
The White speedy car is very expensive.
The Red speedy car runs very fast.
The Pink speedy car is very expensive.
This is a car showroom.
```

For inserting the text:

```
$ sed  '4i\
> This is a car showroom'  vehicle2.txt
```

This command adds or inserts a newline before the fourth line number in the file "vehicle2. txt," as shown below:

```
The yellow speedy car runs very fast.
The White speedy car is very expensive.
The Red speedy car runs very fast.
This is a car showroom.
The Pink speedy car is very expensive.
```

If you want to append multiline text stream at the end of the given input file, then simply use the dollar sign ($), which represents the last line location. In the same way, if you want to add a new multiline text stream at the starting of the given input file, then simply use the insert command with the line addressing 1, i.e., add lines before the first line of input file. In multiline text, you should apply a backslash at the end of each line until reaching the last line.

```
$ sed  '$a\
> This is a car showroom.\
> It is in central city.\
> It also sells motorcycles.'  vehicle2.txt
```

This command appends all lines at the end of the file "vehicle2.txt," as shown below:

```
The yellow speedy car runs very fast.
The White speedy car is very expensive.
The Red speedy car runs very fast.
The Pink speedy car is very expensive.
This is a car showroom.
It is in central city.
It also sells motorcycles.
```

To add the starting of the file:

```
$ sed  '1i\
> This is a car showroom.\
> It is in central city.\
> It also sells motorcycles.'  vehicle2.txt
```

This command inserts all lines at the starting of the file "vehicle2.txt," as shown below:

```
This is a car showroom.
It is in central city.
It also sells motorcycles.
The yellow speedy car runs very fast.
The White speedy car is very expensive.
The Red speedy car runs very fast.
The Pink speedy car is very expensive.
```

- **Double-spacing text:** If you want double spacing between lines, then you can add two blank lines with the append command, as given below:

```
sed 'a\
                        // Give two blank lines
' vehicle2.txt
```

This command provides two blank lines after every line of the file and then prints. Another way of double-spacing text is that by using i, you can insert a blank line before each selected line of the file.

For changing the text:

The change command permits you to replace or change the entire line of text with the supplied text. In contrast to the insert and append commands, the change command may be applied either on a single line address or on a range of line addresses. In the event of range of line address, the entire range of text is replaced with the supplied text, not each line. The change command also replaces the whole line with input text if the line matches the pattern address. Some examples are given below:

Replace the text of line 2 with input text:

```
$ sed '2c\
> Remove the white speedy car.' vehicle2.txt

The yellow speedy car runs very fast.
Remove the white speedy car.
The Red speedy car runs very fast.
The Pink speedy car is very expensive.
$
```

You can use an address range with the change command. It replaces the entire text from line 1 to line 3 with input text or lines:

```
$ sed '1,3c\
> This is a new car 1.\
> This is a new car 2.' vehicle2.txt

This is a new car 1.
This is a new car 2.
The Pink speedy car is very expensive.
$
```

You may also use the pattern address with the change command to replace the text with input text. Here, the change command changes any line of text in the input file with input text that matches the given pattern address, as in the below example:

```
$ sed '/very expensive/c\
> This is a new car 1 with low cost.\
> This is a new car 2 with low cost.' vehicle2.txt


The yellow speedy car runs very fast.
This is a new car 1 with low cost.
This is a new car 2 with low cost.
The Red speedy car runs very fast.
This is a new car 1 with low cost.
This is a new car 2 with low cost.
$
```

```
$ sed '/very expensive/c\
> This is a new car 1 with low cost.' vehicle2.txt

The yellow speedy car runs very fast.
This is a new car 1 with low cost.
The Red speedy car runs very fast.
This is a new car 1 with low cost.
```

6.5.4 The Delete Command and the Use of the Exclamation Sign (!)

Like other editors, sed also offers the feature of deleting the specific lines from the given text stream. So, if you want to delete some lines of text in a file, then you can delete by using the "d" command. As mentioned earlier, the sed editor doesn't alter the original file. All the deleted lines only disappear on the output of the sed editor, but if you wish to keep this output, you must redirect this output into a separate file [4,6]. The original file still comprises all lines including the "deleted" lines. If you forget to mention the addressing pattern, then it will delete all lines from the files, for example:

```
$ cat vehicle2.txt
The yellow speedy car runs very fast.
The White speedy car is very expensive.
The Red speedy car runs very fast.
The Pink speedy car is very expensive.
$

$ sed 'd' vehicle2.txt
                    // Delete all lines of input file
$
```

To delete a specific line:

```
$ sed '2d' vehicle2.txt
The yellow speedy car runs very fast.
The Red speedy car runs very fast.
The Pink speedy car is very expensive.
$
OR
$ sed '2d' vehicle2.txt > New_vehicle.txt
(redirect the output into New_vehicle.txt file)
$
$cat New_vehicle.txt
The yellow speedy car runs very fast.
The Red speedy car runs very fast.
The Pink speedy car is very expensive.
$
```

To delete a specific range of lines:

```
$ sed '3,4d' vehicle2.txt
```

```
        // To delete all lines from the third to the fourth line
The yellow speedy car runs very fast.
The White speedy car is very expensive.

$
```

To delete the lines from a specific line to the last line:

```
$ sed '2,$d' vehicle2.txt
// To delete all lines from the second to the last line
The yellow speedy car runs very fast.
$
```

To delete the lines from the first line to a specific line:

```
$ sed '^,3d' vehicle2.txt
// To delete all lines from the beginning to the third line
The Pink speedy car is very expensive.
$
```

To delete the lines that match the supplied pattern:

```
$ sed '/ very expensive /d' vehicle2.txt
The yellow speedy car runs very fast.
The Red speedy car runs very fast.
    $
```

To delete the lines that match the given range of the supplied pattern (i.e., delete the lines between two specified pattern ranges):

```
$ sed '/Pattern1/,/pattern2/d' input_file  > New_output.txt
```

As mentioned in the above syntax, you are able to delete a range of lines by using two text patterns, namely *pattern1* and *pattern2*. The *pattern1* acts as the starting point of line deletion, and *pattern1* acts as the stopping point of line deletion. It is very important to note that the above syntax must be used very carefully, and must be clear that which part of the sections you want to delete; otherwise, it may give a surprising output and delete the undesired data. The following example helps you to understand the operation of the above-mentioned syntax:

```
$ cat vehicle3.txt
This is car store number 1.
The yellow speedy car runs very fast.
The White speedy car is very expensive.
The Red speedy car runs very high speed.
The Pink speedy car is very expensive.
This car store has good cars.
The Red speedy car is low cost.
```

```
This is car store number 2.

$

$ sed '/very fast/,/high speed/d' vehicle3.txt
This is car store number 1.
The Pink speedy car is very expensive.
This car store has good cars.
The Red speedy car is low cost.
This is car store number 2.
$
```

The above command delete all lines between the first pattern match *"very fast"* and the second pattern match *"high speed"*, in another example as given below:

```
$ sed '/very expensive/,/high speed/d' vehicle3.txt
This is car store number 1.
The yellow speedy car runs very fast.
$
```

In the first incidence of pattern match, it deletes all the lines from the *"very expensive"* pattern match line to the *"high speed"* pattern match line. In the second occurrence of pattern match, the line comprising the *"very expensive"* pattern activates the delete command again, and it deletes the remaining lines of the file, due to not finding the stopping point.

Similarly, it deletes all blank lines because ^ denotes the beginning of the line and $ denotes the end of the line of a file, as given below:

```
$ sed '/^#/d' vehicle3.txt
$
```

```
$ sed '/pattern/,+nd' file.txt
```

To delete the lines that match the pattern along with *n* number of lines after that matched line:
   For example;

```
$ sed '/yellow/,+3d' vehicle3.txt
This is car store number 1.
This car store has good cars.
The Red speedy car is low cost.
This is car store number 2.
$
Note: In a supplied command, if the command address is followed by an
exclamation sign (!), then that command is utilized to all lines except
the lines that match the given address or address range. For example:
```

```
$ sed '2,4!d'   // Delete all lines except lines 2 to 4
```

```
$ sed '/pattern /!d'    // Delete all lines except the lines that comprise
the supplied pattern.
```

6.5.5 The Transform Command

The "y" character is used to represent the transform command. It is the only command of the sed editor that operates on a single character. The function of the transform command (y) is similar to that of the *tr command*. It performs one-to-one or character-to-character position replacement. In the command format, the transform command (y) takes zero, one, or two addresses. The syntax of the transform command is given as follows:

```
'[address]y/input_chars/output_chars/'
```

The transform command performs a one-to-one comparison between the values of *input_chars* and *output_chars*. The first character of the *input_chars* pattern is converted into the first character of the *output_chars* pattern; the second character of the *input_chars* pattern is converted into the second character of the *output_chars* pattern; and so on. This comparison continues till the last character of the mentioned patterns. The *input_chars* and *output_chars* must have the same character length; if not, then the sed editor gives an error message. For example:

```
$ sed 'y/xyz/pqr/' file.txt
```

The above command statement converts every character of file.txt from the character string *xyz* to its corresponding character of the string pqr. It means the "x" character is replaced by the "p" character anywhere on the line, irrespective of whether it is followed by the "y" character. In the same way, the "y" character is replaced by the "q" character and "z" is replaced by the "r" character.

```
$ echo "This is test 1 and 2, include 231 questions in 3 sets of 3
packets." | sed 'y/123/789/'
This is test 7 and 8, include 897 questions in 9 sets of 9 packets.
$
```

The sed editor transforms all occurrences of the matching character irrespective of places. The transform command (y) is also treated as a global command because it performs the transformation operation on every found character in the text line automatically, without looking into the occurrence of the found character, for example:

```
$ sed 'y/abcdefghijklmnopqrstuvwxyz/ABCDEFGHIJKLMNOPQRSTUVWXYZ/'
```

It converts all lowercase characters into uppercase characters.

6.5.6 Pattern and Hold Spaces

sed has two types of internal storage space, namely pattern space and hold space. These spaces are buffers where sed stores data.

1. **Pattern space:** It is an active temporary buffer area where sed run commands and perform execution over a single input line, read from input file. As it is stated that sed executes one line at a time, the current line(s) that are being processed by sed are kept in the pattern space.

2. **Hold space:** The hold space is another supplementary buffer area. The hold space is utilized to hold input text lines temporarily during the execution of other lines kept in the pattern space by sed. It permits you to exchange or append data between the pattern space and the hold space, but you cannot execute and process the typical sed commands on the hold space. The pattern space gets free at the end of every execution cycle. However, the content of the hold space is retained from one cycle to another, not free between the cycles. Table 6.4 shows the five commands that are associated with hold space for various operations.

6.5.7 Quit

There is another functionality of the sed editor where you can stop the operation to a set of lines. This can be done with the help of the *quit (q)* command that instructs sed to stop reading new input lines as well as stop transmitting them to standard output. The basic syntax of the *q* command is given as follows:

```
'[line-address]q'
```

The quit (q) command takes only a single line address, i.e., does not take a range of addresses. Once a line address is matched, the operation will be over and terminated, for example:

```
$sed '50q' hello.txt
```

It quits processing when the 50th line is reached and delivers the first 50 lines of the *hello. txt* file to standard output. This may be very useful to save time only when you need to process some part of the file from its beginning.

6.6 Advanced sed Commands

This chapter highlights the features of various basic sed editor commands that may fulfill your normal daily text-editing requirements and operations [4]. This section discusses some more advanced sed commands with their features. Such commands might not be used frequently, but awareness of their usability is very important.

TABLE 6.4

Various Hold Space Commands

Command	Description
h	Copy data from the pattern space to the hold space
H	Append data from the pattern space to the hold space
g	Copy data from the hold space to the pattern space
G	Append data from the hold space to the pattern space
x	Exchange contents between the pattern space and the hold space

- **Line information:**
  There are three commands to provide and print the line information from the input data stream or file:
- **The print (p) command: to print a line text:**

```
$ cat house.txt
This is a house number 1.
This is a house number 2.
This is a house number 3.
This is a new house number 4.
This is a house number 5.
This is a new house number 6.
```

The p command prints lines that comprise the matching text *"new house"* in the input file:

```
$ sed '/ new house/p' house.txt
This is a new house number 4.
This is a new house number 6.
$
```

To print all lines as specified by the line range:

```
$ sed '3,5p' house.txt
This is a house number 3.
This is a new house number 4.
This is a house number 5.
$
Print all lines from the third to the fifth position line.
```

- **The line number (=) command: to print line numbers of each line, represented as the equal sign character:**
  It is also known as the equal sign command that displays the line number of each line from the input data stream or file.

```
$ sed '=' house.txt
1
This is a house number 1.
2
This is a house number 2.
3
This is a house number 3.
4
This is a new house number 4.
5
This is a house number 5.
6
This is a new house number 6.
$

You can search a particular line and print its line number.

$ sed -n '/new house/{
   > =
```

```
   > p
   > }' house.txt
4
This is a new house number 4.
6
This is a new house number 6.
$
```

- **The list (l) command: to list a line, represented by the l character sign**:

  The list command (l) is used to display both the text and nonprintable characters from a data file. Every nonprintable character is printed either in its octal values, headed by a backslash, or in the typical C-style terminology for common non-printable characters, such as \t for tab characters. For example:

  ```
  $ cat house1.txt
    This house        has     one     address.
  $
  $ sed 'l' house1.txt
  This\thouse\thas\tone\taddress.$
  $
  ```

  The tab character places are displayed with the \t arrangement, and the dollar sign shows the newline character that put at the end of the line.

- **Working with a file:**

  As explained in the s substitution command section, the substitution flag allows you to work with files. But some other sed commands are also available which allow you to work with a file without using the substitution command. These commands are named as the *r (read)* and *w (write)* commands.

- **The *read (r)* command: reading data from a file:**

  The read (r) command permits to include or insert data that is kept in a separate file. The syntax of the read (r) command is given as follows:

  `'[address]r filename'`

  In the above syntax format, the path of *filename* can be mentioned as relative or absolute path from where you are going to take data and insert into the input file. The read command does not provide the facility to mention the range of addresses. It uses only a single line number or text pattern address. For example:

  ```
  $ cat hello1.txt
  This is a hello line.
  This is a hello message.
  $
  $ sed '4r hello1.txt' house.txt
  This is a house number 1.
  This is a house number 2.
  This is a house number 3.
  This is a new house number 4.
  This is a hello line.
  This is a hello message.
  This is a house number 5.
  This is a new house number 6.
  $
  ```

  The sed editor takes the text from the *hello1.txt* file and inserts it into the *house. txt* file. As mentioned in line address, the insertion takes place after the fourth line of the input file, i.e., *house.txt*. Now, type the below command and see the output, and compare with the above:

```
$ sed '/new house/r hello1.txt' house.txt
```

- **The *write (w)* command: writing data to a file**:

    The write (w) command is used to write lines into the mentioned file from the input data stream or file. The syntax of the write command is given as follows:

```
'[address]w filename'
```

In the above syntax format, the path of *filename* can be represented as relative or absolute path into which you are going to write data from the input file. The mentioned address may be defined as a single line number address, a text pattern address, or an address range of line numbers or text patterns. For example, the below command writes the third, fourth, and fifth lines of the house.txt file into the house2.txt file:

```
$ sed '3,5w house2.txt' house.txt
This is a house number 1.
This is a house number 2.
This is a house number 3.
This is a new house number 4.
This is a house number 5.
This is a new house number 6.
$

$ cat house2.txt
This is a house number 3.
This is a new house number 4.
This is a house number 5.
$
```

The sed editor takes lines from the third to the fifth position from the *house.txt* file and writes those lines into the *house2.txt* file.

- **Changing the execution flow**:

    Generally, the sed editor executes all the commands starting from the first line and moves toward the end. Like the other editors, it also provides a way to change the flow of command execution corresponding to structured programming environment.

- **Branching**: The branch (b) command ignores a section of commands based on an address space condition and allows to execute a set of commands within the data stream. Basically, it permits to shift the control to another line in the script file. The basic syntax of the branch (b) command is given as follows:

```
[address]b[label]
```

In the syntax, it is optional to mention the label. If you do not mention the label, then control is shifted to the end of the script; otherwise, execution continues at the line with the mentioned label condition.

- **Testing:** The test (t) command behaves like the branch command, and it is also used to update the execution flow of sed commands stored in a script. The test command jumps to a label based on the successful substitution outcome that must be performed on the current address line. It indicates a conditional branching. The basic syntax of the test (t) command is given as follows:

```
[address]t [label]
```

If the substitution command performs its operation successfully and substitutes a pattern, then the test command jumps to the mentioned label. If you do not mention the label, then control jumps to the end of script if the test condition succeeds. If the label is mentioned, then execution continues at the line followed with label condition.

- **Using the ampersand (&) sign:**
  In the substitution command, the ampersand sign (&) is utilized to indicate the matching pattern. By using the ampersand (&) sign, you can replace it with pattern matched. If you want to use ampersand (&) as a literal, then use the backslash sign with ampersand as "\&" to output a simple ampersand character. For example:

```
$ echo " This is a new house with new car." | sed 's/new/"brand &"/g'
This is a "brand new" house with "brand new" car.
$
```

When the pattern matches the word *new*, "*brand new*" appears in the substituted word. Hence, the *new* pattern is replaced with the "*brand new*" pattern in the input string, because here the ampersand (&) sign indicates the matching pattern, i.e., *new*. So "*brand &*" means "*brand new*."

- **Negating a command using the exclamation sign (!):**
  The exclamation sign (!) is used to negate the operation nature of a command. In such a condition, a command could not perform the operation on the given condition; rather, it performs the operation in the opposite nature of the given condition, i.e., reverse the operation nature of a command. For example:

```
$ sed '/new/!p' house.txt
This is a house number 1.
This is a house number 2.
This is a house number 3.
This is a house number 5.
$
```

Generally, the print (*p*) command prints the lines that match the given pattern, and it would print only the lines that comprise the word *new*. But by adding the exclamation mark (!) to the *p* command, it will reverse the operation of the print command and now it will print all the lines that do not comprise the word *new*.

6.7 sed Advantages

sed is a non-interactive and most prominent text-processing tool. It can be used as an editing filter, permitting to make fast changes in file data. It has some key advantages that make it quite popular among the Linux users, which are given as follows:

- sed is a unique and non-interactive command line text editor able to perform complex editing tasks very simply.
- It uses regular expressions for pattern matching.
- It allows simple programming that permits users to create a separate file that comprises commands, referred to as sed script.
- It is most often used for simple, one-line tasks rather than long scripts.
- It is used to create modified filters for extracting and manipulating data in text files.
- It is used to perform in-a-place editing and selective printing feature of text files.
- sed programs are basically small and simple.
- The s (substitute) command of sed is one of the most important and widely used commands.
- In sed script program, you can describe the function of each line as comment by starting the line with the # symbol.
- It provides the multiline processing feature.
- sed is an open-source editor and you can get the latest version of the GNU sed editor from its GNU page free of cost and install it on computer under GPL.

6.8 sed Drawbacks

The sed editor performs many useful tasks, but on many instances, you require more complex processing. In such a situation, you need a more powerful language like awk or perl as compared to sed. There are some limitations associated with sed in comparison with other advanced filter tools, which are listed below:

- It uses a heavy syntax format that is very difficult to remember and use in some instances.
- There is no provision to go back in the file.
- There is no provision to do forward references such as /..../+2.
- There are no facilities to manipulate numbers.
- In POSIX standard, sed has a limit on line lengths for the pattern and hold spaces. In GNU sed, there is no built-in limit on line lengths due to the use of the malloc() function.

6.9 Summary

sed in Unix/Linux system stands for stream editor, and it is a non-interactive, highly powerful text-editing tool. It can manipulate a stream of data based on the rules that are provided in the command line or stored in the sed script file, which may referred complex program. It can be used as an editing filter that is considerably faster than an interactive editor. The main applications of sed scripts are performing multiple edits to the same file, marking a set of search and replacement edits over a set of files, extracting relevant information from a file, and performing edits in a pipeline. The substitute command replaces matching patterns in several lines, while other commands, for example, append, insert a new text. sed has two types of internal storage space: The first is the pattern space, which is an active buffer; and the second is the hold space, which acts as a supplementary buffer. The sed editor offers a way to change the execution flow by using branching commands or testing commands. The main limitations of sed include a heavy syntax format, no provision to go back in file, and no facilities to manipulate numbers.

6.10 Review Exercises

1. Define the term pattern matching. Give some examples of pattern-matching programming language.
2. Define sed usability. Which programming language is used to implement sed?
3. What are the differences between the pattern buffer and the hold buffer in sed?
4. Define the basic syntax, comment, and addressing mechanism of a sed script.
5. How to distinguish a sed script file from others? Write a sed command to substitute "bestcity" in place of "citybeautiful" in a file.
6. What will be the output of the command "% echo "STRING" | sed 's/[0–9]*/& &/'" if the STRING is (i) 326 fgr and (ii) fgr 326? If the two outputs are not the same, then modify the sed commands for the outputs to be the same.
7. How does the output differ in the command "echo "Second" | sed 'i\First'" as compared to "echo "Second" | sed 'a\First'" in Linux terminal?
8. Write the sed command for the following:
    a. "Sed 's/\/bin\/bash/\/bin\/csh/' /etc/passwd" using "!"
    b. "Sed 's/\/usr\/local\/bin/\/common\/bin/' <old >new" without using the "\" delimiter
    c. To transform characters "a" to "d," "b" to "e," and "c" to "f" in a file
    d. To substitute "1" with "2," "3" with "4," and "5" with "6"
9. The *echo* command to print a string on the terminal is given as follows: $ echo "first second"; then:
    a. Modify the command using sed to obtain the string "second first" as the output.
    b. If "First" and "Second" were present instead of "first" and "second," then write the appropriate command to get the correct output.
    c. If "1st" and "2nd" were used instead of "first" and "second," then modify the sed command from part a) to get the output "2nd 1st."

10. How would you delete a particular line using sed in file.txt file? Give the command.

11. Write the sed command for the following:

    a. To reverse the first four characters in a line of a file

    b. To print the first 10 lines present in a file

    c. To print the lines containing the words "second" and "fourth" from the stream

    d. To modify the third line of a file with "this is the third line"

    e. To delete all the comment lines in a C program containing comment

12. Write the terminal command to substitute "begin" with "START" and "end" with "FINISH" in a file using the following:

    a. Pipelining and sed

    b. Only sed

13. What will be the output of the sed command "sed '3r<newfile>' <myfile>"?

14. You are writing a C program in a team and you have not mentioned the version number and the author name of the program in the code. Write a command using sed to insert the author name and version at the start of the file given below:

```
Version : 1.0
Author_name : sernith
Filename: add_here.c
#include <stdio.h>
int main(){
printf("Add here");
}
```

15. Suppose that you are writing a C program, and after writing it, you realize that from the third line onward, you made a mistake by writing variable 'c' in place of variable 'b'. Write the sed command to replace all the occurrences of "b" with "a" from the third line onward and save the changes made in the file using piping with the *cat* command.

16. How will you use pipeline with the sed command to make the command work like the **read command** in sed?

17. Write the sed command to include vertical tab spaces between every line in the given piece of text in a file named as "input.txt."

    input.txt:

    I am a cse student studying in an engineering college with the various technical clubs such as ACM. ACM is the official technical club of CSE department. A good number of students are part of this club. This technical club organizes various technical activities related to computer science. Some popular activities are OOPs, C, python and data structure programming context and hackathons.

18. Write a command using sed to make a backup of the file before editing it.

19. Write a command to get the list of all the usernames from the **/etc/passwd** file.

20. Highlight some key pros and cons of sed.

References

1. Jan Goyvaerts and Steven Levithan. 2012. *Regular Expressions Cookbook*, O'Reilly Media, Inc.
2. Ken Pizzini, and Paolo Bonzini. 2018. *GNU sed, a Stream Editor*. Open Source under GNU Free Documentation License.
3. Dale Dougherty, and Arnold Robbins. 1997. *Sed & Awk*. O'Reilly Media, Inc.
4. Richard Blum, and Christine Bresnahan. 2015. *Bible- Linux Command Line & Shell Scripting*. John Wiley & Sons.
5. William E. Shotts, Jr. 2012. *The Linux Command Line: A Complete Introduction*. No Starch Press, Inc.
6. Sumitabha Das. 2013. *Your UNIX/Linux: The Ultimate Guide*. The McGraw-Hill.

7

Advanced Filters: awk

For complex text manipulation, Linux users, programmers, and system administrators have promptly used sed and awk tools. You are already familiar with sed filters as elaborated in the previous chapter. Awk is a convenient and expressive programming language that allows to perform easy manipulation of structured data and helps to create formatted reports. This language is utilized and being present on various computing platforms. The awk language is specified by the POSIX standard. The first awk version was written in 1977 at AT&T Bell Laboratories and came into the light with Unix Version 7, around in 1978. Awk also exists as nawk (newer awk) and gawk (GNU awk). The GNU implementation, gawk, was written in 1986 and well suited with the POSIX standard of awk language. It means all the programs written in awk language should be executed with gawk. The gawk is one of the most famous freely available open source, installed and used on Linux distribution, as well as on other freely available Unix systems, like FreeBSD and NetBSD. This chapter not only describes awk programming language in general but also highlights gawk features, referred to as GNU awk. Further, we tried to elaborate all topics and awk features with some key examples that make your learning very interesting and informative.

7.1 Awk Introduction and Concept

Awk is a programming language for text processing and mostly used as a tool for data extraction and reporting. It was written by Alfred Aho, Peter Weinberger, and Brian Kernighan at Bell Labs in the 1970s and launched in 1977. The name of awk is derived from the surnames of its developers. It is a filter and also well known as a pattern matching programming language. The function of awk programming is a set of operations that perform on textual data based on input pattern and generate reports. Basically, it searches files for lines that comprise input patterns. When a line matches with one of the input patterns, awk performs mentioned actions on that line. The awk continues to process on lines until it reaches the end of the input file. One of the very prominent features among few Unix filters that awk can do computation. Awk also accepts extended regular expressions (EREs) for pattern matching and also provides C-type programming constructs along with numerous functions and built-in variables. The familiarity awk may also help you to understand the concept of perl language. Because perl programming uses most of the awk features. Due to the programming language feature of gawk, it brings stream editing one step ahead as compared to sed editor, because sed editors have only editor command features, not programming language feature. The key programming feature of gawk is report generation. Because it helps to extract data from large text files and thereafter formats them into a readable report. With the awk comprehensive programming language feature and concept, you can perform the following tasks:

- Declare and describe variables to store data.
- Frame a text file as comprised of records and fields, like textual database.
- Do arithmetic and string operations.
- Utilize programming construct concepts, such as loop and conditional statements, to incorporate logical control over execution.
- Generate formatted and readable reports by extracting data elements
- Declare and describe user functions.
- Run Unix/Linux commands from a script file.
- Easy interaction and execution with numerous input streams.
- By using regular expressions, you may separate records and fields.
- Perform manipulation over inputs.
- Interactive and dynamic way of adding built-in functions.

7.2 Awk Features over Sed

Every tool has its own specific features with pros and cons. It is totally depending upon you to get the best result from a specific problem. But, this section tries to highlight the requirement of sophisticated processing of tasks to get the required result. As discussed in the preceding chapter, sed, stream editor, works with the stream of character on a per-line basis by using simple conditions like pattern matching and address matching. One of the prompt uses of sed is to search and replace a particular character or pattern in a file. You may write simple program with sed commands, and uses a regular expression such as * (zero and more occurrence of many characters) and q(quit). So sed is preferable for simpler text processing work, but for high and complex text processing, awk is preferred [1,2]. Awk is a tool that uses programming language for processing text, stored in a file. Awk treats a file as a collection of records, and by default, each line treats as a record. Further, each line is broken up into the sequence of fields. So we can say that a line is a collection of fields, and the first word in a line is treated as 1st field, then second word is treated as 2nd field, and so on. Therefore, the number of words in a line is equal to the number of fields (NF) in that line. Awk read a line at a time, scanned each field of line for every pattern, as mentioned in the awk program, and performed action associated with that pattern, if a pattern is matched with a field. That's why awk is also known as line-oriented editor. It offers high robust programming concepts involving if-else, while, do-while, for loop (C style), array iteration, and others. The awk pattern includes regular expression and special symbols, but in the latest version, it also includes multidimensional array features. The programmer may also create relational patterns, group of patterns, range, and other to solve a complicated problem, where user required more control over various operations. The GNU awk (gawk) has several extensions, including true multidimensional arrays in the latest version. There are many other variations of awk including mawk and nawk. The mawk is an interpreter for the AWK Programming Language and written by Mike Brennan. The nawk stands as "new awk" and written by Brian Kernighan.

As you are aware that text files can be roughly or deeply structured. Awk views a file as a tabular format, comprises items in a column that can be considered as a deeply structured. By using the awk program, you can reorder the column data, and even, you may change columns into rows and rows into columns.

- **Summarized features of AWK:**
  - It considers any input file as a database and process using variables.
  - Awk offers various built-in functions and operators for programming, like arithmetic, string, and time functions. These functions ease manipulation over the inputs.
  - It offers many tools to produce formatted reports.
  - It provides regular expressions, loops, conditional statements, arrays, and other programming constructs.
  - It breaks each input line into various fields, i.e., identify individual words that are available for processing with an awk program.
  - Initially, it was created as a programmable editor, but soon after users realized that awk program may perform extensive range of other complex tasks as well.
  - The enhanced version of awk, namely nawk, extends additional support for writing bigger programs and handles general programming problems very easily.

7.3 Structure of an AWK Program

The awk programming language is different from other programming language, due to the data-driven nature of awk programs. The data-driven nature is basically described those data that involve in operation as per the program definition when it found. Due to this, awk is also widely known as pattern scanning and processing language. On the contrary, most programming languages are procedural, and you must describe all steps of the program in detail. Procedural languages are usually much stiffer because it requires a clear description of the program data that are going to be involved in processing. In such condition, awk programs are generally easy to read and write. The key operation of awk programs is to search lines from files that comprise or match any set patterns or conditions, supplied by the user. When a line matches with any of the mentioned patterns, then awk performs stated actions on that line and goes up to the last line of input file. The awk program comprises series of rules. Every rule specifies one *pattern* rule and another *action* rule. The *pattern* rule specifies that which pattern to be searched, and *action* rule specifies that what kind of actions to be performed after finding the mentioned pattern. Therefore, grammatically, awk program is a sequence of one or more *pattern-action* statements, i.e., a rule comprises a pattern followed by an action [3]. The action is enclosed in { } to distinct it from the pattern. It is to note that newlines are typically treated as a separate rule. The awk program structure comprises optional BEGIN statement, pattern-action pair, and optional END statement. Therefore, the basic structure of the awk program is as mentioned below:

An awk program consists of:

```
BEGIN {action}
pattern {action}
pattern {action}
pattern {action}
.........
.........
.........
pattern {action}
END {action}
```

- An optional BEGIN segment
  - execute before reading input
- Pattern-action pairs
  - Process input data
  - For each pattern matched, the specified action is performed
- An optional END segment
  - execute after the end of input data

7.4 Writing and Executing AWK Program

There are many approaches to execute awk command. The command line format of the awk is given as follows:

```
awk options 'program' file(s)
```

As per the above-mentioned syntax, it executes the program, on the given set of input, one or many files; here program is *pattern-action* statements. There are three ways to execute the awk program.

- **Program and input files are provided together as command line arguments:**

  ```
  awk 'pattern-action statement' input-file1 input-file2 ...
  ```
  It read a line at a time from one or more files as mentioned on command line. It is very useful for executing small- or middle-level awk programs in size, directly on command line to avoid creating a separate file for such awk programs.

- **Program is provided as command line argument and input is taken from standard input (keyboard):**
  If no input is mentioned on command line, then awk reads from standard input, as mentioned below:
  ```
  awk 'pattern-action statement'
  ```

In this way, you can omit the input files from the command line and execute awk that comprises only pattern-action statements. Here, awk takes the standard input, i.e., from the keyboard. It will continue until press Ctrl-d, i.e., end-of-file notification. This procedure is suitable when the program is short (a few lines).

- **Program is read from a file**:

  If an awk program is very large, then it is more appropriate to write the whole program into a separate file, known as awk program or awk script. This script file is used by awk to perform various operations on the supplied input files, as mentioned in the given below syntax:

  ```
  awk -f script_file input-file1 input-file2 ...
  ```

  The -f option informs the awk utility to get the awk program from awk script. Any filename may use as a source file. For example, you could place the below program:

  ```
  BEGIN { print "Hello, Welcome You" }
  ```

  into the file named *"xyz."* Then, this command could be executed as given below:

  ```
  awk -f xyz
  ```

But if you wish that this should be identified as an awk program file, then add an extension ".awk" to the filename, such as, *xyz.awk*. It does not affect the execution of the awk program, even it makes more familiar way to manage all awk program files in your system.

7.4.1 To Make Executable awk Programs

Once you are aware of the procedure to execute an awk program. Then, you may wish to write self-contained executable awk program. This may be done by using the "#!" script mechanism in awk programs. For example, in the file *"xyz"* or *"xyz.awk,"* you just add the *"#! /bin/awk -f"* statement as first line of the awk program in file as given below:

```
#! /bin/awk -f
BEGIN { print "Hello, Welcome You" }
```

Then, save file and make it executable by using *chmod* utility. After that, you may simply type *"xyz"* or *"xyz.awk"* at the shell prompt to execute awk programs:

```
awk -f xyz
   OR
awk -f xyz.awk
```

7.4.2 Standard Options with awk

Table 7.1 lists the standard options associated with awk.

TABLE 7.1

List of Option Associated with Awk

Option	Notation Detail	Description
–F *fs*,	--field-separator=*fs*	**Specifies and sets the input FS in a line to fs.** This is similar as a predefined and built-in variable FS. Gawk permits *fs* to be a regular expression. Every record or input line is split into fields by using space bar key or any other user-definable character as FS. Awk refers to each of the fields by predefined variables $1,$2,$3...,$n. $0 variable assigned to access the whole record
-v *var=value*	--value assign	Before starting the execution, allocate the value *value* to variable *var*. It is declared and present in BEGIN block of awk programs
–f *file*	--file=*program_file_ name*	Specifies a *program_file_name* to read awk programs rather than from the first argument
–mf=NNN, –mr=NNN	--Memory limit	To assign different memory limits to the value NNN, where f and r are flags. Flag f sets the maximum NF, and Flag r sets the maximum record size. These lags are joined with –m option. This option is not offered in gawk, because gawk has no limit on memory
–w compat	--compat	Use compatibility mode to allow gawk to perform like awk
-W version -W	--version --copyright	Print gawk version and copyright information on the error output

7.5 Awk Patterns and Actions

The awk is a line-oriented programming language and its program is comprised of pattern-action statements. In some cases, it also referred to as pattern-action statement language. The pattern section comes first and thereafter the action section. The action statements are enclosed in **curly** braces, { }, which separate them from pattern sections.

Therefore, syntactically, a rule comprises a pattern followed by an action. In a pattern section, rules are usually separated by newlines. The basic structure of awk programs is given as follows:

```
pattern { action }
pattern { action }
pattern { action }
.......
.......
```

For each pattern, an action can be mentioned. This action can be executed on each line that matches the pattern. Awk reads one line at a time, and for each input line, it evaluates the rules, mentioned in the pattern section, and if the pattern/rule matches the current input line, then it executes the assigned action. If pattern/rule is missing, the {action}is applied to each single line of input, i.e., all lines. Similarly, if {action}is missing, then matched entire line is printed. Both pattern and {action} are optional, but both cannot be missing. In awk programs, if any line starts with the # character, then that whole line treats as a comment line.

7.5.1 Pattern

Basically, patterns regulate the execution of actions, matched the pattern, and then execute the associated action. Table 7.2 describes various types of pattern structures with a condition under which they do matching.

TABLE 7.2

List of Various Types of Pattern Structure in awk

Pattern Statements	Description
BEGIN { statements }	The statements are executed once before the first input line is processed
END { statements }	The statements are executed once after all input lines or records are read
General expression { statements }	The statements are executed at each input line where the specified expression is true. The general expressions can be comprised of numbers, operators, quoted strings, user-defined variables, functions, or any of the predefined variables
/regular expression / { statements }	At each input line, the statements are executed which comprise a string matched with specified the regular expression. The regular expression uses extended set of metacharacters and generally referred to as string-matching pattern
compound pattern { statements }	A compound pattern can be combined with the Boolean operators || (or), &&(and), and f (not). The statements are executed at each input line where the compound pattern is true
pattern1 ? pattern2 : pattern3 {statement}	If the first pattern, *pattern1*, is true, then the second pattern, *pattern2*, is used for testing; or else, the third pattern, *pattern3*, is evaluated. Only one of the pattern, second or third pattern, is executed
pattern1, pattern 2 {statements }	The *pattern1* and *pattern2* are referred to as matching range patterns. It matches all input records starting with a line that matches *pattern1*, and continuing until a record that matches *pattern2*, inclusive. It does not combine with any other sort of pattern expression

7.5.2 Actions

In a pattern-action statement, action statements are enclosed in curly braces { and }. The pattern tells when the action is to be executed. Sometimes, the assigned action statement is very simple in nature, like single print or assignment. But other times, it may comprise a series of numerous statements separated by semicolons or newlines. The statements in actions may be comprised of usual variables or array assignments, input/output commands, built-in functions, control-flow commands, and user-defined functions, as similar to the most of the programming languages. The action statement may follow the following structure:

```
Common expressions statement, comprising with constants,
variables, assignments, function calls and others.
printf (format, expression-list)
print expression-list
if (expression) statement else statement
if (expression) statement
for (expression ; expression ; expression) statement
for (variable in array) statement
while (expression) statement
do statement while (expression)
break
continue
next
exit expression {statements}
exit
```

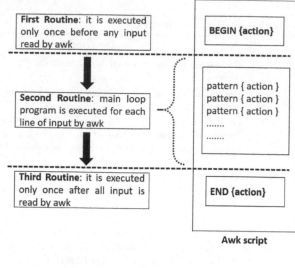

FIGURE 7.1
BEGIN and END pattern in awk script.

7.6 BEGIN and END Patterns

The awk program allows you to add two special procedures or actions along with main awk program, namely BEGIN and END. The BEGIN and END actions do not match any input lines. The statements in the BEGIN procedure are executed before any input reads by awk; subsequently the statements in the END procedure are executed after all input has been read by awk. For example, if you want to print something before starting the execution of main awk program like a heading or the detailed description of program, then the BEGIN section may be used. Similarly, the END section is beneficial to print some concluding remarks or statements of the program after execution is over. Hence, BEGIN and END provide a way to increase control over the program initialization and closing. BEGIN and END do not combine with other patterns, for example:

```
$ gawk 'BEGIN {print "Welcome to the New World!!!!"}'
Welcome to the New World!!!!
$
```

The BEGIN and END sections are optional. Broadly, you can divide an awk script into three major sections: what happens before main program, what happens during main program, and what happens after main program processing [4, 5]. The flow of execution among these three sections is shown in Figure 7.1.

7.7 Awk Variables

A variable is an identifier that references a value. In awk programs, the defined expression may include different types of variables, like user-defined variable, built-in variable, and fields. To declare a variable, you give a name and then assign a value to it. Such variables are referred to as user-defined variables that comprise letters and digits and underscore

TABLE 7.3

AWK Built-in Variables

Variable	Description
ARGC	Number of arguments on the command line
ARGV	An array comprising the command line arguments, indexed from 0 to ARGC-1
ENVIRON	An associative array of environment variables that comprise the values of the current environment variable
FIELDWIDTHS	A whitespace-separated list of numbers defining the precise width of each data field
FILENAME	Name of the current input filename. If no files are specified on the command line, the value of FILENAME is –
NF	Number of fields (NR) in current input record
NR	Number of the current input lines or records.
FNR	Like NR, but relative to the current file, i.e., the input record number in the current input file
FS	Input FS character, by default a space or blank
$0	Complete input record
$n	nth field in the current record; fields are separated by FS
IGNORECASE	Controls the case sensitivity of all regular expression operations. If IGNORECASE is not equal to zero, /rS/matches all the strings rs, rS, Rs, and RS. If with all awk variables, the opening value of IGNORECASE is zero, so all regular expression operations are normally case sensitive
OFMT	Output format for numbers ("%.6g" by default)
OFS	OFS character, by default a space or blank
ORS	ORS character (by default a newline)
RS	The input record separator character (by default a newline)
RLENGTH	Length of the string matched by match () function; –1 if no match
RSTART	First position in the string match by match () function; 0 if no match
SUBSEP	The character is used to separate multiple subscripts in array elements, by default "\034"

but do not start with a digit. Each variable stores a string or a number or both. Since the type of a variable is not declared in awk, so you don't have the provision in awk to define what type of value will be stored in a variable. Therefore, awk uses the suitable value based on the context of the expression, whenever, it is necessary, awk will change a string value into a numeric one, or vice versa. Several predefined variables are available for you to use in the awk program as needed, referred to as "built-in variables." All built-in variables are mentioned in uppercase names and summarized in Table 7.3. Field variables are described the fields of the current input line that are referred to as $1, $2, $3 through $NF; $0 refers to the whole line. Section 7.9 describes in detail field variables.

7.8 Records and Fields

There are numerous ways of providing the input data to an awk program. The most common way to provide the data to awk program through a file. In this arrangement you put all input data in a file, say *data_file* and typed as below:

```
awk 'program file' data_file
```

Awk reads its standard input if no filenames are given. If the filename is given say *data_file* then awk read input from *data_file* to run awk program. Generally, it assumes that input file is structured, not just as an unending string of characters. To executing the awk program, awk reads a line of text from input file, *data_file* . In input text line, every data field is determined by a field separation character. In awk, by default, field separator (FS) character is taken as any whitespace character, like tab or space character. In each input line, the whole line is referred to as a *record* and every word is stated as a *field*. By default, fields are separated by whitespace, like words in a line. The whitespace in awk means any string of one or more space characters, tabs, or newlines. The characters separating the fields are normally referred to as delimiters.

The basic purpose of fields gives more appropriate ways to understand these pieces (fields) of record/line. You can access the information of each field of a record with the help of field variables. These field variables make awk program so powerful to perform fast computation on data file.

You use a dollar sign ($) to refer to a field of a record in an awk program, followed by the numeric number. Thus, $1 refers to the first field, $2 to the second, and so on as given in the below description:

- $0: signifies the complete line of text.
- $1: signifies the first field or word in the line of text.
- $2: signifies the second field or word in the line of text.
- $3: signifies the third field or word in the line of text.
- $n: signifies the nth data word or field in the line of text.
- NF: the number of fields in the current record or in the line of text.
- NR: total number of input records or lines read so far.

For example, suppose the following is a line of input in awk programs:

```
Hello you are welcome to Linux class.
```

In the above-mentioned input line, the first field, or $1, is "*Hello*," the second field, or $2, is "*you*," and so on. Note that the last field, $7, is "*class*." Because there is no space between the "s" and the ".", the dot period is taken and put into the part of the seventh field. NF is a predefined variable whose value is the NF in the current record. Awk automatically updates NF value every time when it reads a record. No matter, the last field in a record can be represented by $NF, regardless of the numbers of fields. Hence, the value of $NF is the same as $7, which is "class." but if you wish to use a field beyond the last one, such as $8 (the record has only seven fields), then you will get the empty string. There is a limit on record length; it is usually about 3,000 characters.

You can modify the FS character with -F option, followed by the delimiter character on the command line. The below example would update the FS by tab character:

```
$ awk -F"\t" '{ print $3 }' data_file
```

In a subsequent part of this chapter, we are going to use *employee.txt* file as an input file for various examples of awk programs. Each line contains the name of the employee, age in years, salary in dollars, and the country where it is currently posted.

```
$ cat employee.txt
Sanrit      29      7700      India
Jone        43      6500      Australia
Smith       33      8700      USA
Joju        25      6900      France
Rosy        37      5780      England

$ gawk '{print $1}' employee.txt
Sanrit
Jone
Smith
Joju
Rosy
$
```

This program uses the $1 field variable to display only the first data field for each input line of the text. If you are reading a file that uses a different FS characters, then specify -F option:

```
$ gawk -F: '{print $1}' /etc/passwd
root
bin
daemon
adm
lp
sync
shutdown
[...]
```

In the above example, it displays the first data field of each line in the password file of the Linux system. In the */etc/passwd* file, it uses a colon (:) to separate the data fields, so it prints the first field of each line in the password file.

7.9 Simple Output from AWK

The basic awk operations are printing field, selecting input, and transforming data. To view the output of various operations on data of the given input file, use inbuilt **print** and **printf** statement. The **print** statement is used for simple output and **printf**, fancier or formatted output.

7.9.1 The *print* Statement

Use the print statement to produce output with simple, standardized formatting. You can mention only the strings or numbers to print, in a list separated by commas, as shown in the below syntax:

```
print item1, item2, item3, item4...
```

The items to print may be constant strings or numbers or field variables (such as $1), or any other awk expressions [4, 6]. Numeric values are converted to strings and then printed. The simple statement, used *"print"* with no items, is equivalent to *"print $0"*: it prints the whole current input record or line. To print a blank line, use "print "," where "" means empty string. To print a fixed piece of text, use a string constant, such as "Hello, Welcome," as single item. The print statement is entirely general for computing. Hence, at many instances, we need to print the desired fields of selected records or the output of some calculation. The following subsequent sections are going to explain various ways to see the output of awk operation.

- **Printing every line**:
  In *pattern-action* sections of awk program, if an action has no pattern, then by default, action is performed to all input lines. The *print* statement, mentioned in action sections, prints the current input line, so the program code is

  ```
  { print }    //print all input lines
  ```
  The field variable, $0, also prints the whole current input line:
  ```
  { print $0} // Similar to the above print statement
  ```

- **To print specific fields**:
  You can print more than one field of each record or line with a single *print* statement. To print the second and fourth fields of each input line of input file *employee.txt*, type the below command:
  ```
  awk '{print $2, $4 }' employee.txt
  29 India
  43 Australia
  33 USA
  25 France
  37 England
  ```
  In *print* statement, if expressions are separated by a comma, then, by default, the output is separated by a single blank space when they are printed. But you may change this default way of printing the fields.

- **Print Number of Fields (NF)**:
  The number of fields is built-in variable to print the total NF in a line. You can use any valid expression after a $ to specify the content of a certain field. Awk counts the NF in the current input line and stores it in variable NF. For example:

  ```
  awk '{print NF, $2, $NF }' employee.txt
  ```
  It prints the total NF, second field, and last field of each input line of input file employee.txt.

- **Compute and print**:
  You can perform some computations among the field values and print the results on the standard output. For example:

```
# cat data_1.txt
```

5	2	2
6	4	6
7	3	4
8	4	5

```
#awk `{ print $1, $1 * $2, $3 }' data_1.txt
```

5	10	2
6	24	6
7	23	4
8	32	5

It prints first field, product of first field with second field, and third field of each line.

- **Printing Line Numbers (NR):**
  Number of Record (NR) or line is another awk built-in variable that counts the number of lines read so far from input file. You can use NR to print the line number as a prefix of each input line of *employee.txt* file as given in the below example.

```
awk {print NR, $0 }' employee.txt
```

1	Sanrit	29	7700	India
2	Jone	43	6500	Australia
3	Smith	33	8700	USA
4	Joju	25	6900	France
5	Rosy	37	5780	England

- **Print text along with output:**
  In the print statement, you can also put some texts along with the field variable and expression. The add-on text must be surrounded by double quotes. For example, by adding some text, you can give some more information in each line of *employee.txt* as given below:

```
awk `{print $1, "is working in", $4, "and getting", $3, "dollars
salary"}' employee.txt
Sanrit is working in India and getting 7700 dollars salary
Jone is working in Australia and getting 6500 dollars salary
Smith is working in USA and getting 8700 dollars salary
Joju is working in France and getting 6900 dollars salary
Rosy is working in England and getting 5780 dollars salary
```
  In the output statement, the text inside the double quotes is printed along with the fields and computed values.

- **Output separators (data field and record)**
  As discussed earlier, a *print* statement contains a list of items, variables, or expressions, separated by commas. By default, the outputs are normally separated

by a single space. But you can use any string of characters as the output field separator (OFS). For this, you can use built-in variable OFS and set your own FS string.
The initial value of this variable is single space string, i.e., ("").

While printing each records of input file, output record separator (ORS) built-in variable
is used to separate the output among the records. Each print statement gives one record
and then ORS string and thereafter prints the next record. The built-in variable, ORS specifies this string. The initial value of ORS is newline character string "\n", so normally each
print statement makes a separate line for each record.

You can set new values to the variables OFS and/or ORS. So that, in output, fields and
records are separated by assigned values. The common place to declare this is in the
BEGIN statement, so that it occurs before any input is read and processed. You may also
assign the values to these variables on the command line, before giving the names of your
input files, by using "-v" option.

For example, in the given below statement, it prints the first and second fields of each
input record, separated by two hash characters (##), with a blank line added after each new
record:

```
awk 'BEGIN {OFS = "##"; ORS = "\n\n"}
      {print $1, $2}' employee.txt

>Sanrit##29##7700##India
>
>Jone##43##6500##Australia
>
>Smith##33##8700##USA
>
>Joju##25##6900##France
>
>Rosy##37##5780##England
```

In another example,

```
$ cat file1
data11,data12,data13
data21,data22,data23
data31,data32,data33

$ gawk 'BEGIN{FS=","} {print $1,$2,$3}' file1
data11 data12 data13
data21 data22 data23
data31 data32 data33
$
```

The print statement automatically puts the value of the OFS variable between each of the
data fields in the output. By setting the OFS variable, you can use any string to separate
data field in the output, as shown in the below examples:

```
$ gawk 'BEGIN{FS=","; OFS="-"} {print $1,$2,$3}' file1
data11-data12-data13
data21-data22-data23
data31-data32-data33
$ gawk 'BEGIN{FS=","; OFS="--"} {print $1,$2,$3}' file1
```

```
data11--data12--data13
data21--data22--data23
data31--data32--data33

$ gawk 'BEGIN{FS=","; OFS="<<-->>"} {print $1,$2,$3}' file1
data11<<-->>data12<<-->>data13
data21<<-->>data22<<-->>data23
data31<<-->>data32<<-->>data33
$
```

The FS and OFS variables describe how gawk program operates data fields in the data stream. Another built-in variable is FIELDWIDTHS that permits you to read the record without using a FS character. In some instances, rather than using a FS character, some data are placed in certain columns within the record. For such cases, you must set the value of FIELDWIDTHS variable to give a desired arrangement of the data in the record. As in the given example, awk ignores FS value and compute data fields based on the assigned field width's size and gives output. In this example, awk uses field width in its place of FS characters"

```
$ cat file2
20212.134245.4437
317+864.26394.024
24946.42682.73365

$ awk 'BEGIN{FIELDWIDTHS="3 5 3 6"}{print $1,$2,$3,$4}' file2
202 12.13 424 5.4437
317 +864. 263 94.024
249 46.42 682 .73365
$
```

The *FIELDWIDTHS* variable declares four data fields, and awk describes the data record accordingly. In each record, the string of numbers is split based on the declared field width value.

7.10 Fancier Output

The print statement is widely used to give fast and simple output. If you want more accurate and control over the output formatting, then use the **printf** statement instead of **print** statement [4]. With *printf* function, you can specify your formatting structure of output, which means to have a full control over the OFS character.

Awk programming tool is widely known for generating formatted output after performing the desired computation over given inputs. So, when you're creating detailed reports, then you need to place data in a desired specific format and place, like the width between each item, formatting choice of numbers, etc. The *printf* statement is catered all such requirements and very similar to the ANSI C programming language library function printf.

The syntax of *printf* statement is;

```
printf ( "format-expression", [item1, item2, item 3, .......... item n])
```

TABLE 7.4

List of Control Character used in Format Specifier

Control-Letter	Description
%c	Displays a number as an ASCII character Like, "printf "%c", 65" output is character letter "A"
%d %i	Both are equivalent and display an integer value. The "%i" description is for compatibility with C language
%e %E	Displays a number in scientific (exponential) notation. For example, `printf "%4.2e\n", 2512` prints "25.12e+02" with a total of four significant figures of which two follow the decimal point. Here "4.2" is a modifier as described. By using "%E", it displays "E" instead of "e" in the output scientific notation
%f	Displays a number in floating-point format For Example, `printf "%4.2e\n", 2512` displays "2512.00," with a total of four significant figures of which two follow the decimal point. Here "4.2" is a modifier as described below
%g	Displays a number either scientific notation or floating point, whichever is shorter or uses fewer characters
%o	Displays an unsigned octal value of a given number
%s	Displays a text string
%x	Displays an unsigned hexadecimal value of a given number, uses a-f from 10 to 15 receptivity
%X	Displays a hexadecimal value, but using capital letters for A through F to represents the decimal number from 10 to 15, respectively
%%	A single % character; no argument is converted

The printf statement syntax is broadly divided into two sections: format-expression and list of items or arguments. The value of format-expression is taken as a string constant in quotes that exactly specify how the formatted output should appear on screen by using both text elements and *format specifiers*. The second part is the list of items or arguments, such as a list of variable names, that correspond to the *format specifications*.

A format specifier is a unique code that uses as a placeholder in output for each mentioned item or argument given with the *printf* command. The first format specifier matches the first item, the second specifier matches the second item, and so on. A format specifier starts with the character "%" and ends with a format control-letter. The format specifier uses the given below syntax:

```
%[modifier]control-letter
```

Here, control-letter is a one-character code that specifies the type of data value to be displayed on output. The modifier is an add-on formatting feature that is optional. Tables 7.4 and 7.5 are mentioned the list of various control-letters and modifiers, respectively, that can be used in the format specifier for formatted output, by using **printf** function.

A format specification may also incorporate modifiers that give more control over the output format of values. The modifiers place between the "%" and the format-control letter. Table 7.5 gives some possible modifiers as listed below.

TABLE 7.5

List of Modifiers

Modifiers	Description
width	This is a number value specifying the minimum width of an output field. Putting any number between the "%" sign and the format control character, then it expands the value to this width. If the value is shorter than the output width, then it covers up with spaces on the left. For example, `printf "%5s", "xyz"` `prints ' xyz'` The width value is a minimum width, not a maximum. If the output is longer than the specified width characters, then it can be as wide as required, like `printf "%3s", "sanrit"` `prints 'sanrit'`
.prec	This is a numeric value that specifies the number of digits to the right of the decimal place for the """, "E", and "f" formats. For the "g", and "G" formats, it species the maximum number of significant digits. For the "d", "o", "i", "u", "x", and "X" formats, it specifies the minimum number of digits to print. But in the case of string output, it denotes the maximum number of characters from the string that should be given on output, like `printf "%.4s", "sanrit"` `prints 'sanr'`
-(minus sign):	The minus sign is used for left justification instead of right justification when placing data in the formatted space. For example, `printf "%-5s", "xyz"` `prints 'xyz '`

7.10.1 Output into Files

By using the redirection operators > and >> , you can store the output into another file instead of printing it on the standard output.

7.11 Arithmetic and Variables

The awk program comprises various expressions in which you can store, manipulate, and retrieve data on the given input data. An expression comprises variables, numeric constants, string constants, operators, functions, and regular expressions. We already discussed regular expressions in detail in Chapters 3 and 5. Like other programming languages, awk programming language also has its own inbuilt variables, operators, functions, arrays, control statements, etc. This section is going to briefly describe the constants, variables, arithmetic operators, and control statements.

7.11.1 Constant

The simplest type of expression is the constant, which always has the same value. There are three types of constants: numeric, string, and regular expressions. Each uses in the appropriate context when you need a data value that isn't going to change. Numeric constants can have different forms but are internally stored in an identical manner.

- **Scalar Constants: Numeric and string constants**:
  - **Numeric constant**: A numeric constant stands for a number. This number can be an integer, a decimal fraction, or a number in scientific (exponential) notation, for example, 2720, 2.720, or 2.720e+2.
  - **String constant**: A string constant consists of a sequence of characters enclosed in double quotation marks, for example, "sanrit."
- **Nondecimal numbers: octal and hex numbers**:
  The awk allows the use of octal and hexadecimal constants in program text. So there is a special notation to represent octal and hexadecimal numbers in binary system as derived from C and C++ languages. In binary string, octal numbers start with a leading "0", and hexadecimal numbers start with a leading "0x" or "0X," so the below example shows the value of binary digits 101 in octal and hexadecimal numbers system:

Decimal 101: decimal value 101 and representation as 101

Octal 101: decimal value 17 and representation as 0101

Hexadecimal 101: decimal value 33 and representation as 0x101

This example shows the difference:

```
$ gawk 'BEGIN { printf "%d, %d, %d\n", 101, 0101, 0x101 }'
101, 17, 33
$
```

- **Regexp Constants: Regular Expression constants**:

Most regexps used in awk programs are constant. A regexp constant is a regular expression description bounded in forward slashes, such as /^beginning and end$/, but some operators are involved in matching and computation operation that cannot be comprised literally in string constants or regexp constants. For such characters, you can represent it in string beginning with another character, namely backslash (\), referred to as escape sequences [2]. Table 7.6 lists some escape sequence characters.

7.11.2 Variable

A variable is an identifier though you can store values at one place in your program and use it later in another part of your program. Variable can be operated completely within

TABLE 7.6

List of Some Escape Sequence Characters

Escape Sequence Characters	Description
\a	Alert or bell
\b	Backspace
\n	New line
\t	Tab
\r	Carriage return
\v	Vertical tab
\\	Literal backslash

TABLE 7.7

List of Arithmetic Operators

Arithmetic Operators	Description
+	Addition
−	Subtraction
/	Division
*	Multiplication
%	Modulo
^	Exponentiation

the program script, and you can also assign value to the variable on the awk command line. To declare the variable, simply you must name it and assign a value. Basically, in awk, variable is divided into three categories: built-in variable, fields, and user-defined variable. We already described the build-in and field variables in earlier sections of this chapter. Those variables that are declared by programmer in program script are referred to as user-defined variables.

Awk variables are case sensitive and take numeric (floating point) or string values. Awk uses suitable variable values based on the context of the expression. Variables do not have to be initialized, by default; user-defined variables are initialized to the null string that has numerical value 0. The following expression assigns a value to x:

```
x = 5
```

x is the name of the variable,=is an assignment operator, and 5 is a numeric constant or the value of variable x. In another example, the below expression assigns the string "Welcome" to the variable z:

```
z = "Welcome"
```

The dollar sign ($) symbol is used to refer fields. In the following way, you can assign the value to the second field ($2) of the current input line to the variable y:

```
y = $2
```

The following expression subtracts 1 to the value of y and assigns output to the variable w:

```
w = y - 1
```

A various number of arithmetic operators and inbuilt functions may be used to evaluate such arithmetic expressions and functions. Various inbuilt arithmetic functions are also described in Section 7.16, and arithmetic operators are listed in Table 7.7.

- **Control-flow statements**
    Like other programming languages, awk also provides several conditional control statements that allow you to make decisions before performing any action. All condition control statement, *if-else* statement, *while* statement, *do-while* statement, and *for* statement, can only be used in action sections of awk programs.

- **A conditional statement**:
    It is introduced by if-else statement that evaluates an expression placed within parentheses, as given in the below syntax:

```
if ( expression )
   action1
[else
   action2]
```

After the expression evaluation, if it is true (nonzero or nonempty), the *action1* is performed. Otherwise else part *action2* is performed.

- **Looping**

    A loop is a concept that allows us to perform one or more actions repeatably. In awk, a loop can be specified by using a **while, do-while,** or **for** statement, as given below:

    - **while Loop**
      ```
      while (condition)
          action
      ```

    - **do-while Loop**
      ```
      The do-while loop is a slight variation of the while loop as shown
      in the below syntax:
      do
        action
      while (condition)
      ```

- **for Loop**

    The *for* loop offers a more compact syntax as compared to *while* and *do-while* loops and produces the same result. The *for* loop syntax is much easier to use because it takes all the required information of a loop within the *for* statement parenthesis as shown in the below syntax:
    ```
    for (set_count; test_count; increment_count)
          action
    ```

    The newline after the right parenthesis is optional. The *for* loop is comprised of three expressions that are given as follows:

    - **set_count**: sets the initial value for a countervariable.
    - **test_count**: gives a condition for loop termination or stop.
    - **increment_count**: sets the value for incrementing the counter each time at the end of each loop, just before testing the test_count condition again.

7.12 Computing with AWK

In awk programs, action part comprises the series of statements separated by semicolons of newlines. In the above sections, you have seen the examples that comprise only a single statement, i.e., *print* statement, in the action section. But this section provides some more examples to show numeric and string computations. This section basically explains the use of inbuilt variables along with how you can create your own variable to perform various numeric and string computations as well as storing data. Some computing operations are mentioned here with the input file *emp_pay.txt*. In this input file, each line contains the name of the employee, age in years, per hour pay in dollars, and the total number of working hours in a day.

```
$ cat emp_pay.txt
```

Sanrit	29	14.5	9
Jone	43	11.3	6
Smith	33	15.0	9
Joju	25	13.0	8
Rosy	37	11.2	7

- **Count the total numbers**:
  To write a program to print the total number of employees who are working more than 7 hours per day:
  ```
  BEGIN
  $4 > 7 { x = x + 1 }
  END { print x, " employees is working more than 7 hours" }
  ```
  For every line in which the fourth field exceeds 7, the previous value of x is increased by 1. So, with the input file *emp_pat.txt* as input, the output of program is:
  ```
  3 employees is working more than 7 hours          // output on screen
  ```
  By default, awk variables, x, used as numbers and started with the value 0, so there is no need to initialize the variable "x".

- **Sums and averages**:
  In another example, to get the sum of total pay of all employees per day, and then average salary per day. For getting the result of the above-mentioned statement, built-in variable NR is used, which holds the numbers of lines that are equivalent to the number of employees. Here is a program that uses to calculate sum and average pay per day:
  ```
  BEGIN
  { for(i=1;1<=NR; i++) ,
      x=x +($3*$4),
  }
  END { print NR, " employees are working"

  print "total pay of all employees per day is", x
  print "average pay all employees per day is", x/NR }
  ```
  The first action collects the total pay for all employees. The END action prints the subsequent outputs.
  ```
  5 employees are working
  total pay of all employees per day is 515.7
  average pay all employees per day is 103.14
  ```

- **Print whole input line**:
  The field variable $0 stores the entire complete line. So use the print function in action part to print $0 value for each input line. It will print the whole input file

lines. If you want to print the last line of input file, then use the print function in END sections of awk programs:

```
{ print $0 }       // For printing all lines
END

END { print $0 }    // For printing , last line only
```

- **Built-in functions:**

  We have already discussed that awk provides numerous built-in variables and built-in functions to perform computing on input values. Section 7.17 described various built-in functions in detail, but apart from these functions, the below program uses three functions, *NR, nc,* and *length* to count the number of lines, words, and characters, respectively, for the input. For ease, we assume that each field as a word for each input line. Here, *length* built-in function calculates the total numbers of characters in the first field, i.e., number of characters in employee names. The below program prints the total number lines, total number of words, and number of characters (only in employee names) of the input file emp_pay.txt.

  ```
  { nc = nc + length($0)
  nw = nw + NF          // NF= no. of field in each input line.
  }
  END {print NR, "lines," , nw, "words,", nc, "Characters" }
  ```

  For the input file emp_pay.txt , the output is:

  ```
  5 lines, 20 words, 23 characters
  ```

- **Some small awk programs:**

  Some small or one-line awk programs are given in tabular form as shown in Table 7.8:

7.13 Handling Text

It is worth mentioning here that most of the programming languages are well capable to manage numbers, but in awk programming language, it can manage both numbers and string of characters. Therefore, awk variables can hold both numbers and strings of characters and suitably translates back and forth as required. The below program evaluates to find the maximum paid employee per hour:

```
$3 > maxpay { maxpay = $3; maxemp = $1 }
END { print "The highest hourly paid rate is", maxpay, "for", maxemp }
```

The output is

```
The highest hourly paid rate is 15 for Smith
```

In the above program, the variable *maxemp* holds a string value, whereas *maxpay* holds a numeric value.

TABLE 7.8

Some One-Line awk Programs

Basic Description	Awk Code
Print the total number of input lines	END { print NR }
Print the eighth input line	NR == 8
Print the last field of every input line	{ print $NF }
print the last field of last input line	{ field=$NF} END { print field }
Print each input line, having more than three fields	NF>3
Print each input line in which the last field is more than 6	$NF>6
Print the total number of lines that comprise "Yellow" pattern	/Yellow/ { nlines++ } END { print nlines }
Print each line that has at least one field	NF>0
Print each line that is longer than 50 characters	length($0)>50
Print the NF in every line followed by the line itself	{ print NF, $0 }
Print the first three fields (in reverse order) of each line	{ print $3, $2, $1 }
Exchange the first two fields of every line and then print the line	{ temp=$1; $1=$2; $2=temp; print }
Print each line with the first field replaced by the line number	{ $1=NR; print }
Print each line after removing the fourth field	{ $4=" "; print }
Print the fields of each line in reverse order	for (i=NF; i>0; i=i - 1) printf("%s ", $i) printf ("\n")
Print the sums of all fields of every line	sum= 0 for (i=1; i <= NF; i ++) sum=sum+$i print sum

Add up all fields in all lines and print the sum	{
	 for (i=1; i <= NF; i ++)
	 sum= sum+ $i }
	 END { print sum }

7.14 String Manipulation

A string constant is formed by enclosing a sequence of characters within quotation marks, like "xyz" or "Welcome, hello everyone." String constants may contain the C escape sequences for special characters as listed in Table 7.8. String expressions are formed by concatenating constants, variables, field names, array elements, functions, and other expressions. For example, the below awk program prints each record headed by its record number and a colon, with no blanks.

```
{ print NR ":" $0 }
```

In awk, you can create a new one string by combining individual input string. This operation is called concatenation. The concatenation operation is performed in the awk program by writing string values one after the other string for the input file *emp_pay.txt* in given below program:

```
{names = names $1 " " }
END { print names }
```

The action part stores the first name of each input line followed by the space. Hence, after all input lines have been read, the names of all the employees comprise in a single string printed on the standard output, as given below:

```
Sanrit Jone Smith Joju Rosy        //output on screen
```

The awk programming language provides numerous built-in string functions that you can use for string manipulation as described in Section 7.17.

7.15 Array and Operators

Array in Awk: An array is a variable but has a special feature to store a set of values. Such values are usually related to some manner. You can access the individual elements by using index value of array. Each index is enclosed in square brackets. In awk, only one-dimensional array concept is provided to store string and number values. There is no requirement to declare arrays, array sizes, and array elements [7]. Like variables, it is considered to be declared when you are assigning a value to an array element or the moment of its use. For example, assigns the string "Welcome" as an element of the array, named x:

```
x[1] = "Welcome"
```

In awk, array concept is slightly different from the other programming languages, like by default, array elements are initialized to zero or an empty string unless initialized clearly. Array increases its size and index automatically that may be any numbers; it can even be a string also. In awk, all arrays are associative arrays. The unique feature of an associative array is that its index can be a number or a string. Therefore, awk arrays are referred to as associative arrays.

You may declare an array variable using a standard assignment statement. Here is the basic syntax of assigning the value to array variable:

```
var[index] = value
```

Here *var* is a variable name, *index* is referred to as associative array index value, and *value* is data element value. Here, there are some examples of array variables, both strings and numbers in gawk:

```
city["Prayagraj"] = "holy-place"
city["Delhi"] = "Capital"
city["Mumbai"] = "Financial Center"
arr[1]=21
arr[2]=20
arr[3]=25
arr[4]=12
arr[5]=04
```

With the above-mentioned variables, you can write the below awk program to understand the array concept:

```
'BEGIN{
> city["Prayagraj"] = "holy-place"
> city["Delhi"] = "Capital"
> print city["Prayagraj"]
> print city["Delhi"]
> var[1] = 21
> var[2] = 20
> var[3] = 25
> sum = var[1] + var[2] + var[3]
> print sum
> }'
```

The output is as given below:

```
holy-place
capital
66
```

- **Iterating through array variables**

    In awk, for every array element, a pair of values is maintained: index of the element and value of element. The problem with associative array variables is that you might not have the information about the index values. The elements are not kept in any particular manner as compared to conventional array. Therefore, an associative array index can be anything. The index value is used to access the element that is saved internally as a string. For example, an array element using arr[1]="San," awk converts the number 1 to a string and stored. If you need to iterate through an associate array in awk, you can use a special format of the *for* statement:

    ```
    for (var in num)
    {
    Statements  //Perform something with num[var]
    }
    ```

    Here *num* is mentioned as the name of array, and *var* is referred to as variable name. The *for statement* loops each time assign the variable *var* the next index value from the *num* associative array. It is significant to remember that the mentioned variable is stored the value of the index and not the data element value. You can simply extract the data element value by using the variable as array index.

7.15.1 Multidimensional Arrays

There is no provision to provide multidimensional arrays directly in awk. However, you can create your own simulated multidimensional array by using one-dimensional arrays. So you can write x[i,j] or x[i][j] by concatenating row and column values with a suitable separator. For this, you can use the nested *for* loops to combine rows and columns for multidimensional array. For example, if you want to create your own simulated two-dimensional array in awk, then you have to use two nested *for* loops, as shown below:

```
for (i = 1; i <= 5; i++)
for (j = 1; j <= 5; j++)
arr[i "," j] = 0
```

The above-mentioned code statements produce an array of 25 elements, where subscripts appear to have the following form: 1,1 1,2 1,3, and so on. But in memory, these subscripts are saved as strings.

- **Deleting array variables**

  In awk, you can delete an array element by removing the respective array index of that element from an associative array. The delete command syntax is as shown below:

  ```
  delete array_Name[index]
  ```

  The delete command is used to remove the associative index value as well as associated data element value from the array:

  ```
  'BEGIN{
  > arr["x"] = 11
  > arr["y"] = 21
  > for (test in arr)
  > {
  > print "The value of index", test," is:", arr[test]
  > }
  > delete arr["x"]
  > print "---"
  > for (test in arr)
  > print "The value of index", test," is:", arr[test]
  > }'
  ```

  The output is as shown below:

  ```
  The value of index x is :11
  The value of index y is :21
  ---
  The value of index y is :21
  ```

  After deleting the index value from array, then you are unable to retrieve it again.

7.15.2 Operators in awk

A variety of operators may be used to produce awk expressions; some key operators are listed in Table 7.9.

7.16 Built-in Functions

The awk programming language offers numerous built-in functions to execute mathematical and string functions [3]. It also provides some built-in time functions to perform the time-related data. In these functions, arguments are passed in C-style, delimited by commas, and bounded by a matched pair of parentheses. But in the case of some functions, awk allows to use functions with and without parentheses, such as print and printf (). The below tables highlighted various awk built-in functions.

TABLE 7.9

List of Operators

Symbol	Meaning
=, +=, -=, *=, /=, %=, ^=, **=	Assignment, both operator assignment and absolute assignment (*var =value*)
?:	C conditional expression. This is the form of *expr1 ? expr2: expr3*. If *expr1* is true, then *expr2* is evaluated; otherwise, *expr3* is evaluated. Only one of the *expr2* and *expr3* is evaluated
||	Logical OR
&&	Logical AND
in	Array membership
"!"	Match regular expression and negation
<, <=, >, >=, !=, ==	Regular relational operators
[blank]	Concatenation
+, -	Addition, subtraction
*, /, %	Multiplication, division, and modulus (remainder)
+-!	Unary plus and minus, and logical negation
^, **	Exponentiation
++ --	Increment and decrement, either prefix or postfix
$	Field reference

TABLE 7.10

List of Arithmetic Function

Arithmetic Function	Description
atan2(y,x)	Returns the arctangent of y/x in radians ranging between $-\pi$ and π
cos(x)	Give cosine of x, with x in radians
exp(x)	The exponential of x
int(x)	The integer part of x, truncated toward 0
log(x)	The natural logarithm of x
rand()	A random floating-point value between 0 and 1
sin(x)	Give sine of x, with x in radians
sqrt(x)	The square root of x
srand(x)	x is new seed for rand(). Use *x* as a new seed for the random number generator. If no *x* is provided, the time of day will be used. The return value is the previous seed for the random number generator

- **Built-in arithmetic functions**: Table 7.10 lists some key mathematical built-in functions as offered in gawk.

  Besides the above-mentioned standard mathematical functions, gawk also offers few functions for bitwise operation of data, which are given in Table 7.11. The bit manipulation functions are very helpful when you are working with data, which are in binary form.

- **Built-in string function**: Table 7.12 lists some key string built-in functions as offered in gawk.

- **Built-in time functions**: The awk programming language also offered some functions to deal with time parameters that are listed in Table 7.13.

TABLE 7.11

List of Bit Manipulation Function

Bit Manipulation Function	Descriptions
and (x1, x2)	Executes the bitwise logical **AND** operation between values of *x1* and *x2*
compl(x)	Executes the bitwise complement of *x*
lshift(x, count)	Shifts the value *x* count number of bits left
or(x1, x2)	Executes the bitwise logical **OR** operation between values of *x1* and *x2*
rshift(x, count)	Shifts the value *x* count number of bits right
xor(x1, x2)	Executes the bitwise logical **XOR** operation between values of *x1* and *x2*

TABLE 7.12

List of Built-in String Function

String Functions	Description
gsub(r,s)	substitute s for r globally in the current record, return number of substitutions
gsub(r,s,t)	For each substring matching the regular expression r in the string t, substitute the string s and return the number of substitutions. If t is not supplied, use $0
index(s,t)	Returns the index of the string t in the string s, or 0 if t is not present
length(s)	return length of the string s, if a string is not provided, then print the length of $0
split(s,a,r)	Splits the string s into the array a on the regular expression r and returns the NF. If r is overlooked, FS is used in its place. The array a is cleared first
sprintf(fmt, expr-list)	Prints *expr-list* according to *fmt* and returns the resulting string
sub(r,s,t)	Similar to gsub(), but only the first matching substring is replaced, i.e., substitute s for first r in t, return number of substitutions. If t is not supplied, use $0
substr(str,i,n)	Returns the *n*-character substring of *str* starting at *i*. If *n* is excluded, then remaining of *str* is used
tolower(s)	Transforms all characters in string s to lowercase
toupper(s)	Transforms all characters in string s to uppercase

TABLE 7.13

List of Built-in Time Function

Time Function	Description
mktime(datespec)	Transforms a date specified in the format YYYY MM DD HH MM SS [DST] into a timestamp value
systime()	Returns the timestamp for the current time of day
strftime(format, timestamp)	Formats *timestamp* into a formatted day and date, using the date()shell function format. If *timestamp* is missing, the current time of day is used

7.17 Summary

Awk is a very useful and communicative programming language for various computing platforms that allow easy manipulation of structured data and help to the generation of formatted reports. You can use the awk program to extract the data element from larger set of data file and give output as per your desired format. Awk patterns regulate the execution of actions, match the pattern, and execute the executed action. In pattern statement actions, actions are enclosed in curly brackets where pattern tells when the action is supposed to be executed. The begin and end functions do not match the input lines and provide a way to increase control over initialization and closing. Awk variables are the identifiers that reference a value. The output may include the statement, specific field, NF, line numbers, text along with output, output operators, or other fancy outputs. Awk program comprises various expressions that consist of constants and variables. Computation with awk consists of counting the total number, sum, average, printing whole output, and built-in functions. With managing numbers, awk is also good at managing strings of characters; we can create new strings by adding individual input strings. All arrays are associative arrays with it is indeed as number or string. For every element of array, a set of value is maintained. However, there is no provision to provide multidimensional array directly in awk so you can create your own simulated multidimensional array using 1-D array. You can delete an array element by removing the respective array index of that element in an associative array. Awk programming languages also provide many standard commands and built-in functions like arithmetic functions, string functions, and in time functions.

7.18 Review Exercises

1. What is the output of the following programs?

   a. What is the output of the program?

   ```
   #!/usr/bin/awk -f
   BEGIN {
     a=int(33.5)
     print (a*63)
   }
   ```

   b. What is the output of this program?

   ```
   #!/usr/bin/awk -f
   BEGIN {
     print log(27)
   }
   ```

   c. What is the output of this program?

   ```
   #!/usr/bin/awk -f
   BEGIN {
     print index("yourself","linux")
   }
   ```

   d. What is the output of the program?

   ```
   #!/usr/bin/awk -f
   BEGIN {
     print toupper("LinUxYoursElf_1_$")
   }
   ```

e.   Will there be an error in program in program execution?

```
    #!/usr/bin/awk -f
BEGIN {
    system("date")
    print "linux yourself
}
```

Let sample file be named ***department.txt.***

Name	Designation	Department	Salary
Daniel	Professor	Biology	78000
Aaditya	Lecture	Biology	50000
Mark	HoD	Biology	100000
Bharat	Professor	Anatomy	75000
Khushdeep	Lecturer	Anatomy	50000
Pawan	Lecturer	Zoology	50000
John	HoD	Zoology	100000
Raman	Professor	Zoology	95000

Write the output for the following commands:

```
awk '{print}' department.txt
awk '/Professor/ {print}' department.txt
awk '{print $1,$2}' department.txt
awk '{print NR,$0}' department.txt
awk '{print $1,$NF}' department.txt
awk 'NR ==4, NR ==8 {print NR,$0}' department.txt
awk '{print NR "- " $1 }' department.txt
awk 'NF > 0' department.txt
awk 'END { print NR }' department.txt
awk 'BEGIN {for(i=1;i <=6;i++) print "square of",i, "is",i*i; }'
```

2. What is the difference between the built-in string functions gsub() and split() in awk programming?
3. Write a sequence of commands to find the sum of bytes (size of file) of all files in a directory.
4. Write a command to print the fields in a text file in reverse order?
5. What will be the output of the following code: awk " $1 >= "K" {print $1}" department.txt
6. Observe the following lines of code and answer the following questions:

```
awk ' BEGIN {print "The number of times biology appears is:"}
   /Biology/ {counter+=1;}
END  {printf("The value of counter is %d",counter)}' department.txt
```

a)   Predict the output of the code.
b)   What will be the output if /Biology/ is replaced with /^Biology/

7. Predict the output of the following:

```
awk 'BEGIN {printf ("The output :");}  {cout+=$4;} END {printf("%d\
n",count);}' department.txt
```

8. Predict the output of the following:

```
awk 'BEGIN {printf("The output :\n");}  /Biology/ {count+=$4;}  /
Anatomy/ {counta+=$4;} END {printf("Biology:%d
Anatomy:%d",count,counta);}' department.txt
```

9. Write a bash script using awk to display the length of each department name in the file department.txt.

10. Predict the output:

```
       awk '{nc=1+length($0)+nc
         nw=NF + nw}
      END { printf (" characters: %d,  Words : %d,   Lines:%d",
  nc,nw,NR");}' Department.txt
```

11. Write a program in awk to find the name of professor having salary more than 75000 in department.txt.

12. Write a program in awk to find the average salary of professor of anatomy department in department.txt.

13. Find the sum of the size of files in your home directory that was modified in the previous month using awk command.

14. Write an awk program to display the total number of professors in each department.

References

1. Dale Dougherty and Arnold Robbins. 1997. *Sed & Awk*. O'Reilly Media, Inc.
2. Jan Goyvaerts and Steven Levithan. 2012. *Regular Expressions Cookbook*. O'Reilly Media, Inc.
3. Arnold D. Robbins. 1996. *AWK Programming Language*. Open Source under GNU Free Documentation License.
4. Alfred V. Aho, Brian W. Kernighan and Peter J. Weinberger. 1988. *The AWK Programming Language*. Addison-Wesley Publishing Company.
5. Arnold D. Robbins. 2018. *GAWK: Effective AWK Programming*. Open Source under GNU Free Documentation License.
6. Sumitabha Das. 2013. *Your UNIX/Linux: The Ultimate Guide*. The McGraw-Hill.
7. Richard Blum and Christine Bresnahan. 2015. *Bible- Linux Command Line & Shell Scripting*. John Wiley & Sons.

8

Shell Scripting

In Chapter 3, the basic features of the Linux shell and command-line execution of various commands have been discussed. As mentioned earlier, various types of shells are developed and available for Linux, such as Bourne Again SHell (BASH), Korn Shell (ksh), C-Shell (csh), T-Shell (tsh), and Z-Shell (zsh). You can choose any one of the shells for your Linux system. Bash is very popular and powerful; it is available on many Unix flavor systems including Sun's Solaris and Hewlett-Packard's HP/UX. Bourne Again SHell (BASH) is an improved form of Bourne shell (sh). Bash is compatible with Bourne shell (sh) that integrates valuable features from the Korn shell (ksh) and C shell (csh). Every shell provides many programming features and tools, which can be used to create a shell program, also referred to as shell script. The bash is flexible and offers a powerful set of programming commands that facilitate you to build complex scripts. Hence, bash has become default shell and shell scripting on most Linux/Unix distributions. This chapter uses bash to elaborate the basics of the shell scripting concept and features, including how to create a shell script, interactive script, define a variable, arithmetic function, conditional control structure, comparison operation, and string operation. You need only one type of shell to do your work. In bash, the shell script filename has the extension of *.sh* or *.bash*.

8.1 Shell Script

A shell is a fundamental and key part of the Linux computing environment. It is known for both a powerful command-line interface and a scripting language interpreter. The command-line interface of the shell provides an environment to execute commands. Shells are user programs that offer a customizable interface between the end-user and Linux kernel. We have covered many shell features in Chapter 3, but some of the key features offered by the shell are listed below:

- Offers an interactive communication or textual user interface between the user and the operating system.
- Provides an environment for the execution of other applications and programs.
- Helps for launching and managing both user and system commands and programs.
- Executes as user program in program space.
- Customizes the shell features according to its operating environment.
- Executes shell in both interactive and noninteractive environments; end-users use interactive environment and system use noninteractive environment.

- Provides an operating environment that can be customized by using configuration files.
- Communicates environment variables to child processes of the shell.
- Offers its own programming language with rich features that allow users to write shell programs, also referred to as *shell script*.

Most simply, a *shell script* is a text file that comprises a series of one or more commands, by assuming that each line of the text file keeps a separate command [1]. On the command prompt, if you are executing a group of commands regularly, save all commands in a file and execute the files as a shell script or shell program. This way, time can be saved and the chance of errors can be minimized during executing such a group of commands.

Simply, if you executed two commands together on the same command-line prompt, separated with a semicolon, you can mention that it is the execution of the simplest and smallest shell script in bash. For example:

```
$date; echo $SHELL
Thu Sep 20  13:15:08 IST 2018
/bin/sh
```

The date command executes first, displaying the current date and time, followed by the output of the SHELL environment variable, which is the path of the currently running shell of the system. This technique is well suited for a small script. The major disadvantage of this technique is that you must type the entire list of commands at the command prompt up to the maximum command-line character count, i.e., 255 characters whenever you run such commands. To avoid this tedious task, you can put all such commands into a text file and run the text file as a shell script file.

8.2 Creating a Script

If you are executing a small or big list of commands regularly, you must store all commands in a text file and execute the file as a shell script or shell program. A single shell script acts as a single command in Linux and typically used to automate routine tasks. Shell scripts are generally used in system configuration and administrative tasks. A decent Linux/Unix system administrator must be reasonably proficient in shell programming. Like booting a Linux machine, every time it executes the shell scripts in */etc/rc.d* to restore the system configuration and setup process. As mentioned, a shell script is a simple text file, so you need a text editor, such as vi or vim, to create a file as a shell script file and thereafter, write the list of commands in the file. This chapter discusses bash, which is the GNU Project implementation of the standard Unix Bourne shell *sh*. It offers functional advances over *sh* for both programming and interactive use. Broadly, you can create the shell script in the following way:

Step1: Use the *touch* command to create an empty file on the command line with the name *script_file.sh*. Thereafter, you can open and enter the list of commands by using any text editor.

```
#!/bin/bash
# Hello.sh: Sample shell script to get the basic details of system.
echo " echo " Hello, Date & Time : `date` "   #display system date & time
echo " Current login shell name: $SHELL"   # Display pth of shell
echo " Name of Operating System" ; uname # Name of operating system
exit 0
```

FIGURE 8.1
Hello.sh shell script file.

```
# touch script_file.sh     // create an empty shell script file
```

Step 2: Use any text editor, such as *Vi/Vim*, to create and open a shell script file with the name *scriptfile.sh* and thereafter write commands and save the script file. The following example elaborates the basic structure of a shell script, for example, writing a simple shell script to get system information such as date, current shell path, and the name of the operating system of system, as shown in Figure 8.1.

When you write a shell script file, you must specify the absolute path of the shell interpreter of the current shell at the first line in the text file. The line should start with the exclamation point character (!) along with hash sign (#), followed by the full path of interpreter of bash, **/bin/bash**. The first line of *Hello.sh* comprises a string beginning with #!. This is not a comment line, called interpreter trail line. The syntax for this is:

```
#!/bin/bash
```

So, all bash scripts often begin with #!*/bin/bash*, assuming that bash is installed in /bin directory. Normally, the hash sign (#) is used for a comment line in a normal shell script. If a line starts with the hash sign (#), the shell ignores the entire line text up to the next NEWLINE character, i.e., the interpreter ignores and not processed the line. After indicating the shell interpreter path and comment lines, commands are entered into each line of the file. You can combine more than one command on the same line by separating them with a semicolon. But in a shell script, the shell prefers to write commands on discrete lines.

After writing the commands in a file, save the file as a script, called *Hello.sh* (Figure 8.1). Now, you created a shell script, but you need to do a few more steps before you run the newly created shell script file. The following section highlights the steps to be done to assign execute permission in the shell script.

8.3 Making a Script Executable: chmod

Whenever you created a new file, the system *umask* value determines the default file permission setting for the newly created file. If the system *umask* value is set to be 022, the system is assigned with read and write permission only. Hence, in this way, the info.sh shell script file has only read and write permission, not the execute permission. Then, run the script using the following command:

```
# sh Hello.sh

Hello, Date & Time: Thu Sep 20  13:15:08 IST 2018
Current login shell name: /bin/sh
Name of Operating System :
Linux
```

The next important step is to assign execute permission to *Hello.sh* script file. To make any file executable, use the *chmod* command with the +x option, as shown below:

```
# chmod +x Hello.sh // Add execute permission along with the
existing file permission.
```

Once the execute permission is done, the script is executable by its owner, group members, and others on the system. With the permissions set, you can now execute your script file by using ./ followed by *Hello.sh*, as given below:

```
# ./Hello.sh
Hello, Date & Time: Thu Sep 20  13:15:08 IST 2018
Current login shell name: /bin/sh
Name of Operating System :
Linux
```

8.3.1 Path of Script File

In the previous way of execution, the '*./filename*' syntax indicates that the file you mentioned is residing in the current working directory.

One of the very important features of the shell is that you can run your script file like any other command at the prompt. In this way, the script's filename becomes a new shell command. The facility to run any executable program, including a shell script, under Linux is solely subject to its path (or location) in the file system. Either you must mention the path of the file to run or it must be located or available in a directory known by the shell. All executable programs or code directories are listed in the PATH environment variable [2]. For this, you need to include the path of your shell script, *Hello.sh*, in the PATH environment variable (see Chapter 3 for details).

```
# echo $PATH             // to see the list of directories listed in
                            PATH environment variable
/usr/local/sbin:/usr/local/bin:/usr/sbin:/usr/bin:/sbin:/bin:/usr/src/
games:/usr/local/rit:/home/abc
```

- Add the directory path of *Hello.sh* script to the PATH environment variable. For example, If you kept your shell script file (Hello.sh) in */home/rit/scripts* directory, then add the command as below:

```
$ PATH=$PATH:/home/rit/scripts    // Add in PATH environment variable
```

```
$ echo $PATH                      // See newly added path in PATH Env
    /usr/local/sbin:/usr/local/bin:/usr/sbin:/usr/bin:/sbin:/bin:/usr/
    src/games:/usr/local/rit:/home/abc:/home/rit/scripts
```

A script becomes very useful when it is executed on the command line, just by typing its name like any other command as follows:

```
#hello.sh          //executed and give the output
#mv Hello.sh XYZ   // rename the script file to any other name without
                   mentioning the .sh extension
 #XYZ              // produce the same result as Hello.sh
```

- **Importance of Shebang (#!)**

  Linux offers various types of shells and its script files on a system. Each shell scripting language comes with a unique and specific command structure. There needs to be a method to communicate Linux to specify which interpreter to be used to execute a script. Therefore, a special line is to be used at the top of each script to identify the specified interpreter program that reads the rest of the lines of that script file. The special line must begin with #!, a concept frequently referred to as *shebang*, which assumed for Sharp (#) Bang(!) [3,5].

  For bash, the shebang line is:

  ```
  #!/bin/bash
  ```

  This command informs that the program, mentioned as bash, is found in the /*bin* directory and assigns bash to be the interpreter for the particular script file. Similarly, the *shebang line for other shells* is given in Table 8.1:

8.4 Interactive Script: Read

After getting the familiarity about how to write a shell script, it is worth mentioning here that you can write and run a shell script in both interactive and noninteractive modes. Interactive scripts require input from the user during the execution of the script. On the other hand, noninteractive scripts can be run in the background without any user intervention. The key advantages of both modes are mentioned in Table 8.2

The subsequent sections elaborate on how to write an interactive script by using *read* built-in command and command-line *positional parameters*.

- **Using *read* built-in command:**

  The *read* is the shell built-in command for taking input from the user to make an interactive shell script. This command is the complement of the *echo* and *printf* commands. It takes one or more variables. The syntax of the *read* command is as follows:

TABLE 8.1

List of Shebang Line Notation

Shebang Line Notation	Description
#!/bin/sh	The Bourne shell
#!/bin/csh	The C-shell
#!/bin/tcsh	The enhanced C-shell
#!/bin/zsh	The Z-shell
#!/bin/ksh	The Korn shell
#!/bin/sed	The stream editor
#!/usr/bin/awk	The awk programming language

TABLE 8.2

Description and Advantages of Interactive and Noninteractive Scripts

Mode	Description	Advantages
Noninteractive script	Scripts run without any interaction from the user at all	• The script executes in a predictable technique every time. • The script may run in the background. • Example: init, startup, reboot scripts.
Interactive script	Requires input from the user	• Higher flexible scripts can be built. • Users can customize the script as per the requirements. • The script can describe its progress as it executes. • For example, any user can create a script using read built-in command or through command-line positional parameters.

```
read [options] Var1 Var2 ... VarN
```

The *read* command is associated with some options as mentioned in Table 8.3.

For example, a shell script, *search.sh*, is created to search a pattern from a given file, as shown in Figure 8.2.

In the *search.sh* script file, read command takes the inputs from user and stores it in respective variable names, *fname* and *pattern*. To print the value of a variable, use $ before the variable name and execute the script:

```
# sh search.sh
Enter the filename: employee.txt
Enter the pattern to be search from employee.txt file: sanrit
Searching for sanrit from file employee.txt
Sanrit : 29: 7700: India
```

The above script stops at two places. First, it reads the filename and stores it in the *fname* variable. Second, it reads the pattern string and stores it in the *pattern* variable. Thereafter, the grep command runs with these two variables as its arguments and gives the output.

TABLE 8.3

List of Options of Read Command

Option	Description
-a array_name	Stores the words in sequential indexes of the array variable *array_name*. The numbering of array elements starts at 0.
-d delim	Assigns the delimiter character to *delim*, which is used to notify the end of the line. The default line delimiter is a newline.
-e	Gets a line of input from an interactive shell.
-n NCHARS	**read** command stops reading after *NCHARS* characters, instead of waiting for a complete line of input.
-p PROMPT	Displays the string *PROMPT*, without a newline, before beginning to read any input.
-r	Due to this option, the read command does not interpret backslash as an escape character, rather backslash is treated as a part of the line.
-s	Silent mode. If read is taking input from the terminal, do not echo keystrokes.
-t Timeout	Time out and return failure if a complete line of input is not read within specified *Timeout* seconds. *Read* command will not read any data if the *Timeout* value is zero. If the timeout is not mentioned, the value of the shell variable TMOUT is used in its place, if it exists.
-u FD	Reads from the file descriptor *FD* instead of standard input.

```
#!/bin/sh
# search.sh: Interactive script by getting the inputs from user

echo "Enter the file name: \c"
read fname
echo "Enter the pattern to be search from $fname file: \c"
read pattern
echo "Searching for $pattern from file $fname"
grep "$pattern" $fname
exit 0
```

FIGURE 8.2
search.sh shell script.

In the above script, each variable is mentioned with individual *read* command, but you can mention one or more variables with single read command as given below:

```
read fname pattern
```

Another example:

```
#!/bin/sh

echo "Enter the Numbers a, b, and c: "
read a b c
echo " The Entered value of a=$a, b=$b and c=$c"
exit 0
```

8.5 Shell Variable

In Chapter 3, we have briefed the concepts of variables, which are referred to as data holders. During the execution of a script, you can include user and other data values in the shell command to process information. You can perform this operation by using variables, which can hold a value/information temporarily within the shell script. The value of such variables can be used with other commands in the script by placing a dollar sign in front of the variable name. As discussed in Chapter 3, variables can be categorized into two: local variable and environment variable. Apart from these variables, one special variable is used to get the value of a variable from the command line during execution, known as the positional parameter [4]. This section shows how to use positional parameter variables and environment variables in shell scripts.

8.5.1 Positional Parameter Variable: Command-Line Arguments

This section elaborates another method to interact with a shell script, i.e., the use of command-line parameters. The command-line parameters permit a user to enter data values on the command line along with the name of the script. When you execute this script, for example:

```
# ./search.sh employee.txt sanrit
```

In the above execution, two command-line parameters, *employee.txt* and *sanrit*, are mentioned along with the script name search.sh. These two command-line parameters act as inputs for executing the script.

Reading parameters

When the Bourne shell runs a script, all the entered command-line parameters are handled using special variables, known as "positional parameters/variables," i.e., you can access within your script. All positional parameter variables are assigned with a standard number, starts with $0 to $9. $1is assigned as the first positional parameter, $2 as the second positional parameter, and so on. The details are mentioned in Tables 8.4.

The name of the script or command itself is specified as the special positional variable $0 (zero). The shell assigns the name of the script, as it appears on the command line to $0. If there is no argument mentioned on the command line, $0 will always comprise the first item looking on the command line, i.e., the pathname of the program being executed. The factorial program is an example of one command-line parameter as given below:

```
<<< Example of one command line parameter>>>

# cat fact.sh
  #!/bin/bash
  factorial=1
      for ((num = 1; num <= $1 ; num++ ))
        do
            factorial=$[ $factorial * $num]
        done
      echo "The factorial of $1 is $factorial"
#
```

TABLES 8.4

List of Positional Parameters

Positional Parameter /Variables	Description
$0	Name of the script or command executed on the command line
$1	First command-line argument
$2	Second command-line argument
$3	Third command-line argument
$4	Fourth command-line argument
$5	Fifth command-line argument
$6	Sixth command-line argument
$7	Seventh command-line argument
$8	Eighth command-line argument
$9	Ninth command-line argument
$*	List of all command-line arguments as a single string (each argument separated with a space)
$$	PID of the current shell
$@	Comprises the value of all command-line arguments or positional parameters excluding $0
$!	PID of last background process
$#	Number of command-line arguments or positional parameters mentioned in command line excluding $0
$?	Prints the exit status of the preceding command

```
# ./fact.sh 6
The factorial of 6 is 720
#
```

<<< Example of two command line parameter>>>

```
# cat Add.sh
  #!/bin/bash
  sum=$[ $1 + $2 ]
  echo "The first input number is $1."
  echo "The second input number is $2."
  echo "The Sum of $1 and $2 is = $sum ."
#
# ./Add.sh 4 7
The first input number is 4.
The second input number is 7.
The Sum of 4 and 7 is = 11.
#
```

It is to recollect that every positional parameter is divided by a space, so that shell understands that the given white space separates two values. But if you want to incorporate a space as a part of the positional parameter value, you should use a single or double quote sign, for example:

```
# cat hello.sh
  #!/bin/bash
  echo "Hello, I welcome $1."
#

# ./hello.sh 'Ritvik and Sanvik'
  Hello, I welcome Ritvik and Sanvik.
#
```

If the shell script or command requires more than nine command-line positional parameters, you can make small changes in the declaration of positional parameter variable names. After the ninth variable, the variable numbers are surrounded with braces, e.g., ${10}, ${16}, ${41}, ${245}, and so on. An example command with more than nine command-line positional parameters is given below.

```
# cat add1.sh
  #!/bin/bash
  sum=$[ ${10} + ${12}]
  echo "The total input provided on command line is $#."
  echo "The Tenth input number is ${10}."
  echo "The Twelve input number is ${12}."
  echo "The Sum of ${10}th and ${12}th input is = $sum ."
#
# ./add1.sh 1 2 3 4 5 6 7 8 9 10 11 12 13 14
The total input provided on command line is 14.
The Tenth input number is 10.
The Twelve input number is 12.
The Sum of 10th and 12th input is = 22.
#
```

- *Shift* **command**

    Another important command offered by the bash is the *shift* command, which helps to manipulate the command-line parameters. The *shift* command exactly shifts the command-line parameters in their relative positions. By default, the shift command moves each parameter variable one position to the left, for example, the value of variable $3 is moved to $2, the value of variable $2 is moved to $1, and the value of variable $1 is discarded because $0 remains unchanged and fixed to store the name of the program/script. The shift command is helpful especially when you are not aware of the total available parameters on the command line. For example:

```
# cat display.sh
#!/bin/bash
        x=1
while [ -n "$1" ]
  do
     echo "Input No. $x = $1"
     x=$[ $x+ 1 ]
     shift
  done
#

# ./display.sh ritvik sanvik kinjal misthi tiya
Input No. 1 = ritvik
Input No. 2 = sanvik
Input No. 3 = kinjal
Input No. 4 = misthi
Input No. 5 = tiya
#
```

8.5.2 Environment Variable

You have already known about the environment variable available in the Linux system, as explained in the Global Variable section in Chapter 3 because the environment variable is also referred to as the global variable. This variable is very important because you can access the values of these variables from your shell scripts as well. Shell provides an operating environment to the various programs/scripts, which comprise numerous variables. Each variable holds a value that is used by programs/scripts and other shells. In a shell script, mostly shells use two types of variables, shell and environment variables.

- **Shell variables:**

    Shell variables, also referred to as local variables, are visible within the current shell in which they are defined. Subshell or child process does not inherit them in contrast with the environment variable. These variables may also be assumed to be user-defined variables. The built-in command **set**, without any option, is used to display all variables, including environment variables. No specific command is available to display only the shell variables. Some shell variables are routinely set by the shell and are presented for use in shell scripts. By convention, shell variables are case-sensitive variables, but generally, these are declared in the lowercase. A detailed description of the variable is given in Chapter 3.

```
$ set        //display all the variables including environment
             variable in sorted order
```

You can declare any defined local variable into environment variable by using the *export* command:

```
$ export MYVAR        // Exporting a local variable
```

While exporting a variable, do not include a preceding dollar sign with the variable name. When a local variable is exported to the environment variable, it passes the information of that variable to all subshell or child processes. It means that such variables will be available to all programs/scripts run by your shell.

- **Environment variables**:

  The environment variables are very useful in running applications/scripts that create subshell/child processes. Environment variables are also known as global variables because their information is shared to all processes started by the shell, including other shells/subshells. When you start your shell session, the Linux system sets various built-in environment variables, which means child processes inherit such variables in their environment. By convention, environment variables are assigned as uppercase names. Every shell keeps numerous environment variables, including the following variables:

  - **BASH**: Display the full path of bash command, generally it is /bin/bash.
  - **HOSTTYPE**: Print the computer architecture on which the Linux system is running. For Intel-compatible PCs, the value is i386, i486, i586, i686, or i686-Linux. For AMD 64-bit machines, the value is x86_64.
  - **PATH**: A colon-delimited list of directories through which the shell looks for executable programs as you enter them on the command line.
  - **HOME**: Your home directory, e.g., */home/sanrit*.
  - **USER**: Your username, e.g., *sanrit*.
  - **PPID**: The parent process ID of the currently executing command or process.
  - **UID**: Every username is assigned with a user identification number, referred to as UID. The user ID number is stored in the /etc/password file.

By using **env** or **printenv** commands on the shell prompt, you can display a complete list of environment variables available.

```
        $ printenv        // display the global environment variables
OR
        $ env
```
You can use *echo* commands to display the content of the variable as given below.
```
        $ echo $HOME              // display current user home directory
        /home/sanrit

        $ echo $USER              // Display current user name
        sanrit

        $echo $SHELL              // Display current shell
        /bin/sh
```

8.6 Shell Arithmetic

Shell provides powerful tools for performing an arithmetic operation in shell scripts. There are two different ways to perform various arithmetic operations in your shell scripts, which are described in the subsequent sections.

- **Using expr command (for integer value operation)**

    The Bourne shell offers a special command *"expr,"* which allows the processing of equations from the command line as well as from the script. **expr** performs the four basic arithmetic operations and the modulus (remainder) function. It handles only integer values, and the decimal parts of values are simply ignored.

    ```
    $ x=4 y=7              //  Multiple assignments without a ;
    $ expr 4 + 6                // Whitespace required
    10
    $ expr $x - $y
    -3
    ```

    Furthermore, converting a string into a numeric expression is comparatively simple by using backticks or double parentheses. An arithmetic expression is generally used in conjunction with *expr* command to perform an arithmetic operation surrounded by backticks (`: character just above the tab key).

    ```
    x=`expr $x + 4`   //The 'expr' command performs the expansion.
    ```

    The expr command, known as an all-purpose expression evaluator, concatenates and evaluates the arguments according to the mentioned operation. It is noted here that arguments must be separated by spaces, otherwise, it will give a syntax error. Operations may be performed by using arithmetic, comparison, string, or logical operators, which are mentioned in the subsequent sections. Some examples of expr command are elaborated here:

    ```
    $expr 5 % 3
    2

    $expr 1 / 0
    returns the error message, expr: division by zero Illegal
    arithmetic operations not allowed.

    $ expr 4 * 2
    expr: syntax error
    $
    To remove the above error, you need to use the backslash character
    (the shell escape character) to recognize any character that may
    be misunderstood by the shell before being passed and processed by
    the expr command. Therefore, the multiplication operator must be
    escaped when used in an arithmetic expression with expr:
    $ expr 4 \* 2
    8

    y=`expr $y + 1`
    Increment a variable, with the same effect as let y=y+1 and
    y=$(($y+1)). This is an example of arithmetic expansion.
    ```

```
Shell script for adding two numbers by using expr command:

#!/bin/bash
# An example of interactive shell script using the expr command:
Read x, y
z=`expr $x + $y`
z1=`expr $x \* $y`
echo "The addition of $x and $y is $z"
echo "The Multiplication of $x and $y is $z1"
exit 0
```

- **Arithmetic expansion**

  The bash permits arithmetic operations to be performed by expansion [5]. The basic syntax of Arithmetic expansion is:

  ```
  $((expression)),
  ```

  where *expression* is an arithmetic expression comprised of arithmetic operators and operands. It permits you to practice the shell prompt as a calculator. Arithmetic expansion supports only integers (no decimal point numbers).

  The shell evaluates expression and replaces $((expression)) with the result of the evaluation. This syntax is as same as the syntax used for command substitution [$(...)]. On a command line, you can use $((expression)) as an argument to a command or in place of any numeric value. For example:

  ```
  x=$(($x+4))          // similar as mentioned with expr command
  x=$((x+4))      // Also correct.
  $echo $((2 + 2))
  4
  ```

  Here is the example of integer division and see the output. It displays only the integer part, i.e., remainder operators:

  ```
  $echo " Eight divided by three is equals = $((8/3))"
  Eight divided by three is equals = 2
  ```

  In arithmetic expansion, spaces are not important, and it may be nested in arithmetic expressions. For example, multiply 4^2 by 5:

  ```
  $ echo $(($((4**2)) * 5))
  80
  ```

  Single parenthesis may also be used to group multiple subexpressions. With this method, you can rewrite the above-mentioned example using a single expansion, instead of using two expansions. You will get the same result:

  ```
  $ echo $(((4**2) * 5))
  80
  ```

  Using bc command (for floating-point value operation)

  The above section elaborated how the shell can handle various types of integer arithmetic, but you cannot operate on floating-point numbers (at least, not directly with the shell). There are several approaches for overcoming this limitation of the bash, for example, taking the help of some programming tools such as Embedding perl or awk programs as a solution. The preceding chapter has already explained awk programming. Another approach to overcome this problem is that you can use a specialized calculator program. One such program found on most Linux systems is the bash calculator [6], known as *bc*. The *bc* built-in program supports quite a few features, including variables, loops, and programmer-defined functions. This section does not cover *bc* in detail; it gives an overview. For more details about *bc*, you can refer to the manual (man page) documented, available with all Linux distribution, as given below:

```
$man bc          // See the complete help manual of bc program.
```

The *bc* is a kind of program or acts as a programming language to design such a program, which allows you to enter floating-point expressions at a command line. The *bc* program interprets the given expressions, executes them, and gives the result. The built-in *bc* identifies the following inputs:

1. Numbers (both floating points and integers)
2. Variables (both simple variables and arrays)
3. Comments (start with "/*" and end with "*/")
4. Expressions
5. Programming conditional statements (if-then statements)
6. Functions

The basic syntax of built-in *bc* is given below.

```
bc [ -hlwsqv ] [long-options] [ file ... ]
```

A file comprising the calculations/functions that to be piped from standard input. The various options associated with the bc command are given in Table 8.5. There are four special variables used in *bc*: **scale, ibase, obase,** and **last**.

- **Scale** variable specifies how certain operations use digits after the decimal point. The default value of scale is 0. For example:
  .000123 has a length of 6 and a scale of 6.
  2021.0001 has a length of 7 and a scale of 4.

TABLE 8.5

List of bc Command Options

Option	Notation	Description
-h	help	Displays the usage and quit
-i	interactive	Forces interactive mode
-l	Math library	Defines the standard math library
-w	warn	Gives warnings for extensions to POSIX bc
-s	standard	Processes exactly the POSIX bc language
-q	quiet	Does not print the normal GNU bc welcome
-v	Version	Displays the version number and copyright information and then quits

- **ibase** and **obase** identify the conversion base for input and output numbers. The default base for both input and output is 10.
- **last** (an extension) is a variable that has the value of the last printed number.

  When you are using *bc* interactively, you simply type the calculations and press enter, and the results are immediately displayed. The *bc* command leaves the interactive session by using **quit.** For example:

```
$ bc
14 * 5.6
78.4
3+5+6-3
11
4.276 * (4 + 7)
47.036
15 / 3 + 4
9
quit        /*To exit the bash calculator, you must enter quit.*/
$
```

As mentioned above, the floating-point arithmetic values are regulated by a built-in variable, identified as scale. You must set this value to the predicted number of decimal places that you prefer to see it in the output, for example:

```
$ bc
3.62 / 6
0               /* by default, scale is 0 */
scale=5         /* Set the scale value is 5 */
3.44 / 6
.60333
quit
$
```

8.6.1 Arithmetic Operator

The supported list of arithmetic operators by *expr* and *bc* commands is given in Table 8.6. The operator's precedence and associativity are similar as in the C language.

TABLE 8.6

List of Arithmetic Operators

Operator	Description	Example with expr Command
+	Addition (a+b)	'expr $a+$b'
−	Subtraction (a−b)	'expr $a−$b'
*	Multiplication (a*b)	'expr $a * $b'
/	Division (with expr command, expansion supports only integer arithmetic; hence, produced results are integers (a/b)	'expr $a / $b'
%	Modulus or remainder (a%b)	'expr $a % $b'
**	Exponentiation (a^b)	$a**$b
++	Increment (a+1)	'expr $a+1'
--	Decrement (a-1)	'expr $a−1'

8.6.2 Logical Operator

Logical operators, also known as Boolean operators, are mostly used in conditional statements as mentioned in Table 8.7. The result of the logical operators is either 1 (TRUE) or 0 (FALSE).

8.6.3 Conditional Operator

Many shell scripts need logic flow control between the commands based on the conditional statements in the script. Conditional statements are used to make decisions and execute some parts of the code while skipping other codes based on conditions. The conditional statement comprises the [[compound command and the *test* and [built-in commands. The *test* and [commands evaluate their behavior based on the mentioned arguments and conditional operators, which are listed in Table 8.8. The subsequent sections highlight the various control structures. For example, a control structure comprises a comparison (ternary) operator as mentioned below:

```
expr1?expr2:expr3
```

In the above execution, if expression *expr1* is evaluated to be true (nonzero arithmetic true), *expr2* will be evaluated, else expr3 will be evaluated.

8.7 Control Structure

In the sequential operations, all the commands are executed in the proper sequential order as mentioned in the program. However, sometimes, programs require a part of the program to be executed based on logical/conditional control between the commands in the

TABLE 8.7

List of Logical Operators

Operator	Description	Example
!=	Not equal to	expr1 != expr2: Result is 1 if both expression results are not equal
&&	Logical AND	expr1 && expr2: Result is 1 if both expressions are nonzero
| |	Logical OR	expr1 | | expr2: Result is 1 if either expression is nonzero
!	Logical NOT	! expr: if result is 1, expr is 0; if result is 0, expr is 1
==	Equal	[$a == $b]
<=	Less than or equal	[$a <= $b]
>=	greater than or equal	[$a >= $b]
<	less than	[$a < $b]
>	greater than	[$a > $b]
Bit Operations *(operators work at the bit level)*		
~	Bitwise negation	Negate all the bits in a number
<<	Left bitwise shift	Shift all the bits in a number to the left
>>	Right bitwise shift	Shift all the bits in a number to the right
&	Bitwise AND.	Perform an AND operation on all the bits in two numbers
|	Bitwise OR	Perform an OR operation on all the bits in two numbers
^	Bitwise XOR	Perform an exclusive OR operation on all the bits in two numbers

script. The bash provides conditional control structures that allow you to choose which part of Linux commands to be executed [7]. Most control structures are similar conditional control structures found in other programming languages but with a few differences. The control structure permits you to change the execution flow of a program.

- **What is Condition?**
  - A condition is an expression that evaluates a Boolean value (true or false).
  - In other words, a condition may be either 1 (true) or 0 (false).
  - In a shell script, a condition is used in flow control and loop statements.

There are two primary methods for flow control execution based on conditional check parameters in the shell script. The first conditional check method is *if statement*, which executes a set of codes if a given condition is true. The second conditional check method is *case statement*, where you can select among multiple available sections of code based on the conditional check. In both cases, only one section of code is executed, and others are completely ignored. The *test* command offers a way to examine different conditions in an if-then or other control statements. The syntax of the *test* command is very simple as given below:

```
test condition
```

The condition is a sequence of parameters and values that the *test* command evaluates. To know more about *text* commands along with various conditional operators, see Section 8.10.

The bash offers another method of testing a condition without using the *test* command in the control structure. For example, in *if-then* statement, the condition can also be evaluated in the square bracket as shown below:

```
if [ condition ]
then
    set of commands
fi
```

It must be noted that we must define the test condition in square brackets [ ] very carefully because space is required **after the first ([ ) bracket** and a space **before the last ( ]) bracket**, otherwise you will get an error message. The subsequent section highlights the syntax of various control structures.

8.7.1 if Statement

The extremely fundamental type of structured command is the *if statement* or *if-then statement*. The basic syntax of the *if-then* statement is given below:

```
If [condition]
then
    command actions
fi
```

It executes command actions statement *if condition* is true. Here is a simple example to explain this concept:

```
$ cat numb1.sh
#!/bin/bash
read -p "Please enter a number:" n
if [$n -gt 5]
then
    echo "Entered number $n which is greater than 5 number."
fi
```

When you run this script from the command line, you get the following results:

```
$ sh numb1.sh
Please enter and confirm number 5 is :5
Thanks for entering 5 number.
$
```

8.7.2 if then else Statement

The if-then-else statement offers another set of commands in the control statement. The basic syntax of *if-then else* statement:

```
if [condition1]
     then
     command list if condition1 is true
     else
     command list if condition1 is false
fi
```

The basic syntax of *nested if-then-else* statement:

```
if [condition1]
     then
     command list if condition1 is true
     elif [condition2]     // if condition 1 is false
     then
     command list if condition2 is true]
     else
     command list if condition2 is false]
fi
```

The conditions are tested with the use of the *test* or [ ] command. For example:

```
#!/bin/sh
if [ $1 -ge 75 ]
then
     echo "passed the exam with First div."
elif [ $1 -ge 50 ]; then
     echo "passed the exam"
else
     echo "fail in Exam"
fi
```

You can continue to string elif statements together by creating one huge *if-then-elif* composite as shown below:

```
if condition1
then
        command set 1
elif condition2
        then
        command set 2
elif condition3
        then
        command set 3
elif condition4
        then
        command set 4
fi
```

8.7.3 while Statement

The while command is a structured command that controls the flow of your shell scripts. The while command permits you to mention a condition to test and then execute the loop through a set of commands until the mentioned test condition returns to a zero exit status. It checks the condition at the start of each execution. If the condition returns to a nonzero exit status, the while command continues the execution of mentioned commands between the do-done statement. The basic syntax of while statement:

```
while [ condition]
        do
                command1
                command2
                ..
                ....
                commandN
        done
```

8.7.4 do-while Statement

Like the other programming languages, the do-while statement behaves similarly to the while statement as discussed in the previous section. In a while statement, the conditional expression is checked at the starting of each execution of a set of commands. But in the do-while statement, the conditional expression is tested at the end of the loop. It means the do-while statement will iterate at least once, whereas a while statement could not be entered at all if the conditional expression is false. The basic syntax of the do-while statement:

```
do
        list of commands;
while[ condition]
```

Hence, you can replace any while loop with a do-while loop only if the script requires the loop to be executed at least once before checking any condition.

8.7.5 Loop (for) Statement

Sometimes, you need to repeat a set of commands until a specific condition is encountered. For this, you need to use *for* loop command that iterates through a series of values.

The *for* loop command is very popular and offered by most computer programming languages. In modern bash, it provides two forms of *for* commands:

- **Traditional format of bash *for* command**:
    The original *for* command's syntax is as follows:

    ```
    for var in list-Values
    do
          list of commands;
    done
    ```

    where *var* is the name of a variable that will increment during the execution of the loop. You can supply the list of values in the iteration from the list-values parameter. You can mention values in the list-values in several ways. In each iteration, the mentioned variable *var* contains the current value in the list-values. The first iteration uses the first item in the list-values, the second iteration uses the second item in the list-values, and so on until all the items mentioned in the list-values have been used. As per shell script logic, you can specify one or more standard bash commands between *do* and *done* statements. For example:

    ```
    $cat test1.sh
    #!/bin/bash
    # basic example of for loop command
    for i in X Y Z W;
    do
    echo "i=$i";
    done
    ```

    After execution, it gives the following result.

    ```
    $./test1.sh
    i=X
    i=Y
    i=Z
    i=W

    $ cat subject.sh
    #!/bin/bash
    # basic example of for loop command
    for S in Math Physics Chemistry Computer Electronics  Arts
    do
    echo "The subject name is $S
    done
    ```

    After execution, it gives the following result.

    ```
    $ ./subject.sh
    The country name is Math
    The country name is Physics
    The country name is Chemistry
    The country name is Computer
    The country name is Electronics
    The country name is Arts
    $
    ```

- **The C-language format of bash *for* command**:

   In the latest versions of the bash, it has a second form of *for-command* syntax, which resembles the format available in the C programming language. The format is also supported by many other programming languages as well.

   In C programming language, *for* command has a particular way to declare the following parameters for the successful execution of the loop.

   1. **Initialization variable or statement**: To initialize the starting value of the loop

   2. **Conditional expression**: To check conditions for further iterations. If the mentioned condition becomes false, the *for*-loop iteration is stopped, and if the condition becomes true, it will continue.

   3. **Increment/decrement statement**: The way of changing variable value after each iteration

The for-loop code in C language is as follows:

```
for(initial statement; condition expression; repeat steps)
{
List of statements;
}
```

For example:

```
for (i = 0; i < 15; i++)
{
printf("The next value is %d\n", i);
}
```

The above C-code gives a simple iteration loop, where i used as a variable. The first statement assigns a default value to the variable i, i.e., i=0. The middle statement mentions the conditional parameter under which the loop will iterate; here, the loop will terminate if the value of i is greater than 14 (loop will terminate when the condition becomes false). The last statement defines the repeat step, which may be an increasing or decreasing step with respect to mentioned conditional statement. After each iteration, the value of variable *i* is incremented by 1.

   The bash also supports a c-style format of *for* loop with a minor change in the syntax as given below:

```
for ((expression1; expression2; expression3));
do
List of commands
done
```

   where *expression1*, *expression2*, and *expression3* are arithmetic expressions, and *list of commands* mentioned between do and done statements is the list of commands that to be executed during each iteration of the loop. The *for* loop behavior may also be constructed with while statement with similar nature as follows:

```
((expression1))
        while ((expression2));
```

```
        do
                List of commands
                ((expression3))
        done
```

The *expression1* is used for loop initialization, *expression2* is used to determine the loop terminate condition, and *expression3* is executed at the end of each iteration of the loop to increment or decrement the loop per the mentioned loop termination condition.

```
$cat loop_test.sh
#!/bin/bash
# simple example of C-style bash for loop command
for (( i=2; i<=6; i=i+1 ));
do
echo "i value is $i"
done
```

After execution, it gives the following result.

```
$./loop_test.sh
        i value is 2
        i value is    3
        i value is    4
        i value is    5
        i value is    6
```

In this example, expression1 initializes the variable i with the value of 2, expression2 permits the loop to continue execution till the value of i remains less than or equal to 6, and after each iteration, expression3 increments the value of i by 1.

8.7.6 switch/case Statement

Another important tool provided by the bash is the *case* statement, which is similar to the switch statement of the C programming language. The *case* statement is also referred to as multiple-choice compound command that helps to match a pattern with a list of alternative options (a series of strings). The *case* statement is referred to as *switch* in the C shell. Each alternative option is related to the corresponding action. One of the key advantages of this case statement is that you can perform multiple different tests with a minimum number of control statements. But the major limitation is that it tests only shell patterns, not the regular expression. However, the case statement is very useful for creating a menu-driven or choice base control operation. The basic syntax of a case statement:

```
    Syntax 1:
            case  $variable-name  in
                pattern1)
                     List of commands;;

                pattern2)
                     List of commands;;
```

```
                        pattern3)
                            List of commands;;
                                ....
                                ....
                                ....
                        patternN)
                            List of commands;;

                    *)
                            List of commands;;
                esac
```

OR

```
Another syntax of case statement is as follow;
 Syntax 2:

        case   $variable-name   in
                pattern1|pattern2|pattern3)
                List of commands;;

                pattern4|pattern5)
                List of commands;;

                pattern6|pattern7|pattern8)
                List of commands;;

                pattern9)
                List of commands;;
                ....
                ....
                ....
                patternN)
                List of commands;;

            *)
                List of commands;;
        esac
```

The properties of *case* statements are given as follows:

- The case statement permits you to simply check pattern (conditions) and then execute the set of command lines if that condition becomes true. You can combine more than one pattern by using OR logical operator (|) for a single check as mentioned in syntax 2 form of the case statement.
- the *$variable-name* is compared against the patterns until a match is found.
- *) acts as the default pattern in *case* statement. It will execute only when no match is found.
- The pattern can include wildcards.
- You must include ;; at the end of each command. The shell executes all the statements up to the two semicolons that are next to each other.

- The *case* statement starts with the "case" word and ends with the "esac" word.
  These two words define the body of the *case* statement.

For example:

```
$cat menu_command.sh
#!/bin/sh
# menu_command: Help to execute 6 Linux Commands as choice menu.
#
echo " COMMAND MENU\n
1. Name of OS\n2. List of files\n3. Current Date\n 4. Current month
Calendar\n5. List of users\n6.Quit to Linux\n Enter your option: \c"
read x
case "$x" in
1) uname ;;
2) ls-l ;;
3) date ;;
4) cal ;;
5) who ;;
6) exit ;;
*) echo "Invalid option" # ;;
Esac

$
```

8.8 String Operators

Bash offers various string operators that to be used in shell scripting. Some of the key string comparison operators are listed in Table 8.9 with suitable examples.

8.9 Functions

Like other programming languages, bash also provides a function feature in the shell. Functions are the best way to reutilize the code in a script. A lengthy program is more complex and harder to design, code, and maintain. Therefore, it would be useful if the overall functionality of a script is separated into smaller and logical subsections. These subsections are referred to as shell functions. Particularly, the function is very useful when you have certain tasks that need to be executed several times. So, rather than writing the same code again and again, it is better to write it once in a function and call the function every time as required.

- **Creating functions:**
  Shell functions are subroutines, procedures, and functions in other programming languages. So, function creation is very simple and uses the following two syntaxes to declare it in the script.

TABLE 8.9

List of String Operators

Operator	Description with Example
= (equal)	This operator checks whether the two strings are equal or not. For example:

```
string1="home";
string2="house";
if [ $string1=$string2 ]
then
  echo "Input strings are Equal";
else
  echo " Input strings are not Equal";
fi
```

Output:
Input strings are not equal

!= (not equal)	This operator performs just opposite of = operator. For example:

```
string1="home";
string2="house";
if [ $string1 != $string2 ]
then
  echo " Input strings are Equal";
else
  echo " Input strings are not Equal";
fi
```

Output:
Input strings are equal

-z (zero-length)	This operator checks whether a string is zero length or an empty string. For example:

```
string1="home";
if [ -z $string1 ]
then
  echo "zero Length"
else
  echo "Not zero length "
fi
```

Output:
Not zero-length

\> (greater than)	This operator checks whether the first string is greater than the second string in comparison with a number of characters. For example:

```
string1="home";
string2="house";
if [ $string1 \> $string2 ]
then
  echo "String1 is Greater String than String2 "
else
  echo "String1 is not Greater String than String2 "
fi
```

Output:
String1 is not greater than String2

(Continued)

TABLE 8.9 (*Continued*)

List of String Operators

Operator	Description with Example
-n (non-zero length)	This operator checks whether a string is a nonzero length. For example:

```
string1="home";
if [ -n $string1 ]
then
  echo "Not zero Length"
else
  echo "Zero length "
fi
```

Output:
Not zero-length

| length *STR* | Returns the numeric length of the string *STR* |
| *STRING* | This operator checks whether a string is not an empty string. If it is empty, then it returns as false. For example: |

```
string1="home";
if [ $string1 ]
then
  echo "Not Empty String"
else
  echo "Empty String "
fi
```

Output:
Not empty string

- **Syntax 1:**

```
function function_name()
    {
        statements;
        statements;
        statements;
    }
```

where function is a shell keyword

- **Syntax 2:**

```
function_name()
    {
        statements;
        statements;
        statements;
    }
```

Both these ways of specifying a function are valid, and there is no specific advantage or disadvantage to one over the other. Generally, in other programming languages, arguments are passed to the function by listing them inside the brackets (). However, in Bash, you need not list anything inside the brackets (); it simply acts as a syntax notation. The name of your function is *function_name*, and by using the *function_name*, you can call it as many as required in your scripts. The function name must be followed by parentheses ( ), followed by a list of commands enclosed within braces { }.

NOTE: The function definition (the actual function declaration) must appear in the script before calling that function in the script. It means that the function should be declared before calling it the script.

For example:

```
$cat test_func.sh
#!/bin/bash
New_function1()
{
     echo "My function1 works very well!"
}

     function New_function2()
{
     echo "My function2 works very well!"
}

New_function1
New_function2
Exit 0
     $
$ ./test_func.sh
My function1 works very well!
My function2 works very well!
$
```

- **Passing arguments:**
   Sometimes, you may have to pass on some arguments and process them within the function. So, you can send user input data to the function in the same way of passing command-line arguments to a shell script. You can provide the arguments right after the function name. These parameters would be represented by $1, $2, $3, and so on.

```
$ cat pass_argu.sh
#!/bin/bash
# Passing arguments to a function
print_hello () {
echo "Hello, My name is $1 and living in $2 city"
}
print_hello Jone London
print_hello Sanrit Delhi
exit 0
$

$./ pass_argu.sh
Hello, My name is Jone and living in London city
Hello, My name is Sanrit and living in Delhi city
$
```

- **Bash function return value:**
   Most of the programming languages have the concept of return value for functions that return to their original calling location or function. However, bash functions do not support this feature and allow to pass the values of a function's local

variable to the main routine by using the keyword *return*. Thereafter, the returned value is stored to the default variable $?. For example:

```
$cat add.sh
#!/bin/bash
addition(){
    sum=$(($1+$2))
    return $sum
}
read -p "Enter the integer X: " x
read -p "Enter the integer Y: " y
addition $x $y
echo "The sum result is: " $?
$
$./add.sh
Enter the integer X:5
Enter the integer Y:4
The sum result is: 9
$
```

We pass the arguments x and y to the *addition* function. The *addition* function executes it through the line sum=$(($1+$2)). Then, the value of the sum variable is transferred to the main routine through the statement line return $sum. By default, the values of $sum will be stored to the default variable $?, and then, the line statement echo "The sum result is:" $? gives the output. *Shell scripts can only return a single value.*

8.10 Advanced Shell Scripting

This chapter along with previous Chapters 5–7 brief out the shell scripting concept in detail. These chapters provide all details so that you can start writing your shell scripting with the bash properties. To fully utilize the power of shell scripting, you need to master RE (regular expressions), commands, and other commonly used tools in scripts, such as grep, expr, sed, and awk, interpret and use regular expressions. This section discusses the *test* command (offers a way to examine different conditions) and arrays in bash; such utilities will help you in writing a more complex shell script.

8.10.1 Array

In the C programming language, an array is a variable to store only the same type of data. Whereas in the bash script, an array is a variable to hold multiple values of the same type or different types of data [6]. Any variable may be used as an array. There is no maximum limit to the size of an array. It is not mandatory that array variables need to be indexed or assigned contiguously or consecutively. The array concept was not supported by the original Bourne Shell, but the newer versions of bash and other newer shells supported the array concept. Newer versions of bash support one-dimensional arrays. By default, in the shell script, everything is treated as a string. There are numerous ways to declare an array in a shell script, which are described as follows:

- **How to declare an array in shell scripting?**
  1. **Indirect declaration**

     In this type, the array declaration should be as follows:

     ```
     ARRAY_NAME[INDEXNR]=value
     ```

     where INDEXNR is a value or treated as an arithmetic expression that must be a positive number. There is no need to declare the array first. For example:

     ```
     ARRAY_NAME[10]= 20
     ARRAY_NAME[34]= 21
     ARRAY_NAME[52]= Mango
     ARRAY_NAME[42]="Hello, My name is San"
     ```

  2. **Explicit declaration:**

     In an explicit declaration of an array, we use the built-in command *declare*, as given below:

     ```
     declare -a ARRAY_NAME
     ```

     First, declare the array and then assign the values.

  3. **Compound assignment:**

     Array variables may also be created using the compound assignment. It helps to declare an array with a group of values, as syntax given below:

     ```
     ARRAY_NAME=(value1 value2  .... valueN)
     ```

     You can also add other value later.

     OR

     ```
     [indexnumber=] string
     ```

     The index number is optional. If it is given, it is assigned as an index, otherwise, the index of the element assigned is the number of the last index that was assigned, plus one using *declare* built-in command. If no index numbers are defined, indexing starts with 0. For example, an entire array can be declared by enclosing the array values in parenthesis as given below:

     ```
     MY_ARRAY=(mango orange banana grapes)
     ```

     ```
     Individual items cab be assigned of above declaration….
         MY_ARRAY[0]=mango
         MY_ARRAY[1]=orange
         MY_ARRAY[2]=banana
         MY_ARRAY[3]=grapes
     ```

- **How to print array value in a shell script?**

  To print the value of an array, you can use the following syntax.

```
echo ${ARRAY_NAME[INDEX_NUM]}
OR
echo ${name of array [index number of a value]
```

In the above given example of array, namely MY_ARRAY, the below command prints the values of MY_ARRAY;
```
    echo ${MY_ARRAY[0]}  ${ MY_ARRAY[1]}  ${ MY_ARRAY[2]}
```

To refer to the content of an item in an array, use curly braces. If the index number of an array is @ or *, all items of an array are referred or considered. Table 8.10 shows some examples to print the array element.

For example:

```
MY _ ARRAY=(mango orange banana grapes)                          // Array
declaration with items

echo ${MY _ ARRAY[*]}
mango orange banana grapes

echo ${MY _ ARRAY[@]}
mango orange banana grapes

echo ${ MY _ ARRAY [0]}
mango

echo ${ MY _ ARRAY}
mango

echo ${ MY _ ARRAY[3]}
grapes

MY _ ARRAY[4]=Apple
// Add item in array

echo ${MY _ ARRAY[*]}
mango orange banana grapes apple

echo ${#MY _ ARRAY[*]}
5
```

TABLE 8.10

List of Some Scripts to Print Array Value

Script to Print Array Elements	Description
${array[*]}	Displays all the items in the array as a single word.
${array[@]}	Displays all the items in the array as each item is a separate word.
${!array[*]}	Displays all the indexes in the array.
${#array[*]}	A number of items in the array.
${#array[0]}	Length of item zero.
${#array[N]}	Length of Nth item.

```
echo ${#MY _ ARRAY[0]}
5

echo ${#MY _ ARRAY[3]}
6

Multiple elements can be added at a time:
MY _ ARRAY+= (guava cherry)

echo ${MY _ ARRAY[*]}
mango orange banana grapes guava cherry
```

- **Deleting array items:**
  To delete an item from the array or delete array, use the *unset* built-in command along with array name and item index number or its key. For example:

  ```
  $ unset MY_ARRAY[2]
  $ echo ${ARRAY[*]}
  mango orange grapes guava cherry
  $ unset MY_ARRAY
  $ echo ${ARRAY[*]}
  <--no output-->
  ```

8.10.2 Test command:

In Section 8.7, various control structures that are used in many programs are described. Such programs need logical flow control among the various commands/statements that allow a script to ignore some part of the command statements based on the evaluated condition. Such a condition is evaluated with the help of *test* command [8].

The syntax of *test* command is given as:

```
test condition
```

The bash offers another method for testing a condition without using the *test* command in the control structure. Such conditions can be evaluated in the square bracket as mentioned in Section 8.7.

There are three classes of conditions to be evaluated by *test* command. They are:

1. **Numeric comparisons (compares two numbers):**
   The numeric comparison is one of the most common ways to perform a comparison between two numbers to check conditions. Table 8.11 lists the various numeric comparison operators used for testing two numeric values. These operators always start with -, followed by a two-character word and bounded on either side by whitespace.

2. **String comparisons (compares two strings):**
   Another use of the *test* command is to perform comparisons between string values to check conditions. Table 8.12 lists another set of string comparison operators used with *text* command; the same operator is also listed in Table 8.9 with examples.

   NOTE: The *greater-than* (>) and *less-than* (<) symbols need to be escaped, otherwise the shell uses them as redirection operator symbols, with the string values as

TABLE 8.11

List of Numeric Comparison Operators Used with *test* Command

Operator	Numeric Comparison Statement	Description
-eq	X -eg Y	Check, if X is **equal to** Y
-ne	X -ne Y	Check, if X is **Not equal to** Y
-le	X -le Y	Check, if X is less than or equal to Y
-lt	X -lt Y	Check, if X is **less than** Y
-ge	X -ge Y	Check, if X is greater than or equal to Y
-gt	X -gt Y	Check, if X is **greater than** Y

X and Y are two numeric values.

TABLE 8.12

List of String Comparison Operators Used with *test* Command

String Comparison	Description
string1 = string2	*string1* is the same as string *string2*
string1 != string2	*string1* is not equal to *string2* or str2
string1 < string2	string1 is less than string2
string1 > string2	string1 is greater than string2
-n string1	*string1* is not a null string, it means the string has a length greater than zero
-z string1	*string1* is a null string, it means the string has a length of zero

string1 and string2 are two input strings, and the comparison operator compares the strings based on their length.

filenames. Hence, the correct representation of the *greater-than* symbol is (\>) and (\<) for *less-than*.

3. **File comparisons (checks files attributes):**

The third and important use of the *test* command is to check the status of files and directories on the Linux file system. This is a relatively powerful and widely used comparison method in shell scripting. Table 8.13 lists the file comparison attributes.

8.11 Examples of Shell Scripting Program

- Example: Create a shell script using *case* statement:

  Shell Script Name: vehicle_rental.sh

```
$ cat vehicle_rental.sh
#!/bin/bash
# Script using command line argument.
# if no command line argument is given
echo "Available Vehicle for Rent: car , van, jeep, motorcycle"
# if no command line argument is given
if [ -z $1 ]
then
   echo " Sorry, Input"
```

TABLE 8.13

List of File Comparisons

File Comparison	Description
-f file1	True if *file1* exists and is a regular file
-d file1	True if *file1* exists and is a directory
-r file1	True if *file1* exists and is readable
-w file1	True if *file1* exists and is writable
-x file1	True if *file1* exists and is executable
-s file1	True if *file1* exists and has a size greater than zero (nonempty file)
-O file1	True if *file1* exists and is owned by the current user
-e file1	True if *file1* exists (Korn and Bash only)
-G file1	True if *file1* exists and the default group is similar to the current user
file1 -nt file2	True if *file1* is newer than *file2*
file1 -ot file2	True if *file1* is older than *file2*
file1 -ef file2	True if *file1* is lined to *file2*

```
  fi
  # use case statement to make decision for rental
 case $1 in
 "car") echo "For $1 rental is Rs.30 per k/m.";;
 "van") echo "For $1 rental is Rs.15 per k/m.";;
 "jeep") echo "For $1 rental is Rs.25 per k/m.";;
 "motorcycle") echo "For $1 rental Rs. 5 per k/m.";;
  *) echo "Sorry, You may enter wrong input or vehicle is not available
for rental services!";;
  esac
$
```

- Example: Create a shell script using *array* statement:

  Shell Script Name: array_add.sh

```
$ cat array_add.sh
  #!/bin/bash

  echo -n "Enter the array1 : "
  read -a var
  echo -n "Enter the array2 : "
  read -a var1
  i=0
  while ( test ${var[$i]} )
  do
  array[$i]=$( expr ${var[$i]} + ${var1[$i]})
  i=$( expr $i + 1)
  done

  echo -n "Whole array:"
  echo ${array[*]}

  echo -n "First two elements:"
  echo ${array[*]:0:2}
```

```
    echo -n "All elements starting from 1 index:"
    echo ${array[*]:1}
$
```

- Example: Create a shell script using *test* statement:

  Shell Script Name: result.sh

```
$ cat result.sh
#!/bin/bash
# input subject marks
echo -n "Subject 1 : "
read sub1
echo -n "Subject 2 : "
read sub2
echo -n "Subject 3 : "
read sub3
echo -n "Subject 4  : "
read sub4
avg= `expr $sub1 + $sub2 + $sub3 + $sub4`
avg= `expr $avg/4`

# condition check..

if test $avg -le 50
then
   echo " Fail: Your result is $avg "
else
   if test $avg -gt 50 -a $avg -le 60
   then
     echo " Passed with Second Div.: Your result is $avg"
   else
     if test $avg -gt 60 -a $avg -le 75
     then
       echo " Passed with First Div.: Your result is $avg "
     else
       if test $avg -gt 75 -a $avg -le 100
       then
         echo " Passed with Distinction: Your result is $avg"

       fi
     fi
   fi
fi
#
```

8.12 Summary

A shell provides numerous utilities that can be used to create and manipulate files as well as programs, referred to as a shell script. Bash has become default shell and shell scripting on most Linux distributions. Shells are user program that permits customizable interface between the Linux kernel and the user. A shell script executes a small or big list of commands, written in a text file. You can execute this text file as a shell script or shell program to execute all the programs mentioned in it. Before executing any such script file, you must make it an executable script by using the *chmod* command. A script may be run in interactive or noninteractive modes. In interactive mode, user intervention is required by using the READ command. Shell variables can be categorized into local and environment variables. The environment variable is also known as the global variable because its information is shared to all processes started by shell including other shells and subshells. Shell provides a powerful tool for performing arithmetic, logical, and control operations by using built-in arithmetic and logical operators along with control statements. Looping is an integral part of shell scripting. Bash also provides shell function features, which help to break a lengthy and complex program into separate smaller and logical subsections. The test command and array feature help to write advanced shell scripting. The test command in the control structure evaluated three classes of conditions, which are numeric comparisons, string comparisons, and file comparisons.

8.13 Review Exercises

1. Write down the key feature and benefits of shell scripting programming over other programming languages.
2. Describe how to make variables unchangeable and the procedure to create a shell script, and write down two ways to make it executable.
3. In a shell script, explain the importance of the following terms: shebang (#!), environment variable, positional parameter, and interactive script.
4. Write a shell script with a number as input and check whether the input number is even and divisible by 7 and 3.
5. Write a bash script with an input of four subjects and display the average score up to 4 decimal digits.
6. Explain positional parameters with a shell script that comprises 14 positional parameters.
7. Write a shell script to get partition details of the disk, current date, and absolute path of the current working directory.
8. Write a shell script to find out all filenames that are comprised of special characters.
9. Write a shell script to print the total number of files and subdirectories under the given directory name.
10. Write a shell script with a three-digit number as the input and check whether it is a palindrome or not.
11. Write a bash script with three sides of a triangle as the input and display whether the triangle is a scalene or isosceles or equilateral.
12. Write an interactive bash script to make a simple calculator to perform the basic arithmetic operations using the *expr* operator.
13. Write a shell script to find a person's age as on date in years, months, and days from the provided date of birth as inputs.

14. Write a shell script with a 5-digit number as input and generate the reverse of the input number. Also, check whether the reverse and original numbers are the same or not?

15. Write a shell script to display the total number of logged users with their login name and terminal/system name.

16. Write a shell script to find whether the given character is a special character or a number or an alphabet and, further if it is an alphabet, whether it is uppercase or lower case?

17. Write a bash program with a number as input and replace all 2's with the 5's in the given number.

18. Write a bash script with two numbers as input to find GCD (greatest common divisor) of input numbers and display.

19. Write a script using a while statement to copy the content of any given file as the input to a variable and display the individual count of total characters of a special type, number type, and alphabet type.

20. Write a shell script with an integer number as the input and display its octal and binary equivalent.

21. Write a shell script with a filename as an input argument and find whether the file is existing or not. If the file exists, the script should display the absolute path-names of the input file along with its size, type, and permission.

22. Write a script with a *pattern* and *filename* as the input and count the total number of occurrences of the pattern in the file. Further, take another input *pattern1* from a user and replace the first 10 matching patterns with *pattern1* in the given filename.

23. Write a shell script to delete the lines comprising a word <red> if it appears between the 3rd and 15th line of a given file.

24. Write a bash script with 10 numbers as the input and store it in an array. Find the minimum, maximum, and average numbers in the array.

25. Write a bash script to store the result report of students whose names and marks are given using two arrays, one for students and one for marks.

26. Write a bash script to implement the following problems.

    a. Implement stack
    b. Merge sort on an array of numbers
    c. Implement binary search tree
    d. Implement the *grep* command
    e. To clear the hard disk partitions of your machine
    f. To predict the day of a given date
    g. To display all the cores of the system and how many of them are allocated to applications
    h. To make a calendar of the current month

References

1. Machtelt Garrels, Bash Guide for Beginners in the World Wide Web https://tldp.org/LDP/Bash-Beginners-Guide/html/index.html.
2. Richard Petersen. 2008. *Linux: The Complete Reference*. The McGraw-Hill.
3. Sumitabha Das. 2013. *Your UNIX/Linux: The Ultimate Guide*. The McGraw-Hill.
4. Steve Parker. 2011. *Shell Scripting: Expert recipes for Linux, Bash, and More*. John Wiley &Sons, Inc.
5. Mendel Cooper. 2014. *Advanced Bash-Scripting Guide*. Open Source Documents.
6. Richard Blum Christine Bresnahan. 2015. *Linux Command Line and Shell Scripting Bible* John Wiley & Sons, Inc.
7. Eric Foster-Johnson, John C. Welch, and Micah Anderson. 2005. *Beginning Shell Scripting*. Wiley Publishing, Inc.
8. Carl Albing, JP Vossen, and Cameron Newham. 2007. *Bash Cookbook*. O'Reilly Media.

9

Linux System Administration

Linux has hit the mainstream in computing technology, empowering everything from desktop and mobile to the computer-intensive system and servers. Due to part of open-source software, Linux is taking part in various technological innovation and open-source projects like KDE, GNOME, Android, Unity, LibreOffice, Samba, Apache, Mozilla, OpenOffice, and Scilab. Due to this, high-skilled Linux system administrators and engineers are extremely required. This chapter presents useful tips for Linux administration in addition to Chapter 1. Chapter 1 describes the Linux features, distribution, how to install and start with various run levels, creation of user accounts, and much more information related to system administration and maintenance.

Linux administration is a specialized work and covers all the essential things that you must do to keep a computer system in the best operational order. Therefore, every Linux learner or user must understand the important concepts related to Linux administration including:

- To know the login as root user/superuser into the system and understand the power and feature of it.
- To know the concept of run levels, start-up and shutdown processes.
- To know Linux boot loader (LILO, GRUB) and user interface (KDE and GNOME).
- To create, modify, and delete user accounts and directories.
- To learn how to back up or restore system and user important files.
- To check disk usage limit and disk usage space.
- To install and upgrade the software package of Linux system.
- To know the file system functioning and its feature.
- To check and repair file systems and other troubleshooting problems.
- To mount and unmount file system with main Linux file system structure.
- To know about RPM packets and its installation procedures.

9.1 Checking Space

When you are working with Linux system and wish to install or upgrade software packages in the system, you require a fair amount of memory space on your disks. Therefore, it is significant to check memory space from time to time to ensure the availability of free space before the installation of any software package on the system. There are plenty of tools for checking disk space consumption in Linux, but Linux provides a strong built-in command, known as "df" command. The **"df"** command stands for **"disk file system"** or **"disk free space"** [1]. It provides full details of available and used disk space of file system on Linux system. With the help of df command, you can check the memory space of individually mounted disks. The df command displays the lists of all file systems with its device names, utilized disk space (used and available), percentage of the disk space used, and mount point, as in below example:

The basic syntax for df (summarize free disk space) is as follows:

```
                    df [options] [devices]
```

```
$df
Filesystem    1K-blocks     Used         Available    Use%   Mounted on
/dev/sda1     103117024     45928204     52976480     47%    /
devtmpfs      7425908       0            7425908      0%     /dev
tmpfs         7435612       185228       7250384      3%     /dev/shm
tmpfs         7435612       0            7435612      0%     /sys/fs/cgroup
/dev/sdc1     31440900      11684316     19756584     38%    /home
/dev/sdb      10475520      3590868      6884652      35%    /common_pool
$
```

In the above output of df command, all field names are displayed, but you can change the output format by using "valid field name" as mentioned in Table 9.1 as follows:

For example, you can also limit the number of fields in the output of df command by providing the valid field name format with –output option as follows:

```
            Syntax: $ df --output=field1,field2,...
```

```
$ df -H --output=source, pcent, target
File system   Use%   Mounted on
 /dev/sda1    47%    /
 /dev/sdc1    38%    /home
 /dev/sdb     35%    /common_pool
$
```

The df command is a very safe way to see the space information of all partitions, instead of using fdisk command. The fdisk command can also be used to erase partitions. The df command displays only mounted partitions, while fdisk shows all partitions. By default, disk space is shown in 1K blocks, but the popular option is -h, which displays the disk space in human-readable style, generally M for megabytes or G for gigabytes, as shown below:

TABLE 9.1

List of Field Name Format of df Command

Field Display Name	Valid Field Name Format Used with --output Option)	Description
File system	source	Mount point source, generally a device.
1K blocks	size	Total number of blocks.
Used	used	Number of used blocks on a drive.
Available	avail	Number of available blocks on a drive.
Use%	pcent	Percentage of USED space divided by total SIZE.
Mounted on	target	The mount point of the drive

TABLE 9.2

List of df Command Options

Option	Notation	Description
-a	--all	Shows all file systems, including the list of file systems that have 0 block (dummy file systems), which are omitted by default.
-B	--block-size=SIZE	Uses SIZE-byte blocks
-i	--inodes	Lists inode usage information instead of block usage.
-h	--human readable	Displays output size in one-kilobyte blocks by default, -m (output size in one-megabyte), -G (output size in one-gigabyte)
-H,	--si	Like -h option, but uses powers of 1000 not 1024
–T	-print–type	Prints the type of each file system listed such as ext4, btrfs, ntfs, ext2, and nfs4
-l	--local	Limits listing to local file systems
–x	exclude–type=*fstype*	Displays the file systems except for type *fstype*. Multiple file system types can be removed by giving multiple –x options. By default, all file system types are listed.

–t,	--type=*fstype*	Displays only the file systems of type *fstype*.
		$ df -t ext3 #display only ext3 file system
		$ df -t ntfs #display only ntfs file system

```
$ df -h
File system   Size   Used   Avail   Use%   Mounted on
/dev/sda1     99G    45G    50G     48%    /
/dev/sdc1     30G    13G    18G     42%    /home
/dev/sdb2     99M    19M    76M     20%    /boot
/dev/sdb      10G    3.6G   6.5G    36%    /common_pool
tmpfs         1.5G   0      1.5G    0%     /run/user/0
```

See information about specific file system.

You can also see the information of a particular file system, for that you provide a device or mount point as an argument. For example, the following command displays information only for the partition /dev/sda1:

```
$ df /dev/sda
$ df /home/data
$ df .            //for current directory
```

The associated options with df command are listed in Table 9.2.

9.2 Disk Usage Limit

Another useful built-in command, known as du (disk usage), is used to check the disk usage information of files and directories on a system [2]. It is a quick way to check the amount of disk space used by each subdirectory and file of directory arguments. By default, the disk usage space is measured in 1K blocks, unless the environment variable POSIXLY_CORRECT is set. The basic syntax for du (summarize disk usage space) is as follows:

```
du [options] [directories and/or files]
```

By default, the du command displays all the files, directories, and subdirectories of the current working directory, and it displays the total disk blocks occupied by each file or directory. The **du** command also displays the sizes of files and directories in a recursive manner. So the output of du command for the present working directory is given below:

```
$        du
8        ./sanvik
72       ./sample.txt
20       ./sonu.txt
16       ./shyam.txt
8        ./xyz/sksingh
8        ./ritvik
40       ./bin
32              ./.config
128      ./Documents
16       ./ccet
4        ./.vim
200      .
$
```

The number at the left of each line is the number of disk blocks that each file or directory takes. You can also simply specify the directory path as an argument to the command to know about its space usage details, as given below:

```
$ du /home/san/
40       /home/san/downloads
8        /home/san/color/plugins
4        /home/san/programs/hello.txt
12       /home/san/.chrome
24       /home/san/.ssh
```

Table 9.3 lists all the options associated with du command.

9.3 Kernel Administration

In the Linux operating system (OS), kernel is the core part and is responsible for the allocation of system resources and scheduling of various executing processes. It takes the charge to perform essential responsibilities to manage the various parts of system resources, like memory management, disk management, process management, file management, network management as well as interfacing with the hardware components that empower Linux system to work perfectly without bugs. Linux system supports multitasking and multiuser standard feature due to kernel functionality. Hence, proper mechanism must be in place to manage the various responsibilities of the kernel, referred to as kernel administration. When you installed Linux, then it is the responsibility of the kernel to configure the various connected devices to the system. Whenever you want to add a new device, kernel provides such support and tools. Hence, the existing kernel must be reconfigured or updated regularly to support the new devices or software through a defined procedure. Such procedures or steps are generally referred to as building or compiling the kernel. The updating kernel is distinguished through its kernel versions.

TABLE 9.3

List of Options of du Command

Option	Notation	Description
–a	-all	Displays disk usage for all files, not just directories.
-h	**-Human Readable Format**	Displays disk usage in **Bytes, Kilobytes, Megabytes, Gigabytes**, etc.
–b	-bytes	Prints sizes in bytes.
–c	-total	Writes a grand total of all the arguments after all arguments have been processed. This can be used to find out the disk usage of a directory, with some files excluded.
– k	-kilobytes	Prints sizes in kilobytes. This overrides the environment variable POSIXLY_CORRECT.
–l	-count-links	Counts the size of all files, even if they have appeared already in another hard link.
–s	-summarize	Summarizes each argument
–x	-one-file-system	Skips directories that are on different file systems from the one that the argument being processed is on.
–D	-dereference-args	Dereference symbolic links that are command line arguments. Does not affect other symbolic links and very helpful for finding out the disk usage of directories like /usr/tmp where they are symbolic links.
–L	-dereference	Shows the disk space used by the file or directory that the link points to instead of the space used by the link.
–S	-separate-dirs	Counts the size of each directory separately, not including the sizes of subdirectories.

The kernel may be divided into two main categories based on their design, namely monolithic kernel and the microkernels as described in Section 1.9. The latest versions of the kernel are made available at kernel home page site [3] that provides advanced support for the latest software and hardware devices. The kernel version numbering is often mentioned as an important subject. The kernel version is to determine the status, feature, and revision factor of kernel release. The version number for a Linux kernel comprises four segments: the major, minor, revision, and security/bug fix numbers, as described in Section 1.9. You can use the following command to see the detailed information of kernel version and other information:

```
#uname -r // Display kernel version
```

The Linux kernel is being worked on continuously; latest versions are announced and posted in the public domain when they are ready. Linux distributions may include different kernel versions. Table 9.4 lists some weblinks that provide resources and information related to Linux kernel:

Before we discuss kernel compilation and installation, let us make sure that you are familiar with a concept of what a kernel is and the importance in the system.

Most often, when people say "Linux," they are usually referring to a "Linux distribution"; for example, Debian is a type of Linux distribution. As discussed in Chapter 1, a distribution comprises the same kernel (but the kernel version may differ among Linux distributions) and other necessary software packages to make a functional OS. The kernel of any OS is the core of all the systems' software. The kernel has many jobs. The essence of its work is to abstract the underlying hardware from the software and provide a running environment for application software through system calls.

TABLE 9.4

List of Some Popular Weblinks for Linux Kernel and Resource

Weblink	Descriptions
http://kernel.org	The official website of Linux kernel and all new kernels announces from here
http://linuxhq.com	Linux headquarters, kernel sources, and patches
http://en.tldp.org	Linux Documentation Project that comprises all Linux resources
http://vger.kernel.org	The Linux Kernel Development Community
http://kernelnewbies.org	Linux kernel sources and information

At the time of writing this chapter, Linux kernel (version 4.x, where x represents the complete version number of the kernel) contains more than 6 million lines of code (including device drivers). By comparison, the sixth edition of Unix from Bell Labs in 1976 had roughly 9000 lines. All the source codes for the Linux kernel can be found as well as downloaded by anyone from its official home page site [3].

The following key tasks are performed by the kernel:

- **Process scheduling**: Enables multitasking feature of OS.

- **Memory management**: Proper management of memory space that allows more processes to be held in RAM simultaneously leads to better CPU utilization.

- **File system management**: Helps to manage file system on disk, allowing files to be created, retrieved, updated, deleted, and so on.

- **Creation and termination of processes**: Better resource allocation to all executing processes.

- **Access to devices**: Allows and simplifies communication between the system and all peripheral devices.

- **Networking**: Allows to connect remote system, transmits and receives network messages (packets) on behalf of user processes.

- **System call provision for application programming interface**: Allows processes to make a request to the kernel, to perform various tasks using kernel entry points known as system calls.

- **Multiuser environment**: Allows multiple users to interact with the system simultaneously and independently with other users.

- **Multiprocessor support**: The system can use multiple processors and each processor may handle any task, and there is no discrimination among them.

9.3.1 Listing Kernel Modules with *lsmod*

As discussed in Chapter 1 (Section 1.8), the Linux kernel provides various OS features, including support for various devices such as DVD, sound card, Bluetooth, and network cards. For better kernel administration, the functional parts of the kernel are broken out into distinct units with strict communication mechanisms between them. These function parts of the kernel are referred to as modules. In other words, you can say that kernel is a collection of various modules, which can be added (loaded) or removed (unloaded) as required. After the system boots up, kernel modules can be dynamically loaded or linked into the kernel as you need. It can be unlinked or removed from the kernel when it is no longer required. This mechanism extends the functionality of the kernel without the need to recompile (the kernel) or reboot the system. Mostly, Linux kernel modules are known as device drivers like network drivers, or file systems. A device driver enables the kernel to access a hardware component/device connected to the system.

In Linux, the loadable kernel module uses certain kernel commands to perform the job of loading or unloading modules. The *modprobe* command is a general-purpose command that calls *insmod* to load modules and *rmmod* to unload them. These commands are listed in */lib/modules* directory and have *.ko* (kernel object) extension since kernel version 2.6, prior kernel version *.o* extension. To see the list of loaded kernel modules in Linux, use the following command:

```
#lsmod              //Display all loaded modules in kernel
Module              Size        Used by
xfs                 985347      2
libcrc32c           12644       1 xfs
ebtable_filter      12827       0
ebtables            35009       1 ebtable_filter
ip6table_filter     12815       0
ip6_tables          26901       1 ip6table_filter
devlink             15525       0
......
......
#
```

The *lsmod* (list modules) command reads the contents of *proc/modules*.

To load kernel modules—insmod

To load a kernel module, use the insmod (insert module) command with specifying the full path of the module, as follows:
```
$insmod <Absolute path of module>
```

To unload kernel modules—rmmod

To unload a kernel module, use the rmmod (remove module) command with specifying the full path of the module, as follows:
```
$rmmod <Absolute path of module>
```

To get the information about module
```
$modinfo    //Displays information about a module by using
            -a(author),-d(description), -p (module parameters),
            -f(module filename), -v(module version).

$man modprobe    // Complete information about modprobe command
                 for listing, inserting, and removing modules
                 from the kernel
```

9.4 Compiling and Installing

To install a new kernel, you need to download the new kernel packages from www.kernel.org. to your system. The following steps are required to compile and install the Linux kernel from the downloaded source:

1. **Download the latest kernel source from the kernel home page [3]:**
   The filename would be *linux-x.y.z.tar.xz*, where *x.y.z* is the actual Linux kernel version number. At the time of writing this section, one of the latest stable kernel version file is linux-5.2.8.tar.xz, which represents Linux kernel version 5.2.8. Use the wget command to download Linux kernel source code in your current working directory:

   ```
   $wget https://git.kernel.org/pub/scm/linux/kernel/git/stable/linux.
   git/tree/?h=v5.2.8
   ```

   One advantage to compiling the kernel is that you can enhance its configuration, adding support for certain kinds of devices such as Bluetooth devices.

2. **Unpacking the kernel source code (tar.xz file):**
   The kernel source is in the form of compressed archives (.tar.gz). They have the prefix linux with the version name as the suffix (linux-5.2.8.tar.xz). The kernel source is typically installed in /usr/src/linux. But it should not be unpacked or placed in this directory. Generally, the new kernel source code file is downloaded in the current working directory. In this case, we suggest creating a local directory, called *mylinux,* in the current working directory and unpack the archive file by using the tar command. Let's assume that the current working directory is your home directory, then use the following commands to unpack the source file:

   ```
   Create a local directory, namely mylinux, in home directory
   $mkdir ~/mylinux

   Now move the source code into this directory:
   $mv ~/linux-5.2.8.tar.gz ~/mylinux/

   go to the mylinux directory:
   $cd ~/mylinux
   $ls
   linux-5.2.8.tar.gz
   $

   Now here, unpacked the source code
   $tar -xzvf linux-5.2.8.tar.gz

   See the content of mylinux directory
   $ls
   linux-5.2.8.tar.gz
   linux-5.2.8/
   $
   ```

   In the above example, the local directory of the new kernel source would be ~/mylinux/linux-5.2.8. It should comprise the folders, such as *arch, block, fs, init, ipc, mm, kernel, and tools.*

3. **Configure the Linux kernel features and modules:**
   Before compiling the kernel, you need to configure kernel features by specifying the list of included modules (drivers) as required for your system. You can also mention modules which ones are to be left out. There are many ways to do this. A simple and easy way to do this is to first copy your existing kernel config file and then use the configuration tool to make changes (if necessary).

```
$ cd linux-5.2.8
$ cp -v /boot/config-$(uname -r) .config
```

You can configure the kernel using one of the various offered configuration tools, which are *config, menuconfig, xconfig(qconf),* and *gconfig(gkc)*. The configuration options are stored in a file, namely *.config*, kept in the kernel directory.

```
$ make config // Text-based command line utility
OR
$ make menuconfig       //Curses-based graphical utility
OR
$ make gconfig          //X-window (gtk)-based graphical utility,
                        works best under GNOME desktop.

OR
$make xconfig           // X-window (Qt)-based graphical utility,
                        works best under KDE desktop
OR
$ make defconfig        // Creates a default configuration based on
                        the system architecture
```

The kernel is now successfully configured, but it should be modified according to your system, so that it will work correctly (see Section 9.5).

4. **Compiling Linux kernel:**

Now, the kernel configuration is ready, and you can compile your kernel. The only command that is needed here to compile the kernel is the *make* command:

```
# make //Create a compressed kernel image
```

The make command without arguments will create the *bzImage* and *the modules*. But, if you use *bzImage* option with *make* command, then it simply produces a kernel file called *bzImage* and places it in the *arch* directory. For Intel and AMD systems, you find *bzImage* in the *x86/boot* subdirectory, *arch/x86/boot*. For a kernel source, this is in *arch/x86/boot*.

```
# make bzImage              // Simply creates a kernel file, called
                            bzImage but not the modules
```

The above command produces kernel image file in the path *<kernel-source-tree>/arch/x86/boot/bzImage*.

To compile your modules separately, use the *make* command with the module argument.

```
# make modules
```

To speed up the kernel compilation, the *make* command offers a feature to split the built process into several parallel jobs. Each of these jobs then executes separately and simultaneously. This is expressively speeding up the built process on multi-processing systems to enhance processor utilization. Hence, to build the kernel with multiple make jobs on a multicore processor system, use the following command syntax:

```
# make -j n // where n is the number of cores/threads
```

where n is the number of cores/threads that can execute maximum n jobs concurrently. For example, if your CPU has a Quad (4) core, then you can type:

```
# make -j 4
```

If you don't know CPU core count or thread, then use nproc command to get thread or CPU core count, as follows:

```
# make -j $(nproc)
```

5. **Install the Linux kernel modules**

   We compiled parts of the kernel as modules; therefore, we need to install the modules by using *make* command with the *modules_install* option, as follows:

```
# make modules_install
```

The above command installs the modules in the */lib/modules/version-num* directory, where *version-num* is the version number of the kernel. Please note that you should keep a backup copy of the old modules before installing the new modules.

6. **Installing the Kernel:**

   So now you have a fully compiled kernel and need to install it by using the following command:

```
#make install
```

It will install three files into */boot* directory as given below:

- Creates RAM disk image file
  `/boot/initrd-5.2.8.img`

- Compiled kernel file
  `/boot/vmlinuz-5.2.8`

- Creates System.map file
  `/boot/System.map-5.2.8`

7. **Modifying the Boot loader for the New Kernel**

   There are two common Linux kernel boot loaders: GRUB and LILO. After installing the new kernel, you need to amend the boot loader configuration file. For systems running with GRUB, the configuration file will be */boot/grub/grub.conf* or */boot/grub/menu.lst*. But for systems running with newer versions of GRUB2, the configuration file will be */boot/grub2/grub.cfg*. On the other hand, the configuration file for LILO is */etc/lilo.conf*.

   In these boot loader configuration files, we must make a new entry point for the newly installed kernel, so that the new kernel can be seen in the boot menu after rebooting the system. For Fedora Linux system, modify the GRUB2 configuration file as follows:

```
#grub2-mkconfig -o /boot/grub2/grub.cfg
#grubby --set-default /boot/vmlinuz-5.2.8
```

8. **Reboot and start the new kernel in your system:**

   Hope you did everything as prescribed to install a new kernel and then reboot the system and select the new kernel from the boot loader menu during system bootup.

```
# reboot
```

After the system boots up, you can check the kernel version of the newly installed kernel by using the following command:

```
# uname -r
5.2.8-custom
```

Congratulations! You successfully configured and installed the new kernel.

9.5 Modifying

As suggested in the previous section, if you want to modify your kernel configuration, you are advised to retain a copy of your current kernel. In case, if something wrong happens to the modified kernel version, you can always have the option to boot from the saved copy of the previous kernel. It is always advisable that during the installation of the new kernel, you must give a different name to the new kernel, so that the older kernel is not overwritten. You always have the option to modify your system in order to make sure that it will operate correctly with the latest feature.

The newly installed kernel makes use of numerous files in the */boot* directory. Each filename is followed with kernel version's number. The main file is vmlinuz-*kernel_version* (vmlinuz-*5.2.8*), which is the actual kernel image file, including other support files like System.map-*kernel_version* (System.map-5.2.8), config-*kernel_version* (config-5.2.8), and initrd-*kernel_version*.img (initrd-5.2.8.img). The System.map-5.2.8 file comprises kernel symbols required by modules to start kernel functions. Before modifying your current kernel, you must create a separate copy of the current working kernel version (5.2.8) files with another name, as given below here:

```
#cp /boot/vmlinuz-5.2.8 /boot/vmlinuz-5.2.8.old
```

The filename, *vmlinuz-5.2.8.old*, is a separate copy of the current working kernel. Further, you also create backups of the *System.map* and *config* files along with the backup of current modules situated in the *lib/modules/ kernel_version* directory. Otherwise, when you modify or install new kernel, you will miss the modules already working with the original kernel, which means the previous kernel. Now, if you are compiling and installing an updated or new kernel version, then these files are stored in a new directory named with the new kernel version number. In such a situation, the previous kernel system files are kept in a safe place, if required you can run the previously installed kernel.

The kernel configuration is saved in a file, known as *.config*, in the top directory of the kernel source path. After downloading the source code of the selected kernel version into a local directory, the first instance is to configure the kernel with the appropriate options. The most basic method of configuring a kernel is to use the *make config* method:

```
#make config
```

Every kernel code comes with a "default" kernel configuration file; use the following command for default configuration:

```
#make defconfig
```

Now, we have created a basic configuration file, but this file should be modified or upgraded to support the available hardware packages to run the system effortlessly. Basically, three different interactive kernel configuration tools are available to configure or modify kernel [4], which are as follows:

- **A terminal-based (*menuconfig*):** The *menuconfig* is a way of configuring a kernel using console-based program, which provides a technique to move around the kernel configuration menu lists using the keyboard arrow keys. To start up this configuration mode, use the following command:

  ```
  #make menuconfig
  ```

- **A GTK+-based graphical (*gconfig*):** The *gconfig* method of configuring a kernel uses a GNOME-based graphical interface, which allows you to modify the kernel configuration. The *gconfig* is written using the GTK+ toolkit and has a two-pane screen.

  ```
  #make gconfig
  ```

- **A QT-based graphical (*xconfig*):** The *xconfig* method of configuring a kernel uses a KDE graphical interface, which allows you to modify the kernel configuration. The *xconfig* is written using the QT libraries and has a three-pane screen

  ```
  #make xconfig
  ```

9.6 LILO and GRUB

Linux and Unix OSs are known as multiboot OSs that support dual boot installation by having some boot loader program or manager. In Linux distribution, popularly two most common boot loaders are widely in use: LILO and GRUB.

GRUB is one of the more commonly used boot loaders in modern distributions, but LILO is still in use as well. This section briefly describes how to configure and modify the boot loader (GRUB and LILO) files. In Linux system, the below-mentioned folder contains the information of the installed boot loader.

```
/boot—files used by the bootstrap loader (LILO or GRUB)
```

To determine which boot loader is installed in your system, use the following command:

```
$ls -F /boot
```

In the above command output, search GRUB or LILO files.

9.6.1 LInux LOader

LILO is a boot loader for Linux and comes as the standard default boot loader for most Linux distributions. The LILO is a small utility program that can load Linux kernel (or the boot sector of another OS) into memory and start it. When the system BIOS initiates, LILO presents you with the following prompt command:

```
# lilo [options]
```

The *lilo* command is used to map the installer to read the configuration file *"/etc/lilo.conf,"* which contains the information required by the boot loader to locate the kernel or other OSs as default during the installation process [5, 6]. During installation, LILO can be placed either in the MBR of the first boot sector of the disk or in your root partition. Any modification in *lilo* configuration file (/etc/lilo.conf) requires the lilo command to be rerun to make effective all changes. Some commonly used options with *lilo* command are as follows:

```
-C config _ file
            // Read the config _ file instead of the default /
            etc/lilo.conf
 -m map _file
            // Write map _file in place of the default as
specified in the configuration file.
-q          // Query the current configuration
-v          //Increase verbosity
```

Some other key functions associated with **/sbin/lilo** are as follows:

```
1.  /sbin/lilo -A-activate/show active partition
2.  /sbin/lilo -I-inquire path name of current kernel
3.  /sbin/lilo -M-write a master boot loader on a device
4.  /sbin/lilo -R-set default command line for next reboot
5.  /sbin/lilo {-u|-U}-uninstall LILO boot loader
```

lilo configuration file:
The boot= instruction in the configuration file *etc/lilo.conf* specifies the master boot record (/dev/hda) or the root partition of your Linux installation (generally, it is /dev/hda1 or /dev/hda2). If a new Linux kernel is installed, you must modify the lilo configuration file (*/etc/lilo.conf*) and then execute the lilo command to apply the modification done in the configuration file. The /etc/lilo.conf file is used by the /sbin/lilo command to determine which Linux kernel or OS to load and where it should be installed. To see the complete structure of the LILO configuration file, execute the given LILO man page command:

```
#man lilo
```

After executing the above manual page command, mostly the lilo installations use a configuration file as similar to the below structure:

```
boot = /dev/hda     # or your root partition
delay = 10          # delay, in tenth of a second (so you can interact)
vga = 0             # optional. Use "vga=1" to get 80x50
#linear             # try "linear" in case of geometry problems.


boot=/dev/had       # installed on the first hard disk or your root partition
map=/boot/map       # locates map file, not be modified in normal use.
install=/boot/boot.b     # boot sector, not be altered in normal use.
prompt
timeout=30          # time that LILO waits for user input before proceeding
                    with default booting
```

```
message=/boot/message        # LILO displays message to select the operating
                             system or kernel to boot.
default=linux        # default operating system for LILO to boot

<<<<<section to load Linux>>>>>
  image = /boot/vmlinux-version      # your zImage file
  root = /dev/hda1                    # your root partition
  label = Linux                       # or any fancy name
  read-only                           # mount root read-only


<<<<<section to load Window OS>>>>>
  other = /dev/hda4                   # your Window partition, if any
  table = /dev/hda                    # the current partition table
  label = WindowOS                    # or any other name
```

You can have numerous "image" and "other" sections. These sections help you to configure the various kernel images in your *lilo.conf* file. After you save this file, execute the */sbin/*
lilo program to write the configuration updates to the boot section of the disk, as follows:

```
#/sbin/lilo
```

Now, you can safely reboot the system to see the newly configured kernel in LILO boot
loader menu list.

9.6.2 GRand Unified Boot loader

GRUB is a multistage boot loader. It is much more flexible and powerful than LILO. It
recognizes and supports various types of file systems and kernel executable format. With
this feature, GRUB allows loading and boots more than one different OS, both open source
and propriety. When you boot the system, GRUB displays a menu-based option that is
generated by the configuration file **/boot/grub/grub.conf**. Here, a user can select which OS
or kernel to boot, or GRUB will boot the default OS as specified [2][6].

The grub-install script is used to install GRUB. For example, to install GRUB on the
master boot record of the first hard drive in a system, then call grub-install as follows:

```
# grub-install '(hd0)'
```

grub-install looks for a device map file, by default */boot/grub/device.map*, to establish the
mapping from BIOS drives to Linux devices. The device map file includes any number of
lines in the below structure:

```
(disk) /dev/device
```

For example, for a system with a floppy and a single SCSI disk with two partitions (pari-
tion1 and parition2), the file would be seen as given below:

```
(fd0) /dev/fd0 //floppy disk
(hd0) /dev/sda1 //for sda disk partiton1
(hd0) /dev/sda2 //for sda disk partiton2
```

By using the *grub* command, GRUB can also be installed. Let us assume that */boot* directory is on the first partition of the hard disk, then:

```
# grub
grub> root (hd0,0)
grub> setup (hd0)
```

For booting GRUB, if there is no configuration file or the configuration file does not have specific information about which kernel to load, then the below prompt will display:

```
grub> //Provide the root, kernel, initrd, and boot information to
      set up a Linux kernel.
```

- **GRUB configuration file:**
  The GRUB configuration file (/boot/grub/grub.conf), which is used to generate the list of OSs to boot in GRUB's menu interface, basically permits the user to select a particular predefined set of commands to execute.

  For example, the following GRUB configuration file comprises three entries, two for Linux kernels (one old kernel-2.6.16.1 and another installed kernel-2.6.21.1) and one for Microsoft Windows OS, as given below:

```
boot=/dev/hda
default=0
timeout=20
splashimage=(hd0,2)/boot/grub/splash.xpm.gz

<<<<<section to load old Linux kernel>>>>>

title Old Linux (2.6.16.1)
root (hd0,2)
kernel /boot/vmlinuz-2.6.16.1 ro root=/dev/hda2
initrd /boot/initrd-2.6.16.1.img


<<<<<section to load another Linux kernel>>>>>

title New Linux (2.6.21.1)
root (hd0,2)
kernel /boot/vmlinuz-2.6.21.1 ro root=/dev/hda2
initrd /boot/initrd-2.6.21.1.img


<<<<<section to load Microsoft Windows >>>>>

title Windows
rootnoverify (hd0,0)
chainloader +1
```

Further, you can edit the above configuration file structure and make more multiple sections to add the newly installed kernel versions and other information. After you save the file, reboot the system to see the newly configured kernel in GRUB boot loader menu list. Use the up/down keyboard arrow keys to highlight and select the new kernel version for booting. Unlike LILO, the amendment in the GRUB configuration files does not require the *grub* command to be executed again and again after each change.

9.7 Root User (add sudo)

Multiuser feature is one of the key features of Linux and Unix OSs that have a major difference from other OSs. Due to multiuser feature, Linux allows multiple users to use a computer system at the same time without affecting each other. In Linux, basically, there are two types of users—root users or superusers and normal users. A root user or superuser or administrator has the authority to access all the files and has complete control over the system. On the other hand, the normal user has limited privileges toward accessing files or systems. A root user or superuser, also known as an administrator, can add, delete, and modify a normal user account (see Section 1.13).

The superuser or root user is a special user account used to perform system administration operations. If you have the correct password of superuser, then it may enable you to log in as the root user, making you the superuser. A superuser has the power to change practically anything on the system. The command line interface for the root user uses a special prompt, the hash sign, #. If you logged in as the root user, then you will see the # prompt, as follows:

```
login: root
password:
#
```

- **Change root user password**

  During the installation of Linux, you set the root user password. But later you can change the root user password by using *passwd* command, as follows:

```
# passwd root
New password:
Re-enter new password:
#
```

- **Root user access: su**

  The **su (switch user)** command is used to switch the currently logged user to another user account. Sometimes, it becomes necessary to login into another normal user account or superuser account to perform some specific tasks without logging out from your current user login session.

```
Command Title:
su - change user ID or become superuser

Command Syntax
su [options] [username]
```

If *su* command is used with *username,* then it switches from the current user login session to another user login session as mentioned by *username.* On the other side, if *su* command is invoked without a *username,* then by default, *su* command switches from the current user login session to the superuser account. It is highly recommended to use *su* command with (-) argument. See the below examples to understand the difference between *"su"* command and *"su -"* command:

```
Example 1: "su" command with username (sanrit)
```

When you invoke *su* command with username, here *sanrit*, then it is important to note that the user *sanrit* keeps the environment from user *ritvik*, i.e., original login session. Therefore, the current working directory and the path to executable files also remain the same, so *cd* command (without any option) is used to change the current working directory to the *sanrit* home directory, as shown below:

```
[ritvik@Linux ~] $ su sanrit
Password:                    //Enter the password of sanrit user account
[sanrit@Linux ritvik]$
[sanrit@Linux ritvik]$ ls
ls: cannot open directory.:permission denied
[sanrit@Linux ritvik]$
[sanrit@Linux ritvik]$cd    //Return to home directory of logged user account

[sanrit@Linux ~]$
[sanrit@Linux ~]$ls
 bin data1
[sanrit@Linux ~]$
[sanrit@Linux ~]$ exit    // Exit from current user and return to previous
                           login session
 [ritvik@Linux ~] $
```

Example 2: "su-" command with username (*sanrit*)

When you invoke su command with a "-" or "-l" or "--login" options along with username, it provides a login interface like stand-alone login environment when you are logging on normally from login prompt. All the commands below are equivalent to each other.

```
        $ su - sanrit
        OR
        $ su -l sanrit
        OR
        $ su --login sanrit
```

```
[ritvik@Linux ~]     $ su - sanrit
Password:                     //Enter the password of sanrit user account
[sanrit@Linux ~]$
[sanrit@Linux ~]$ls
 bin data1
[sanrit@Linux ~]$
[sanrit@Linux ~]$ exit    // Exit from current user and return to
                           previous login session
[ritvik@Linux ~] $
```

Example 3: "su-" command without username

When you invoke *su* command without username with "-" option (highly recommended), then you simply switch to root user or superuser. To start a shell for the superuser, type the below command:

```
[ritvik@Linux ~] $ su -
```

```
Password:              //Enter the password of root user account
[root@Linux ~]#pwd
/root
[root@Linux ~]#
[root@Linux ~]#exit        //Return to the previous user login shell
[ritvik@Linux ~]$
```

- **sudo command: provide controlled administrative or root user access/privilege**
  In Linux, sudo (**S**uper **U**ser **DO**) command allows you to run commands as another user, by default as the superuser; i.e., it allows a user to have administrative privileges without logging in as root user. To authenticate the use of sudo command, a user must authenticate themselves with the user login password (not the root user password) and confirms the user request to execute a command by checking a file, called /etc/sudoers file or sudoers file. The system administrator configures *etc/sudoers* file to identify users or groups who can use the sudo facility.

  Once a user has been authenticated themselves, a time stamp is recorded and the user may use sudo privilege without a password for a short period of time (5 minutes, unless you changed in /etc/sudoers file). This time stamp can be modified if the user runs sudo with –v option. To edit the sudoers file, use the *visudo* command; for further details, refer to its *visudo manual*.* To see the complete information regarding sudo command, execute man command as follows:

  ```
  $man sudo
  ```

  The basic syntax of sudo command, at the command prompt, is as follows:

  ```
  $sudo command
  ```

  For example, you can run a system_script file by a user login 'ritvik' with the help of sudo command, as given below:

  ```
  [ritvik@Linux ~] $ sudo system_script
  Password:
  System Script Starting...
  ```

  In the above command execution, it prompted the password of ritvik user account (not the superuser's), and once the authentication is validated, the mentioned command is carried out. One key difference between *su* and *sudo* is that *sudo* command does not start a new shell, not load another user's login environment. To see what privileges are granted by sudo command to the current logged user, use the -l option to list them as follows:

  ```
  [ritvik@Linux ~] $ sudo -l
   User ritvik may run the following commands on this host:
   (ALL) ALL
  [ritvik@Linux ~] $
  ```

  Some key options associated with *sudo* command are given in Table 9.5.

---

* Visudo Manual: https://www.sudo.ws/man/1.8.13/visudo.man.html.

TABLE 9.5

List of Some Key Options of *sudo* Command

Option	Description
-v	Prints the version number and exit.
-l	Displays the list of commands permitted (and forbidden) by the user on the current login.
-h	Prints a usage message and exit.
-v	With –v (validate) option, **sudo** will update the user's time stamp, prompting for the user's password if required. This extends the **sudo** timeout for another 5 minutes, or user-defined value for timeout is set in **sudoers**
-b	The **-b** (background) option tells **sudo** to run the given command in the background.
-c	The **-c** (class) option helps **sudo** to run the mentioned command with resources limited by the specified login class.
-a	The **-a** (authentication type) option helps **sudo** to use the mentioned authentication type when validating the user, as permitted by **/etc/login.conf**.
-u	The **-u** (user) option helps **sudo** to run the mentioned command as a user other than root.
-H	The **-H** (HOME) option sets the **HOME** environment variable to the home directory of the target user (by default **root** user) as specified in **passwd**. By default, **sudo** does not modify **HOME**.
--	The -- flag indicates that **sudo** should stop processing command line arguments. It is most useful in conjunction with the **-s** flag.

9.8 Additional Packages

The key fact to measure the quality of any Linux distribution basically depends upon the presently available/offered software packages like applications for desktops, office suites, Internet packages, programming packages, and multimedia applications along with the strength of the distribution's support developer community. In the present scenario, software growth is extremely dynamic and constantly changing. Generally, the renowned and widely used Linux distributions announce new versions after every 5–6 months and many individual programs' updates every day. To keep up the Linux system with this growth of software packages, we must have good package management tools for managing software packages, frequently referred to as package manager or desktop software manager.

Package management is a method of installing and maintaining software on the system. All Linux software packages are currently distributed from online repositories. Software packages' downloads and updates are handled automatically by your desktop software manager or package manager. Nowadays, many Linux users prefer to install software by installing package manager along with required software from their Linux distributor, rather than to download *standard tar archive*. Every software is packaged in an archive using a different file extension format, as compared to standard tar archives (see Table 9.7), to be recognized and managed by a package manager. A package manager archive comprises all required files, like program files, configuration files, data files, and documentation files that create a software application. In one simple operation, the package manager installs all such files in your system. In other words, the package manager helps to install the software packages automatically with minimum user intervention.

- **Packaging systems**

    Different distributions use unlike packaging systems, and usually, a package projected for one distribution is not compatible with another distribution. Therefore, Linux distribution makes their own software packages, but sometimes,

TABLE 9.6

DEB and RPM Packaging Description

Packaging System Manager	Distributions (Some Key Names)	Package Extension	Package Tools
Debian style package (DEB)	Debian, Ubuntu, PureOS, Kali Linux, Linspire	.deb	dpkg, apt, apt-get, aptitude
Red Hat package (RPM)	Red Hat Enterprise Linux, CentOS, Fedora, openSUSE, Mandrake Linux, SUSE, Oracle Linux	.rpm	rpm, yum

TABLE 9.7

File Extensions of Linux Software Package

File Extension	Description
.rpm	A Red Hat software package
.deb	A Debian Linux package
.src.rpm	Source code versions of software packages of Red Hat package
.gz	A gzip compressed file
.bz2	A bzip2 compressed file
.tar.gz	A gzip compressed tar archive file
.tar.bz2	A bzip2 compressed tar archive file
.tar	A tar archive file
.conf	Configuration file
.bin	A self-extracting software file
.Z	A file compressed with the compress command

these packages are also adopted by other distributions to improve the system functionality. Presently, two software packages are widely used on several distributions, namely Red Hat Package Manager (RPM) and Debian Package Manager (DEB). The Red Hat packages are widely adopted by most distributions as compared to Debian packages. The standard compressed archive file has an extension like **.tar.gz** or **.tar.Z**, whereas RPM packages have an **.rpm** extension and DEB uses a **.deb** extension; for more details, see Table 9.6. Further, Table 9.7 lists the various software packages with its file extension.

9.8.1 Red Hat Package Manager

Red Hat Package Manager (RPM) is a command line-driven package management system that was originally written in 1997 by Erik Troan and Marc Ewing. Various popular Linux distributions, like Red Hat, CentOS, Fedora, Oracle Linux, SUSE, openSUSE, and Oracle Linux, used RPM packages and the RPM Package Manager is the default application package manager on the running system [6]. RPM organizes various Linux software packages that provide the service of installation, uninstallation, querying, verifying, and updating software packages installed on Linux systems. An RPM software package works as its own installation program for a software application that must be installed in different directories. The installed program is most likely located in a directory called */usr/bin*.

RPM software package works equally to the Windows Installation Wizard, which automatically installs software with configuration along with any other files (application may be used during execution) in appropriate directories on your system. The RPM software package performs all these automatic tasks for you. RPM also maintains a database to keep track of all the installed software packages. RPM packages may use *yum* tool (Yellowdog Updater Modified) to automatically download, install, and update software from online RPM repositories [8].

The RPM Package Manager offers the following:

- It is free and released under **General Public License (GPL)**.
- It installs, updates, and uninstalls packaged software.
- Keeps the information about the installed packaged software, i.e., whether installed or not under **/var/lib/rpm** database.
- Verifies the integrity of software package and follow-on software installation.
- **rpm command**

  With the rpm command, you can install and uninstall RPM packages including query, build, and verify packages. You can use the *rpm* command right from a shell prompt. There are five basic key modes of rpm command as follows:

  - **Install**: Used to install any RPM package.
  - **Remove**: Used to erase, remove, or uninstall any RPM package.
  - **Upgrade**: Used to update the existing RPM package.
  - **Verify**: Used to verify an RPM package.
  - **Query**: Used to query any RPM package.

  To perform a particular action, *rpm* command uses some set of options, which are listed in Table 9.9. The syntax for the *rpm* command is as follows:

  ```
  #rpm options rpm-package-name
  ```

  A broad description of *rpm* and its features is provided on the manual page which you can see by using the following command:

  ```
  #man rpm // Provide a detailed list of rpm options
  ```

  You may also visit the RPM home page* for further brief details about RPM. Further, some more weblinks are provided in Table 9.8 that offer up-to-date versions and documentation for RPM.

TABLE 9.8

List of RPM Package Weblinks

RPM package weblinks
http://rpmfind.net
http://www.redhat.com
http://freshrpms.net/
http://rpm.pbone.net/
https://help.sonatype.com
https://rpmfusion.org

---

* RPM home page: https://rpm.org.

TABLE 9.9

List of RPM Options

Option	Syntax	Description
-i	rpm -i *rpm_package-file*	Installs a package
-e	rpm -e *rpm_package-file*	Uninstalls (erases) a package
-q	rpm -q *rpm_package-file*	Queries a package
-U	rpm -U *rpm_package-file*	Upgrades, similar to install, removed the previous version.
F	rpm -F *rpm_package-file*	Upgrades, already installed package
-V	rpm –V	Verifies whether a package is correctly installed
-qa	rpm –qa	Displays list of all installed packages

- **RPM package query**

   The -q option informs you if a package is already installed, and the -qa option displays the list of all installed packages and software packages [9]. Further, you may also combine the -q option with the other options to display more information related to the package as given in Table 9.10.

   In the following command, use *more* command to see the output in pager forms, as follows:

   ```
   # rpm -qa | more // Display the list of all installed packages
   ```

   In the next example, the user checks to see whether *mozilla-mail* is already installed on the system or not by using the following command:

   ```
   # rpm -q mozilla-mail
   mozilla-mail-1.7.5-17
   #
   ```

NOTE: **YUM (Yellowdog Updater Modified)**, represented as yum, is an open-source command line as well as graphical-based package management tool for **RPM (Red Hat Package Manager)**-based Linux systems. It was developed and released by **Seth Vidal** *under* **GPL (General Public License)** as an open-source code. It permits users and system administrators to effortlessly install, update, remove, or search software packages on a system.

TABLE 9.10

List of rpm Query Options with –q

Option	Description
-*q* application	*Checks whether the mentioned application is installed or not.*
-*qa*	*Lists all installed RPM packages.*
-*qf* filename	*Lists applications that own the given filename.*
-*qR* application	*Lists applications and capabilities on which this application depends.*
-*qi*	*Displays all package information including description*
-*qc*	*Lists configuration files*
-*qd*	*Lists documentation files*
-*qpl* RPM-file	*Displays files in the mentioned RPM package.*
Qpi RPM-file	*Displays all package information in the mentioned RPM package.*
-*qpc* RPM-file	*Displays only configuration files in the specified RPM package*
-*qpd* RPM-file	*Displays only documentation files in the specified RPM package*

9.8.2 Installation and Uninstallation

You can use -i option with rpm command to install new packages, and with -U option, you can update currently installed packages with new versions. With an -e option, rpm command uninstalls the mentioned package.

- **Installation**
  Only root user can perform rpm installation in Linux system. There are two ways to install an rpm package by using the following commands:

```
#rpm -ivh package_name.rpm    //Install rpm package using -i option
#rpm -Uvh package_name.rpm    //Update rpm package using -U option
```

- **Uninstallation**
  Only root user can perform rpm uninstallation in Linux system. To remove a software package from your system, you must ensure that the package is already installed in your system by using -q option. Thereafter, you can use -e option (erase) to uninstall the mentioned package. It is not essential to use the full name of the installed file with -e option, you simply give the name of the software as follows:

```
# rpm -q software_name //Check whether the software is installed or not
# rpm -e software_name //Uninstall the mentioned software package
```

9.9 GNOME and KDE

Linux OS that offered graphical desktop environment comprises a workspace, menus, and icons. Unlike Microsoft Windows, the look and feel of the graphical environment varies extensively across Linux systems. Linux systems have several alternatives of desktop GUIs, but two GUIs are installed on most Linux systems: GNOME and KDE. Each desktop GUI has its own style and appearance. GNOME and KDE are two dissimilar desktop interfaces with distinct tools for selecting preferences, see Section 1.10. In this section, we briefly explain the various system directories and files associated with GNOME and KDE.

- **GNOME directories and files**
  You cannot say that there is no "best desktop," but there is a desktop that is best for you as per your need, preference, and hardware. GNOME is a powerful and easy-to-use desktop interface, comprising a desktop, a panel, and a set of GUI tools. It also offered a flexible platform for the development of powerful applications. With the help of several GNOME desktop tools, you can customize your desktop, like desktop background, screen saver, and themes.
  Presently, GNOME is supported by several distributions and is the main interface for Red Hat Linux and Fedora. GNOME is free and released under the GNU Public License. You can download the source code, documentation, and GNOME software, directly from the GNOME webpage [10].
  The binary files of GNOME are usually installed in the */usr/bin* directory on your system. GNOME libraries files are residing in the */usr/lib* directory. The detailed list of the GNOME configuration directories and files is given in Table 9.11.

TABLE 9.11

List of GNOME System Directories in Linux

System Directories (GNOME)	Description
/usr/bin	GNOME packages
/usr/lib	GNOME libraries files
/usr/include/libgnome-2.0/libgnome	Header files help in compiling and developing GNOME applications
/usr/include/libgnomeui	Header files help in compiling and developing GNOME user interface parts
/usr/share/gnome	Files used by GNOME applications
/usr/share/doc/gnome*	Documentation file for numerous GNOME packages including libraries
/etc/gconf	GNOME configuration files
.gnome, .gnome2	GNOME configuration files for the user desktop and applications

- **KDE directories and files**

  KDE is another popular graphical desktop that includes the standard desktop features like a window manager, a file manager, and a wide set of applications to support various Linux tasks. The KDE desktop is developed and distributed by the KDE project, supported by a large open group of programmers around the world. KDE is entirely free and open software provided under a GNU Public License, and its source code is freely available for users. The KDE development is managed by a core group, referred to as the KDE core team. Anyone can apply for the core team through membership, at the KDE official website page [10].

  Numerous applications written specifically for KDE are easily accessible from the desktop. Such applications usually have the letter K as part of their name, like KCalc, KNotes, and KMail. You can see a complete set of applications offered by KDE from its home page [11]. A variety of tools are provided with the KDE desktop. These include calculators, console windows, notepads, and even software package managers. On a system administration level, KDE offers numerous tools for configuring your system. When KDE is installed on your system, it is installed in the same system directories as other GUIs and user applications. On Red Hat and Fedora OSs, KDE is installed in the standard system directories with some variations, such as */usr/bin* directory for the KDE program files. The detailed list of the KDE configuration directories and files is given in Table 9.12

TABLE 9.12

List of KDE System Directories in Linux

System Directories (KDE)	Description
/usr/bin	KDE programs
/usr/lib/kde3	KDE libraries
/usr/share/apps	Files to be used by KDE applications
/usr/share/icons	Icons to be used in KDE desktop and applications
/usr/share/mimelnk	KDE desktop files used to build the main menu
/usr/include/kde	Header files help in compiling and developing KDE applications
/usr/share/config	KDE desktop and application configuration files
/usr/share/doc	KDE help document file

9.10 Installing and Managing Software on RPM-Based Systems

As discussed in Section 9.8, in most Linux distributions, basically two major software packages, RPM and DEB, are used. These software packages are to be managed by various tools to manage installation, uninstallation, updating, searching, and many more. In this section, we briefly describe the task of package management successfully. For that, you must be aware of available packing tools. The available packing tools for Debian and Red Hat are divided into two categories: low-level tools and high-level tools, which are described as follows:

- **Low-level tools**: dpkg (Debian) and rpm (Red Hat) that perform installation, upgrade, and removal of package files.
- **High-level tools**: *apt, apt-get, aptitude (Debian), and yum (Red Hat) that perform* the tasks of dependency resolution and metadata searching; i.e., data about the data is performed).
  - **dpkg**: It is a low-level package manager for Debian-based systems. It can install, remove, provide information and build *.deb packages but is unable to download and install automatically with their respective dependencies.
  - **apt-get**: It is a high-level package manager for Debian and derivatives. It offers a simple way to retrieve and install packages, including dependency resolution from multiple sources.
  - **Aptitude**: It is another high-level package manager for Debian-based systems, same as apt-get tool. It may be used to perform management tasks, including installing, upgrading, and removing packages. It also handles dependency resolution automatically in a fast and easy way.
  - **rpm**: It is a low-level package manager for Linux distributions like RHEL, Fedora, and CentOS to perform low-level handling of packages like query, install, verify, upgrade, and remove packages.
  - **yum**: It is another high-level package manager that provides the functionality of automatic updates with package dependency management to RPM-based systems. It also works with repositories, like apt-get or aptitude tools.

Table 9.13 summarizes the various operations that may be performed with the command line package management tools for Debian or Red Hat software package-based system.

9.11 Installing Programs from Source Code

As you are familiar with package management tools, like *dpkg, aptitude, yum, apt-get, and rpm*, they are very handy to install various packages that are already compiled and available on various repositories. If the application or package is not available on a repository or does not have an installer script, then how you can install an application from source code is summarized here with some steps.

TABLE 9.13

List of Operations Performed Associated with Debian and Red Hat Package Management Tools

Management Tasks	Packaging System (Linux Distributions)	Command Structures
Searching for a package	Debian Linux and its derivatives	# aptitude update #aptitude search *package_name*
	Red Hat Linux and its derivatives	# yum search *package_name* # yum search all *package_name*
Installing a package from a repository	Debian Linux and its derivatives	#aptitude update #aptitude install *package_name*
	Red Hat Linux and its derivatives	#yum update #yum install *package_name*
Installing a package from a package file	Debian Linux and its derivatives	#dpkg --install *package_name*
	Red Hat Linux and its derivatives	#rpm -i *package_name*
Removing a package	Debian Linux and its derivatives	#apt-get remove *package_name*
	Red Hat Linux and its derivatives	#yum erase *package_name*
Updating packages from a repository	Debian Linux and its derivatives	#apt-get update
	Red Hat Linux and its derivatives	#yum update
Upgrading package from a package file	Debian Linux and its derivatives	#dpkg --install *package_name*
	Red Hat Linux and its derivatives	#rpm -U *package_name*
Displaying installed packages	Debian Linux and its derivatives	#dpkg --list
	Red Hat Linux and its derivatives	#rpm -qa
Listing information of installed package	Debian Linux and its derivatives	#aptitude show *package_name*
	Red Hat Linux and its derivatives	#yum info *package_name*
Identification of package file	Debian Linux and its derivatives	#dpkg --search *file_name*
	Red Hat Linux and its derivatives	#rpm -qf *file_name*

Many software packages that are developed, designed, or being in the developing phase for cross-platform implementation may not be in the form of an RPM format. Rather, it may be archived and compressed. A compressed archive is an archive file, which is created with *tar* utility and then compressed with a compression tool like *gzip*. As you are aware that the filenames of compressed archive files are followed with the extension like *.tar.gz*, *.tar.bz2*, or *.tar.Z*, all these various extensions specify various compression approaches to compress files and similarly use other decompression commands to decompress files, like *gunzip* utility for *.gz*, *bunzip2* utility for *.bz2*, and *decompress* utility for. Z, respectively.

The following steps are required to install an application software whose source code is available in the form of *tarball* packages (*tar* files). These *tarball* packages are available at different websites, from where you can download and install. For example, you downloaded the file *mysoftware-x.y.z.tar.gz or mysoftware-x.y.z.tar.gz2*.

- **Download the source code package:**

Use wget command to download the tar file, as follows:

```
#wget <web link of the tarball file>
```

The above command will download the *tarball* (e.g., *mysoftware-x.y.z.tar.gz or mysoftware-x.y.z.tar.gz2)* into the current directory. The *wget* command is very flexible and has a lot of options, see help file for more information about it.

- **Decompressing and extracting *tarball* file:**

The below command is used to unpack, decompress, and extract *tarball* file to get the access of source code and other files, as follows:

```
#tar -xvfz mysoftware-x.y.z.tar.gz
(or)

#tar -xvfj mysoftware-x.y.z.tar.gz
```

The above command creates subdirectory with the name *mysoftware-x.y.z.* in the current working directory and stores all extracted files in it.

- **Change directory:**

Use *cd* command to change the directory where the software's source code has been extracted, as in the previous step. The directory name comprises the name and release of the software. In the preceding step, the extraction created a subdirectory *mysoftware-x.y.z.*

```
#cd mysoftware-x.y.z.
```

- **Configuration, compilation, and installation:**

Most software can be configured, compiled, and installed in three simple steps, as follows:
  - **Configuration:**
    Mostly, the packages comprise a configuration script that can be used for configuring the environment. The name of the configuration file is generally *"configure."* To start configuration, run the following command:

```
#./configure // Generate makefile
```

The above command will check and/or create the built environment, and if everything goes well, then it creates a file, known as *"makefile."* The file *"makefile"* is used in the compilation of the software.
  - **Compilation:**
    Once the *makefile* is produced, then in the same directory, execute the following command:

```
# make // Compile all the source codes
```

The above *make* command compiles all the source codes related to the software.

- **Installation:**
  Once the compilation is done perfectly, then all the required binaries files are generated. Thereafter, install all such binary files in the standard defined paths, so that they can be invoked from anywhere in the system. To install, execute the following command:

  ```
  # make install //Execute all binaries files.
  ```

The above steps elaborate on how to fetch, unpack, configure, compile, and install the software from source code. If everything went well, you are able to run the application program. All the software packages or programs or commands are generally installed in numerous standard system directories, such as /bin, /usr/bin, /usr/local/bin, /usr/sbin, or /usr/local/sbin, to access the installed software from any part of the system. Use $PATH environment variable to see the path of system directories.

9.12 Network Management: telnet, rlogin, and rdesktop Commands

Computer network provides a communication facility among connected computers that are frequently improving and allowing faster and economical connections in the present days. Due to multitasking and multiuser key features of Linux OS, the networking concept awareness is always very important for Linux users. Linux is used to build all sorts of networking system tools and applications. To manage the Linux system over network, you must have a reasonably strong understanding of the network and the protocols, which are used to communicate over the system network. In this section, we simply highlight the various key command concepts for remote system access over network in Linux environment.

- **telnet command:**
  You use the *telnet* command (also referred to as telnet protocol) to login remotely or interact with the remote system on your network. The system can be on your local area network or available through an Internet connection.

  ```
  Syntax
  #telnet [option] [host] [port]
  ```

  As per the given syntax, telnet sets up a connection to a host (either a system name or IP address) using port. If a specific port is absent, the default port is assumed to be 23. If telnet is invoked without the *host* argument, it enters in command mode, where it prints a telnet command prompt, i.e., *telnet>*. In command mode, it accepts and executes the commands listed in Table 9.14, for example:

  ```
  $ telnet sanrit
  Trying 192.168.10.154...
  Connected to sanrit.
  Escape character is '^]'.
  GNU/Linux
  Kernel 3.10.0-514.16.1.el7.x86_64 on an i686
  login: san
  ```

TABLE 9.14

List of Command Options of Telnet

Options	Description
-4	IPv4 address resolution.
-6	IPv6 address resolution.
-a	Tries for automatic login. It sends the username via the *USER* variable of the *ENVIRON* option if supported by the remote system. The username is recovered through the *getlogin* system call.
-b *address*	Uses bind on the local socket to bind it to a mention local *address*.
-d	Sets the initial value of the debug toggle to TRUE.
-S *tos*	Sets the IP type-of-service (TOS) option for the telnet connection to the value tos.
-l *user*	Mentions *user* to log in as the user on the remote system. By sending the mentioned name as the *USER* environment variable, it needs the remote system to support the TELNET ENVIRON option.
-n *tracefile*	Opens *tracefile* for recording trace information.
open	Tries to set up connection with specified *hostname*
close	Closes current connection
logout	Forcibly log outs remote user and close the connection
display	Displays operating parameters
host	Indicates the official to specify a hostname or the Internet address of a remote host over the network
port	Indicates a port number (address of an application) or service name to contact. If a number is not specified, the default telnet port (23) is used.

```
Password:
Last login: Fri Apr 12 15:10:23 from 192.168.10.253
...
$logout                  //Connection closed by remote host.
$
```

The following example shows the "telnet>" command prompt, if no host and port provided with *telnet* command:

```
$ telnet
telnet> open sanrit
Trying 192.168.10.154...
Connected to sanrit.
Escape character is '^]'.
GNU/Linux
...
telnet> close
connection closed.
$
```

- **rlogin command:**

  The *rlogin* command permits you to start a terminal session on the remote host. Firstly, the *rlogin* command appeared in 4.2BSD. The -l option allows you to mention the login name of the account. The basic syntax of rlogin is given below:

```
$rlogin system_name //Remote login on system_name host
$rlogin [-8EKLdx] [-e char] [-l username] host
```

TABLE 9.15

List of Some Options of rlogin Command

Options	Description
-8	Always permits an eight-bit input data path; otherwise, parity bits are stripped except when the remote side's stop and start characters are other than ^S/^Q.
-E	Stops any character from being recognized as an escape character. When used in conjunction with the -8 option, it offers a totally transparent connection.
-K	The -K option switches off all Kerberos verification.
-e	Allows user specification of the escape character, which is the tilde (˜) by default. This condition may be as a literal character, or as an octal value in the form of *nnn*.
-d	Turns on socket debugging on the TCP sockets used for communication with the remote host.
-l *username*	Allows you to log in as mentioned *username* on the remote system. This is a very useful command option that enables you to log in as someone else on a remote machine.

For example:
```
#rlogin -l sanrit domain.com //Login as user sanrit to the remote
                          system domain.com.
#rlogin domain.com //Access remote system domain.com for user login
```
The list of key options for rlogin command is described in Table 9.15.

- **rdesktop command:**

    The rdesktop is a client for Remote Desktop Protocol (RDP), released under the GNU GPL. The *rdesktop,* open-source software, available at http://www.rdesktop.org/. *rdesktop,* permits you to connect and manage the number of Microsoft products desktops remotely from your Linux computer. In this section, we brief about the installation of **rdesktop** in Linux system to access the remote desktop of **Windows** computer system using the **hostname** and **IP address**.

    - **Windows settings for rdesktop**

        To allow rdesktop to connect with any given Windows system, you require to make few changes in Windows system itself, which are as follows:

        1. Enable remote desktop under Windows OS.

        2. See and enable RDP port in firewall.

        3. Require user with a password to access the system.

        After making the above changes in Windows configuration settings, then you install rdesktop on your Linux system to access your Windows desktop.

    - **Installation of rdesktop in Linux**

        It is always desirable to use a default package manager like *yum, dnf,* or *apt* to install software to handle dependencies automatically during installation. The following commands are used for various Linux distributions:

```
# yum install rdesktop    // For CentOS/Red Hat Linux
# dnf install rdesktop    // For CentOS/Red Hat and Fedora Linux
# apt install rdesktop    // For Debian/Ubuntu Linux
```

        If **rdesktop software** is not available on your system to install from the default repositories, then you can download the latest release of rdesktop tarball from GitHub link [12] and install it by following the subsequent steps:

```
# wget https://github.com/rdesktop/rdesktop/releases/download/
v1.8.6/rdesktop-1.8.6.tar.gz
# tar xvzf rdesktop-1.8.6.tar.gz
# cd rdesktop-1.8.6/
# ./configure
# make
# make install
```

9.12.1 Connect Windows desktop from Linux system:

After the installation of rdesktop software in your Linux system, now you can connect with the remote Windows system desktop through two ways: *hostname* and *IP address*.

- **Connect using hostname**:
    If you know the hostname of remote Windows system, then you can connect Windows system from Linux desktop by typing the **following** command using -u option as username (sanrit) and (XYZ) as the hostname of remote Windows system:

```
#rdesktop XYZ
Connect Windows system, and then, you will see the Windows login
prompt to enter username and password to access the Windows desktop.
You can also provide the username during the connection of remote
host as follows:
#rdesktop -u sanrit XYZ
```

- **Connect using IP address:**
    If you know the IP address of remote Windows system, then you can connect Windows system from Linux desktop by typing the subsequent command using -u option as username (sanrit) and (192.168.10.153) as IP address of remote Windows system:

```
# rdesktop 192.168.10.153
# rdesktop -u sanrit 192.168.10.153
```

The rdesktop command associated with some key options is listed in Table 9.16.

9.13 Summary

Linux is widely used in computing technology, empowering open-source projects, desktops, and monitors to the computer-intensive system and servers. To work with Linux, you require memory space on your disks; the built-in command *du* is used to check the disk usage information of files and directories on a system. Kernel is a good part of Linux, and it is responsible for allocating system resources and scheduling of various parts of system resources. Kernel performs process scheduling, memory management, file system management, access to devices, networking, etc., and losing kernel module uses kernel commands to perform their job of loading or unloading modules. To install a new kernel, you need to download the new kernel package from kernel home page [3] to your system. After compiling, we need to install the module by using the make command. After retaining a

TABLE 9.16

List of Some Key Options of rdesktop Command

Options	Description
-u <username>	Username for authentication on the remote host.
-d <domain>	Domain for authentication of remote host.
-s <shell>	Start-up shell for the user, run a particular application instead of explorer.
-c <directory>	The initial working directory for the user. Generally used in combination with -s to set up a fixed login environment.
-p <password>	Provides password on the command line to authenticate on host desktop after connection.
-n <hostname>	Client hostname. Normally, rdesktop automatically obtains the hostname of the client.
-f	Allows fullscreen mode. Fullscreen mode can be switched at any time using Ctrl-Alt-Enter.
-t	Disables use of remote control like seamless connection sharing.
-E	Disables transmission encryption from client to remote host.
-m	Do not send mouse motion events.
-a <bpp>	Sets the color depth for the connection (8, 15, 16, or 24).
-T <title>	Sets the window title. The title must be specified using an UTF-8 string.
-z	Enables compression of the RDP data stream to save network usage.
-a <bpp>	Sets the color depth for the connection (8, 15, 16, 24, or 32).
-4	Uses version 4 RDP.
-5	Uses version 5 RDP (default).

copy of the current kernel, you can modify your kernel configuration in case something wrong happens to the modified kernel version. In Linux distribution, the most common boot loaders are LILO and GRUB.

LILO is the default boot loader for most Linux distributions, and GRUB is the multistage boot loader. GRUB is more flexible and powerful than LILO. In Linux, there are two types of users: root users and normal users. The root user is a special user account used to perform system administration operations. In Linux, sudo command allows you to run commands as another user. Package management is a method of installing and maintaining software on the system. With rpm command, you can install and uninstall RPM packages including query, build, and verify packages. Most Linux systems have two GUIs installed: GNOME and KDE. These two are dissimilar desktop interfaces with distinct tools for selecting preferences. GNOME and KDE have different directories and files. telnet command, rlogin command, and rdesktop command are the various command concepts for remote system access over network in Linux environment.

9.14 Review Exercises

1. Write a command to check whether you are using KDE or GNOME.
2. Write a command to add a new user on the command line.
3. How to execute a command as another user in Linux?
4. How to limit file system disk usage per user?

5. Write various Linux commands to check disk space.

6. Write commands in Linux {Distro: Linux, Red Hat;} to compile and install a package.

7. Write steps:

    a. to start a terminal session on a remote host.

    b. to find all the regular files in a directory.

    c. to run rdesktop through an ssh tunnel in one command.

8. Discuss the key tasks performed by the kernel.

9. Write steps:

    a. to see the memory usages of the system.

    b. to print all the internal modules (modules that are already loaded) in the kernel.

    c. to download the latest kernel source and from where.

    d. to determine the boot loader that you are currently using.

10. Discuss **RPM** package and procedure install packages in detail.

11. How to check whether a package is installed on your system or not?

12. Compare GNOME and KDE.

13. Write the procedure to change the GNOME theme on your Linux system.

14. Discuss network management in Linux.

15. Login into your Gmail account using rlogin.

16. Write steps to install rdesktop on your Linux machine.

17. Try to communicate between your PC and your friend's PC using < **telnet command.**

References

1. Gareth Anderson. 2006. *GNU/Linux Command-Line Tools Summary*. Open Source under GNU Free Documentation License.
2. Richard Petersen. 2008. *Linux: The Complete Reference*. The McGraw-Hill.
3. Linux Kernel Organization, Inc., The Linux kernel archives in the World Wide Web. https://www.kernel.org/.
4. Greg Kroah-Hartman. 2007. *Linux Kernel in a Nutshell*. O'Reilly Media.
5. Adam Haeder, Stephen Addison Schneiter, Bruno Gomes Pessanha, and James Stanger. 2010. *LPI Linux Certification in a Nutshell*. O'Reilly Media.
6. Christopher Negus. 2008. *Linux Bible*. Wiley Publishing, Inc.
7. Mark G. Sobell. 2012. *A Practical Guide to Fedora and Red Hat Enterprise Linux*. Prentice Hall.
8. The Red Hat Package Manager, RPM repository available in the World Wide Web https://rpmfind.net/linux/RPM/.
9. Trevor Kay. 2002. *Linux+ Certification_Bible*. Hungry Minds Inc.
10. The GNOME Project, The GNOME's technologies in the World Wide Web. https://www.gnome.org/technologies/.
11. The K Desktop Environment (KDE) Software Open Communities, The KDE Announcements in the World Wide Web. https://kde.org/announcements/.
12. The rdesktop Tarball File, The GitHub download link in the World Wide Web. https://github.com/rdesktop/rdesktop/releases.

Section II

Linux Programming

10

File Management

Everything in Linux is based on the concept of Files, or we can simply say that everything in Linux is a file; if it is not a file, then it is a process. Hence, to understand Linux, it becomes crucial to know about the Linux file system and the basic operations associated with it. When you install Linux on a system, then all the data of the installed operating system (OS) are organized and kept in various files. So, you spend most of your time working on files. All files are kept and organized into various directories. These directories are linked and organized into a tree-like layout or structure, which is referred to as a file system or file structure. A file's existence is not just with its name, but also its location in the tree-like file hierarchical structure. You can create as many files and directories as you want and add more directories to the file structure as required. This chapter will help you to understand the organization and terminology of the Linux file system. It also briefs you on how you can create and remove files and directories and explains the way for naming them. It also shows how to copy and rename them, archiving and compressing them, move them through the file system, and many more. Furthermore, it explains the important file and directory attributes as well as setting and resetting the various file access permissions.

10.1 Filename and Type

A filename is a collection of text string (i.e., a sequence of characters), which is used to recognize a file by the user. In other words, a file is a collection of contiguous blocks of data that are saved in storage devices such as the hard disk drive (HDD), floppy disk, optical disk, or magnetic tape. You can assign names to files on any Linux-like OS for easy recognition and as a reference to find them again by the user in the future. You can allocate a name to the file by using any characters including letters, underscores, and numbers. You may also include dots (.), underscore (_), and some other special characters. You should never begin a filename with a dot character (.) along with some other characters, such as question marks, slashes, or asterisks. Such special characters are reserved for some other system-related tasks. It is worth mentioning here that while assigning filenames, you should assign some meaningful name related to the stored information in that file. something. Briefly, you can use the following characters in the filename:

- Uppercase letters (A–Z)
- Lowercase letters (a–z)
- Numbers (0–9)
- Underscore (_)

- Dot or Period (.)
- Comma (,)

Mostly, the filename's length may be assigned up to 256 characters, but it depends on the respective file system. Some file systems limit filenames to fewer characters, not up to 256. Filenames must be unique within a directory. However, numerous files and directories with the same name may reside in different directories due to having different *absolute pathnames* (i.e., the relative path from the root(/) directory in the file system structure). Through this way, the system will be able to differentiate them.

In some OS, filenames may be divided into two parts: a user-allocated name and an extension, separated by a period. The extension part is very helpful for categorizing the various files. So that you can find the right application to operate such files. You are very much aware of such standard extensions that have been used by various application programs. For example, Microsoft word application files have an extension of .doc (or .docx); similarly, C++ source code files have .cpp; c code files have .c; media player files have .mp3(or mp4 etc.); compressed files have .zip ; graphical information files have .gif (.png or .jpeg), and many more. Yes, you can also make your own file extensions to categorize your files from others.

The following examples are all valid Linux filenames:

- content
- chapter5
- 97roll
- New_Address
- abc.tar.gz   //Compressed archive
- abc.sh   //Bourne shell file
- sum.cpp
- spng.mp3
- myimage.jpg
- New roll number assign (Must enclose it in quotes such as "New roll number assign")

You can use uppercase and/or lowercase alphabet within filenames. Linux OS is case sensitive, so it recognizes the uppercase and lowercase alphabet separately. Files named MARCH, March, and march represent three different files. Generally, files and directories are referred to by their names rather than absolute pathnames. But in Linux, the first level of directories, which reside in the root (/) directory, are generally referred to by their absolute pathnames, for example, */home, /bin, /root, /boot, /etc, /usr,* and so on. There various way to change the name of a file or directory. On the command line, use the *mv* (i.e., *move*) command as given:

```
$mv file_1 file_2 //change the name of file, named file_1 to file_2.
```

On the graphical user interface (GUI), a name can be changed by using the right mouse button, then clicking on the file *icon* (image to represent a file or a directory with its name), and then selecting *Rename* Application Programming Interface (API) in the opened menu bar that appears after pressing right-click. Thereafter, the cursor is moved to the filename label, and the new name is entered by the user.

10.1.1 Hidden Filenames

If a filename starts with a dot(.), then such a file refers to a hidden filename (or a hidden file or occasionally an invisible file). The *ls* command does not display such files. However, by using the -a option with ls command, you can view all files including the hidden files. Every directory in the file system has two special hidden entries that begin with a single dot (.) and double dots(..). The dot (.) represents the pathname of the current working directory, and the double dots(..) represents the pathname of the parent directory of the current working directory.

10.1.2 File Type

When you are navigating the Linux file system, you may encounter different file types. Most files are just files, referred to as regular files, which comprise standard data, such as the text file, image file, word document file, music file, executable file, program file and so on. A directory is another widely used file type. The category of the file type may differ according to the specific implementation of files in Linux. How can you determine the type of file in Linux? For that, you must use the ls command with the -l option (long format) to see the file type[1], as follows:

```
$ ls -l
-rw-rw-r--    1     root    root    04 Sep 17 2018   test1.gz
-rwxrw-r-x    1     root    root    24 Aug 7  2018   hello.c
drwxrwxr-x    2     root    root    13 Apr 24 2018   ccet
drwxrwxr-x    2     root    root    26 Aug 3  2018   sample
$
```

the ls -l display the content of the current working directory in a log format. In the above output of ls -l, the first column shows the file permission of a file, comprising ten-character fields. The first character of the file permission part denotes the file type. For instance, file *hello.c* has "-", which means a regular file type. In the same way, the file *sample* has "d", which means a directory file type. **It is important to point out that Linux file types are not to be mistaken with file extensions**. The names of all seven different file types of the Linux system are listed below along with their character notation:

- Regular file   -
- Directory   d
- Character device file   c
- Block device file   b
- Local socket file   s
- Named pipe   p
- Symbolic link   l

10.2 Linux File System Architecture

Linux organizes and keeps its files and directories in a hierarchical or tree-like structure. It means all the files are organized in a tree-like pattern of directories, also referred to as folders, which may contain files and /or other directories. The first directory in the Linux

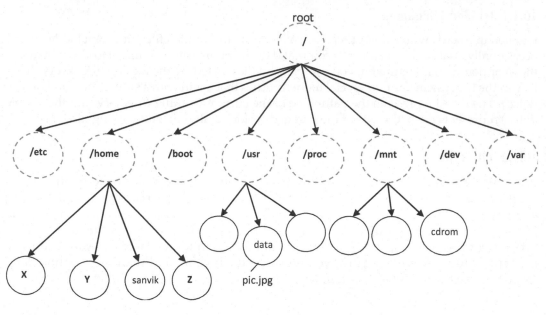

FIGURE 10.1
The hierarchical view of the Linux file system.

file structure is called the root directory or beginning directory, denoted as "/". The root directory further contains files and subdirectories, which again comprises files and sub-directories, and so on.

Within the root directory (/), various system directories contain files and programs that are created during the installation of the Linux system, which keeps system information and features, see Figure 10.1. The root directory also contains a directory called /root and /home. The /root directory is the home directory of the root user or the super user, and the /home directory contains the home directories of all the users on the system. In the /home directory, each user has his/her own home directory with their name that comprises the files and directories of that user. A file system can be created using the *mkfs* command, as described in Chapter 2 (Section 2.3). The hierarchical or tree-like structure organizes files and directories in the file system as shown in Figure 10.1.

10.3 File and Directory Structure

The subsequent section describes the various important directories and pathnames of the Linux file system.

10.3.1 Root (/) Directory

In the Linux file system, the topmost directory or the beginning directory of the file system is root(/). In Linux, you cannot use a forward slash (/) character while assign-ing the name of a file. A forward slash (/) character is reserved for use as the name of the root directory (i.e., the directory that contains all other directories and files of the

TABLE 10.1

List of Subdirectories of the Root(/) Directory

Directory	Content
/	Root directory, the starting directory structure of the Linux file system.
/bin	Directory to hold the program files shared by the system, the system administrator, and the users.
/boot	The startup files, kernel(vmlinuz) files, and boot loader files.
/dev	Device files of all the CPU peripheral hardware.
/etc	System configuration files.
/home	Home directories of all common users.
/lib	Library files of all programs required by the system and the users.
/misc	For miscellaneous purposes.
/mnt	Usual mount point for external file systems of various devices such as CD-ROM, eternal hard drive, digital camera, and so on.
/opt	Files of add-on application software packages (third party).
/proc	It is a Virtual File System (VFS), containing useful information about running or active processes and others system resources information.
/root	The home directory of the *root* user or administrator.
/sbin	Programs files for the system and the system administrator.
/tmp	Temporary files of the system that are removed after rebooting the system.
/usr	All user-related files such as programs, libraries, documentation, and so on.
/var	Keeps all variable files and temporary files generated by users, such as log files, mail queue, store image files before writing on the CD, printer spooler file, download file from the network, and so on.

whole Linux system. It also acts as a directory separator while defining the pathname, see Section 10.3.3.

As we all know, in the Linux file system, the root directory comprises all system and user directories, subdirectories, and files of the whole Linux system [2]. Some directories and files that are generated during the installation of Linux on the system are responsible for the proper functioning of the entire system, for example, the /bin directory contains all essential command binaries of Linux. Table 10.1 shows some system-generated directories that reside directly under the root (/) directory.

10.3.2 Home Directories

There is a subfolder, referred to as */home* , under the root directory (/) that comprises all user's home directories. A home directory is also referred to as the *login directory* of an ordinary user that keeps the user-created personal files, directories, program files, and other data files. When a user logs into the Linux system, that is the first directory after logging into the system. A home directory is created automatically for every ordinary user. By default, the name of a user's home directory is always the same as the *loginname* of that user. Therefore, for example, a user with a username of *sanrit* would have a home directory of also name *sanrit*, and the location of that directory would be */home/sanrit*. The location of this user home directory is referred to an as *absolute pathname*, which is discussed in the next section.

We already discussed that only the superuser can create a normal user by using the *adduser* command. After the successful creation of a normal user, like *"sanrit,"* then a home directory of that normal user is created under */home* directory, as */home/sanrit*.

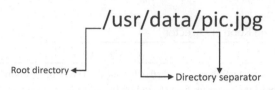

FIGURE 10.2
Pathname format of a file.

10.3.3 Pathnames

A pathname is the route of all directories and their subdirectories, separated with a forward slash (/) from the root or top beginning directory (/) in the file system to the designated directory or file, as shown in Figure 10.2. In the pathname of a directory or file, the first forward (/) is referred to as the root directory, and the remaining forward slashes act as directory separators. For example, as per figure 10.1, the path of *pic.jpg* is shown in figure 10.2.

The path of each directory can be specified in one of the two ways: absolute pathnames or relative pathnames (for more details, see Section 2.3).

10.3.4 System Directories

The Linux file hierarchy contains numerous system directories under the root directory (/) as described in Table 10.1. These system directories comprise files and programs responsible for the proper running and maintaining of the system. Several directories also comprise some other subdirectories with programs for executing a particular feature of Linux as mentioned in Table 10.2. For example, the directory */usr/bin* contains the various Linux commands that users execute, such as *ls, cd, lpl*, and so on.

As you see, the directory */var* keeps some files that change frequently, such as mailboxes, log files, spoolers, lock files, and so on. In the same way, all system files related to devices or peripherals attached to the system (except CPU) are stored or have an entry in the /dev directory. Some command devices system files are listed in Table 10.3.

Similarly, /etc directory stores most of the system configuration files as listed in Table 10.4.

TABLE 10.2

List of Some System Directories

Directory Name	Description
/usr/bin	Keeps user-oriented commands and utility programs, not required for booting purposes.
/usr/sbin	Keeps system administration programs and tools.
/usr/lib	Keeps libraries for programming languages.
/usr/share/doc	Keeps Linux help documentation.
/proc/cpuinfo	Comprises CPUs information.
/etc/fstab	File systems information of the system.
/etc/passwd	Comprises information on each user account.
/etc/hosts	Information on IP addresses with hostnames of the system.
/etc/skel	Comprises files and directories that are routinely copied to new users' home directories.
/usr/sbin	Keeps system administration tools and commands for booting the system.
/usr/share/man	Comprises the online manual help files.
/var/spool	Keeps spooler files, such as printing jobs and network transfer packets.

TABLE 10.3

List of Common Devices

Name	Device Description
cdrom	CD drive
console	Entry for the currently used console.
cua*	Serial ports.
dsp*	Devices for sampling and recording.
fd*	Floppy disk entries, by default, it is /dev/fd0, a floppy drive is the size of 1.44 MB.
hda1, hda2, hda3.... hdb1, hdb2, hdb3.... hd* hdc1, hdc2, hdc3...	Standard support for IDE (Integrated Drive Electronics) hard disk drives (HDD) with their partitions.
sda1, sda2, sda3.... sdb1, sdb2, sdb3.... sd* sdc1, sdc2, sdc3...	Standard support for Small Computer System Interface (SCSI) HDD with their partitions.
ir*	Infrared devices
isdn*	ISDN connections
lp*	Printers
mem	Memory
modem	Modem
mouse (also msmouse, logimouse, psmouse, input/mice, psaux)	All kinds of mouses.
par*	Parallel port entries
ram*	boot device
usb*	USB card and scanner
video*	graphics and video card support

TABLE 10.4

Most Common Configuration Files

Configuration File	Description of Configuration File
bashrc	Configuration file for the Bourne-Again shell.
default	Default options for some commands.
file systems	Known file systems: ext3, vfat, ntfs, iso9660, and so on.
fstab	Lists device partitions and their *mount points*.
ftp*	Client-server Configuration file.
group	Configuration file for user groups: groupadd, groupmod, and groupdel.
inittab	Booting information table.
aboot.conf	Boot loader information.
modules.conf	Module configuration to enable a particularly special feature, such as drivers.
passwd	Lists of local users information: useradd, usermod, and userdel.
profile	System-wide configuration of the shell environmental variables: default properties of new files, limitation of resources, and so on.
rc*	Directory comprising the active service information for each run level.
ssh	Directory comprising the configuration files for the protected shell between the client and the server.
sysconfig	System configuration files including the mouse, keyboard, network, desktop, system clock, power management, and so on.
xinetd.* or inetd.conf	Internet services configuration files.

10.4 Inodes

The Linux file system comprises four main blocks: Boot block, superblock, inode block, and data block (see Section 11.4). The superblocks hold information about the file system but the *inode* blocks comprise information about individual files, and data blocks contain the information stored in the individual files. A user recognizes a file through its name, but the kernel uses *inodes* to represent a file.

In a file system, the inode table comprises one inode, referred to as a unique identification number, for each file residing in the file system, including the physical location of the file data. The inodes are recognized numerically by their sequential location in the inode table [3]. At the time a new file is created, it gets a free inode. Each inode contains information about a file including the following information:

- An inode number (a unique identification number),
- The owner and the group related to the file,
- The file type (regular, directory, pipe ...),
- The file's permission list (read, write, and execute),
- The date and time of creation, last read, and change,
- The file size,
- A disk address mentioning the real location of the file data.

The inodes have their own separate space on the disk. You can see inode numbers of a file by using the –i option with the ls command.

```
$ls -i

75784578 test1.gz
38833142 hello.c
40497221 ccet
39139118 sample
```

The inode number of a file is the first field displayed by the ls –li command in a long format output, as follow:

```
$ls -il
75784578 -rw-rw-r--  1     root    root   04 Sep 17 2018   test1.gz
38833142 -rwxrw-r-x  1     root    root   24 Aug 7  2018   hello.c
40497221 drwxrwxr-x  2     root    root   13 Apr 24 2018   ccet
39139118 drwxrwxr-x  2     root    root   26 Aug 3  2018   sample
$
```

In the above output, the first column represents the inode numbers of each file in the respective row.

10.5 File Operation

One of the key operations of the Linux OS is the organization of files. You may require to do some basic operations on your files, such as printing them on your output screen or the printer. The Linux system offers some set of commands that perform basic file management operations, such as creating files, listing files, displaying files, printing files, searching files, moving files, copying files, renaming files, deleting files, and many mores [4,5]. These are some basic operations that are described in the subsequent subsections.

10.5.1 Creating files: *touch, cat*

- *touch* **command**: It is used to create a file without any content. The file created using the *touch* command is empty. This command can be used when the user wants to create a file without storing data at the time of file creation. The syntax of the *touch* command is as follow:

      $touch [option]file_name(s)

  If you used the *touch* command without any options, it creates new files with names as you are provided as *arguments* (i.e., input data). The *touch* command can create any number of files concurrently. For example, the following command would create three new, empty files named `File _ name1`, `File _ name2`, and `File _ name3` as follows:

      $touch [option] File_name1 File_name2 File_name3 ......

  Table 10.5 listed some options that are associated with the touch command.

- **cat command**:

  The *cat* command is one of the most regularly used commands on the Linux / Unix OS. It has three operations related to the text files that are displaying files, combining files into one file, and creating a new file. The basic syntax of the cat command is:

      cat [options] [filenames] [-] [filenames]

- **Displaying files:**

  The cat command is used to display the content of a text file, by using the following commands:

TABLE 10.5

List Touch Command Options

Option	Notation	Description
-a	--time=atime,	Change the access time only. Used to change or update the last access or modification times of a file.
-m	--time=mtime,	change the modification time only, used to update only the last modification time of a file.
-d	--date time	Use time instead of the current time. It may contain month names, time zones, am and pm, and so on.
-c	--no-create	To check whether a file is created or not. If not created, then don't create it.
-t	YYMMDDHHMM	Use the argument (Years, months, days, hours, minutes, **and** optional seconds) instead of the current time.

```
$ cat file1
  Hello.

  This is a test file1.

  Sanrit is owner of file1

$
```

The -n option display with the numbering of all the lines of the file is shown below:

```
$ cat -n file1
1  Hello.
2
3  This is a test file1.
4
5  Sanrit is owner of file1
6
$
Display with more and less command
$cat file1 | more    // Displays the lines from the starting of
                     the text file but stops after it displays
                     each page of data. The more command is a
                     pager utility.
$cat file1 | less    // Displays the file's contents before it
                     finishes reading the entire lines, i.e.,
                     displays the file from the ending side.
$cat file1 file2 file3  // Displays all files one after another
                        in one go.
```

Table 10.6 lists some key options associated with the cat command for displaying the content of a file.

Now, in another example, you can redirect the output of the *cat* command to another file, namely, *new_file1* by using the *output redirection operator* (>), as follows:

```
$cat file1 > new_file1
```

If a file named *new_file1* already exists, it will be *overwritten* (i.e., all its contents will be erased) by the redirection operator that copies the output of the above command.

- **Combining files**:

  The second role of the *cat* command is concatenation, i.e., combing the contents of the input files into the output file (along with the existing content). By using

TABLE 10.6

cat Command Option for Displaying a File

Option	Description
-n	Display all output lines (including blank lines), starting with numbering 1.
-b	Display all output lines (excluding blank lines), starting with numbering 1.
-s	Replace multiple adjacent blank lines with a single blank line.
-E	Display a $ after the end of each line.
-T	Display tab characters as ^I

the *append operator* (>>), you can copy the contents of the files into one file. This operation only copies, and there is no effect on the content of the original output file.

For example, the following command will concatenate copies of the contents of the three files *file1*, *file2*, and *file3* into a *new _ file1*

```
#cat file1 file2 file3 // Displays the contents of file1, file2, and
                             file3 on the output screen
#cat file1 file2 file3 >> new_file1 // Not display the contents of
                             file1, file2, and file3 on the output
                             screen, rather it is stored in the new_
                             file1 along with the existing content.
```

If a file named, *new _ file1*, already exists, it will not be overwritten, and the new text will be added to the end of the existing file.

- **Creating files**:

The third use of the *cat* command is file creation. For creating small files, it is easier to create new files on a command prompt rather than using any editor such as vi, vim, q, and so on. This can be done by typing the *cat* command, followed by the output redirection operator and the name of the file, which you want to create, and then pressing the Enter key. After this, input the contents into the file. When content writing is over, then finally press the Control key along with *d* key to save and exit. For example, a new file named *new _ file1* can be created by typing

```
$cat > new_file1    // <Press <enter key> to start writing content
 --------
 --------
 --------
 <Ctrl +d>              // Save and exit
 $
```

If a filenamed *new_file1 is* already exists, it will be *overwritten* (i.e., all its contents will be erased) by creating the new, empty file with the same name. Therefore, you must be cautious and prefer to use the append operator rather than the output redirection operator in order to prevent accidental removal. Hence to avoid overwritten of data, type the below command:

```
$ cat >> new_file1        // Add content along with the existing content
                             of file
```

10.5.2 Listing Files: ls:

The most widely used Linux command is *ls*, which is used to list the files and subdirectories of a directory. For example, to display the contents of the current working directory, use only *ls* command without any argument on the command line, as shown below:

```
$ ls
Desktop Documents Music Pictures Public Templates Videos
$

$ls -la       //Displays all files in a long format of the current
                 working directory.
```

You can specify any other directory path to see the content of that directory, besides the current working directory, for example, to see the content of a /usr directory, type the below command syntax:

```
$ ls /usr
    bin     games         kerberos      libexec      sbin     src
    etc     include       lib           local        share    tmp
```

You can specify more than one directory path (absolute path) to see their contents; for example, you may typed both /usr and /var directory to see its content as follow:

```
$ls /usr /var
/usr:
bin games  lib  libexec sbin  src
. . . . .

/var:
account crash games   lib  log  opt  spool yp
. . . . .
```

The ls command has a huge number of possible options, some widely used options are listed in Table 10.7.

10.5.3 Displaying Files: cat, more, head, and less:

- *cat* **command:** The cat and more commands display the contents of a file on the screen. The name cat stands for concatenate.

```
$ cat my data .txt  // displays the content of mydata.txt file
```

The cat command displays the entire text of a file on the screen at a time. It gives a problem when the file size is large, and it could not display all content at once on the screen. If the file is large, then its text goes so fast on the screen that you can see only the last few lines that fit on the screen. This is a major drawback of the cat command to display a large file. It is useful for small files. The *more* and *less* commands are designed to overcome this limitation by displaying one screen of text at a time. Cat command is also used to create a file with contents as described in Section 5.3.1.

- *more* **command:** If the content of a file is very large and the whole file could not be viewed on a single screen, then the more command is very helpful. The more command is used to view the contents of a text file one screen at a time, and thereafter, press the Enter key to scroll down line by line or press the Spacebar key to go to the next page, or press 'b' key, to go back one page and so on. The other options are associated with the more command as described in Section 5.3.9. For example:

```
$more /etc/passwd  // View the content by scrolling on the screen
```

- *head* **command:** The head command displays the first few lines on the screen or the standard output of an input. The basic syntax of the head command is

```
head [options] [file(s)]
```

TABLE 10.7

List of ls Command Options

Option	Meaning (Option)	Description
-a	--all	Display all files, including hidden files that start with a period (.). Normally, hidden files are not listed.
-d	--directory	Display directories such as other files, rather than listing their contents.
-l	--Long	Display the output in a long format, i.e., for each file, it prints file type, permissions, number of hard links, owner name, group name, size in bytes, and timestamp.
-F	--classify	This option will append an indicator character to the end of each listed name (for example, a forward slash if the name is a directory).
-t	--time	Output sort by modification time.
-S	--sort	Output sort by file size.
-h	--human-readable	In long format listings, display file sizes in human-readable format rather than in bytes.
-r	--reverse	Display the results in reverse order. Normally, ls displays its results in ascending alphabetical order.
–c,	--time=ctime, --time=status	Sort directory contents according to the files' status change time instead of the modification time.
-f		Do not sort directory contents, display in the same order as they are stored on the disk. This is similar to enabling –a and –U and disabling –l, –s, and –t
–i,	--inode	Print the index number of each file to the left of the filename.
–k,	--kilobytes	If file sizes are being listed, print them in kilobytes.
–m	--format=commas	List files horizontally separated by commas to accommodate maximum files on each line.
–p		Attach a character to each filename representing the file type.
–u	--time=atime--time=access, --time=use time	Sort directory contents according to the file's last access time instead of the modification.
–x,	--format=across, --format=horizontal	List the files in columns, sorted horizontally.
–R,	--recursive	List the contents of all directories recursively.
–S,	--sort=size	Sort directory contents by file size instead of alphabetically, with the largest files listed first.
–U	--sort=none	Do not sort directory contents; list them in whatever order they are stored on the disk.

By default, the head command prints the first ten lines of each input filename that is mentioned on the command line along with the head command, for example:

```
$head -5 file1.txt file2.txt file3.txt file4.txt
```

The above command displays the first five lines of each input file. The details of various options associated with the head command are described in Section 5.3.7.

- *less* **command**: *Less* is a program like *more* command that allows backward movement in the file as well as forward movement. The *less* command is used to view text files, and if the file is longer than one page or screen, we can scroll up and down to view the content. To exit from the less command view, press the Q key. The *less* command is more powerful and has better configurable display effectiveness when compared to the more command. You can call the *more* or *less* command by tying the command name followed by the filename on the command line to view the file content, as given below:

```
$ less myfile1.txt
```

You can refer to Section 5.3.8 for details along with various options associated with *less* command.

10.5.4 Printing Files: *lpr, lpq, and lprm*

In this section, we describe key commands that are used to print files and control printer operation. Usually, Linux systems use two software packages to manage and perform the printing operation. These packages are a common Unix printing system(CUPS) and Ghostscript. The Ghostscript is a PostScript interpreter, which acts as a raster image processor (RIP) for printing files using connected printers.

Moreover, CUPS offers both the SystemV and Berkeley(BSD) printing commands for printing files. Besides, it also provides a rich number of printer-specific options that permit you to control various printing operations such as how and where the files are to be printed. Besides, CUPS also has the ability to recognize various forms of data and may convert files into a printable form. Table 10.8 shows the sets of printing commands of SystemV and Berkeley(BSD).

- *lpr*: **Print Files (Berkeley-BSD):**

  lpr program can be used to submits files for printing. Files named on the command line are sent to the named printer; if no printer is specified, then it goes to the default destination printer. If no files are provided on the command line, lpr accepts the print file from the standard input, maybe, to make it in pipelines.

  The basic syntax of *lpr* is given as

```
$ lpr [ options ] [ filename ... ]
```

If the filename is not specified, *lpr* assumes that the input is coming from the standard input (usually, the keyboard or another program's output). This permits the user to redirect a command's output to the print spooler, as the example given below:

```
$ cat project.txt | lpr
```

If you wish to print many files at once, then you can mention more than one file on the command line after the *lpr* command, such as

```
$ lpr datafile1 datefile2 datafile3
```

TABLE 10.8

System V and BSD Print Commands

Berkeley (BSD) Command	SystemV Command	Description
lpr	lp	Submit to print a file.
lpm	cancel	Cancel a print file.
lpq	lpstat	Check the status of a print file.

TABLE 10.9

List of lpr Command Options

Option	Description
-p	Print each page with a shaded header with the date, time, job name, and page number. This option is equivalent to "**-o prettyprint**" and is only useful when printing text files.
-P *printer_name*	Specify the name of the printer used for printing files. If no printer is specified, the system's default printer is used.
-U *username*	Specifies an alternate username.
-# *number*	Sets the number of copies to the *number* for print. *number* range is from **1** to **100**. For example: `cat hello.c file1.txt welcome.cpp | lpr -#3`
-h	Disables banner printing. This option is equivalent to "**-o job-sheets=none**".
-l	Specifies that the print file is already formatted for the destination and should be sent without filtering. This option is equivalent to "**-o raw**".
-m	Send an email when printing is completed.
-r	After printing files, the named print files should be deleted.

All printing files are placed in a queue and are printed one at a time in the background. Table 10.9 lists all common options for lpr commands.

- *lpq*: **show printer queue status**:
  Use the *lpq* command to view the contents of the print queue. When lpq is invoked without any arguments, it returns the contents of the `default print queue`. For example, an empty queue for a system default printer named *printer1*:

```
$ lpq
Printer1 is ready and printing
Rank    Owner   Job  Files        Total Size
active  sanrit  31   project.txt  76548 bytes
```

Table 10.10 listed all common options for lpq commands.

- *lprm*: **remove print jobs:**
  Another important feature of any printing system is the capability to cancel a job. So, lprm command is used to cancel print jobs that have queued for printing.

```
$ lprm -
```

TABLE 10.10

List of lpq Command Options

Option	Description
-P *destination*	Mentions an alternate printer or the class name.
-U *username*	Specifies an alternate *username*.
-a	Reports jobs on all printers.
-h *server*[: *port*]	Specifies an alternate server.
-l	Requests a more verbose (long) reporting format.
-W	Displays a wide version of status information with longer queue names, device names, and job numbers.

TABLE 10.11

List of lpq Command Options

Option	Description
-P *printer*	Specify the queue associated with a specific *printer*; otherwise, the default printer is used.
-	If a single hyphen (-) is given, lprm will remove all jobs that a user owns. If the superuser employs this flag, the spool queue will be cleared completely.
user	Remove any jobs queued belonging to that *user* (or *users*). This type of invoking *lprm* command is useful only to the superuser.
job #	A user may remove an individual job by specifying its job number. This number may be obtained by executing the *lpq* command.

The above-mentioned command cancels all the print jobs that are owned by the user who invoked the command. You can also cancel a single print job by using the job number. You can find the job number for a printing job by using the *lpq command,* and thereafter, use that number with *lprm to cancel a job.* You can type the below syntax to cancel job 31 as used in the *lpq* section:

```
$ lprm 31
```

The above command will cancel job 31 (project.txt) on the default printer. Table 10.11 lists all common options for *lprm* commands.

10.5.5 Searching and Linking File: *find, ln*

In this section, we describe the search utility command, *find,* and create links between the files by using the *ln* command.

(a) *find:* **search for files in a directory hierarchy:**

If your file system has a large number of files in various directories, then you require a search utility to locate them. The *find* command offers you to perform the search operation using various search criteria. Some key search criteria are name, type, owner, last update time. and so on. The basic syntax of find gives as

```
$ find directory_pathname -option criteria
```

In the above syntax format, *directory_pathname* is the directory from which the search should start. The directory name is the argument for find commands along with various options that notify the type of search and criteria for the search. Some frequently used key options with the find command are described below:

- **Searching a file by name:**
  If you want to search a file by name, then you use the *-name* option with the *find* command. The *-name* option instructs the find command to search for the file-name that matches with the input pattern, as given below:
  With the directory name followed by the -name option and the name of the file

```
$ find directory_pathname -name filename
```

or

```
$ find directory_pathname -name input_pattern
```

- **Searching the working directory:**
   If you want to search your working directory, then you must use the dot (.) in the directory pathname to denote your working directory. The double dots(..) signify the parent directory. To find all files in the current directory, use the following command input

```
$find .
```

- **Locating directories:**
   Find command is very useful to locate other directories. In Linux, a directory is specified as a special type of file, denoted by *d* in file type categories. Though all files have a byte-stream format, some files, such as directories, are used in different ways. The find command uses an option known as *-type* that locates a file with respect to its given file type. The -type option takes a character modifier that denotes the file type.
   For the directory, the modifier is represented as *d*. In the given examples, the directory name and the directory file type is used to search for the directory known as "documents":

```
$ find /home/sanrit -name documents -type d -print
/home/sanrit/letters/ documents
$
```

You can also search for files by ownership, access permission, and other search criteria. This command has several options and criteria to match that specify search criteria and actions, which are listed in Table 10.12.

(b) *ln:* **make links between files:**
   The *ln* command is used to create *links* between files, either hard or symbolic links. By default, the ln command creates *hard links,* which are basically additional names for an existing file. Hard links cannot be made to directories. Similarly, they cannot cross the limits of the file system and partition. Thus, for making a hard link file named *linkfile1* to a file named *file1,* use the below syntax:

```
Syntax: ln [options] original_file new_file
```
For a hard link file,

```
$ln file1 linkfile1
```

to create a soft link file, use the -s option with the ln command, as given:

```
$ln -s file1 linkfile2
```

Here, *linkfile2* file is the soft link file of *file1.* Table 10.13 lists all common options for *ln* commands.

TABLE 10.12

List of Options and Commands Associated with the *find* Command

Option or Command	Description
-name *pattern*	The search pattern comprises the filename that has to be searched.
-user *name*	Searches files belong to a specified username.
-type *fileType*	Searches for files with the specified file type. *fileType* can be: b—block device c—character device d—directory f—file l—symbolic link
-group *group_name*	The group name will be the group for which files will be searched.
-exec *command*	Executes the command when the files are found.
-print	Displays search result on the standard output. The result comprises the list of filenames, including pathnames.
-ls	Give a detailed listing of each file, with owner, permission, size, and date information.
-lname *pattern*	Searches for soft or symbolic link files.
-gid *name*	Searches for files belonging to a group (refer to as group id).
-size n	Match files of size n.
-type c	Match files of type c.
-group *name*	Searches for files attached to the mentioned group name.
-perm *permission*	Searches for files with a certain permission set.
-context *scontext*	Searches for files according to the security setting (SE Linux).
-size *numc*	Searches for files with the size num in blocks. If c is additional after num, the size in bytes (characters) is searched for.
-uid *name*	Searches for files belonging to a user with respect to its user id.
-mmin *n*	Match directories or files whose contents were updated *n* minutes ago.
-mtime *n*	Match files or directories whose contents only were last updated n*24 hours ago.
-cmin *n*	Match files or directories whose content or attributes were last modified exactly *n* minutes ago.
-delete	Delete the currently matching file.
-empty	Match empty files and directories.

TABLE 10.13

List of Various Common Options for *ln* Commands

Option	Notation	Description
-s,	--symbolic	Used to create symbolic (soft) link files instead of hard links. Produces an error message, if systems do not support symbolic links utility.
-P,	--physical	Makes hard links directly to symbolic links.
–d, –F,	--directory	Permits the superuser to create hard links to directories.
-v,	--verbose	Prints the name of each linked file.
-b	--backup	Backups the files to be removed.
-i	--interactive	Confirms whether to remove the existing destination files or not.
-f	--force	Removes the existing destination file immediately.

10.6 Directories

In Linux, everything is kept and configured as a file. This comprises not only text files, program file, images files but also include directories, partitions, and hardware device drivers. To the user view, the Linux file system looks like a hierarchical structure of directories that comprise files and other directories or subdirectories. Directories and files are recognized by their names. This hierarchy structure starts from a single topmost directory known as the *root directory*, which is denoted by a "/" (forward slash), (don't confuse with superuser or root user directory (i.e.,/root), which is kept under "/". All directories and subdirectories are arranged under the root directory (/). This hierarchical structure of directories is also known as the tree structure of directories.

The Linux file system structure allows you to organize files in these directories so that you can easily find and manage any file and directory. In the Linux system, each user starts with one directory and then subsequently adds subdirectories. You can expand the directory structure to any desired multiple levels of subdirectories as required. You can take the advantage of this strength and organize your files. This is one of the most suitable and useful ways to store your data. The Linux hierarchical structure "grows" downward from the root(/) with paths connecting the root(/)as shown in Figure 10.1

At the last or end of each path is either a directory file or an ordinary file (directory is a special kind of file, having filetype *d*, as compared to an ordinary file, having file type – (hyphen)). Ordinary files, or simply files, look at the ends of paths that cannot start any other paths. On the other hand, directory files, also known as folders or directories, are the points from where you may or may not generate other paths. If you refer to Figure 10.1, the top directory is the root (/), and all other directories and files are away from the root and move downward. Each directory has a parent directory except the root (/) directory. A parent directory is directly connected by a path to the said directory closer toward the root(/). Subsequently, the child directory is directly connected by a path (see Section 10.6.2) to the said directory away from the root(/).

The subsequent subsections describe various operations related to directories such as locating path, creation, deletion, moving, copying, and others [5].

10.6.1 Special Directories

After the successful installation of the Linux system including the Linux file system, the file system creates some main directories and stores Linux OS information, bootable programs, and other system-related contents. All files and directories look under the root directory, even if they are stored on different physical devices such as different disks or different computers. Some key special directories defined by the Linux file system are **/bin** (comprises all binary programs needed for booting), **/boot** (stores the boot loader files such as the GRand Unified Bootloader (GRUB) and Linux Loader (LILO)), **/home** (home directories for local users), **/mnt** (for mounting external peripheral devices such as a CDROM or floppy disk), **/root** (the home directory for the super or root user), **/etc/fstab** (keeps the information about all file systems on the system), /usr (the place where most application programs get installed), **/etc** (keeps server configuration data, generally users can read but not write to this file), **/var** (holds log files, print spool files, and possibly database files) and many more directories. You can see (Section10.3.4) for more details about the system and special directories that are essential to setup Linux OS.

10.6.2 Paths and Pathnames: Absolute, Relative

A path is a way you need to follow in the tree structure to reach from a given source to the designation file. A pathname is a series of directory names that you need to follow to reach the destination file in a tree structure. Pathnames can be defined in two ways: as absolute pathname or as a relative pathname. An absolute pathname starts from the root (/) directory, and a relative pathname starts from the current working directory. For more details, see Section 2.3 of Chapter 2.

For example, the absolute path of pic.jpg file in Figure 10.1 is as follows:

```
/usr/data/pic.jpg
```

10.6.3 Creating and Deleting Directories mkdir, rmdir

- **mkdir: creates a directory**:
  The *mkdir* command is used to create a directory with each given name. The basic syntax of *mkdir* command is

```
$mkdir directory_path
```

By default, the mode of created directories is 777 (octal number) minus the bits set in the umask value (see 10.9 section for more detail). For example:

```
# mkdir New_Dir1
# ls -old New_Dir1
drwxrwxr-x 2 root root 4096 Feb 22 09:48 New_Dir1
#
```

The system creates a new directory named *New_Dir1*. In the long listing output of *ls* command, the directory's record begins with a "d", which indicates that *New_Dir1* is a directory not a file. A frequently used option with *mkdir* command is the *-p* option, which will tell *mkdir* to create not only subdirectories but also any of its parent directories if they do not already exist. For example, if you need to create a hierarchy of directories */home/sanrit/subdir1/mydir1/mydir2* starting from the current working directory (i.e., the directory in which the user is currently working). If the current working directory of the user is */home/sanrit*, then by using the -p option, *mkdir* will automatically create *subdir1* and *mydir1* directory along with *mydir2* directories.

To create a single directory called *mydir1*, then use the following command:

```
$mkdir mydir1
```

To create a hierarchy of directories such as *subdir1/mydir1/mydir2* in your present working directory (i.e., */home/sanrit*), then use the following command with the -p option:

```
$mkdir -p subdir1/mydir1/mydir2
```

A multibranched directory hierarchy can be created with the help of a single command rather than using a separate command on the command line to create each branch. For example, the tree beginning with *sanrit* directory and having the two branches *CSE* and *ECE* can be created as follows:

```
$mkdir -p sanrit/ECE/schemes/subjects sanrit/CSE/schemes/subjects
```

Several subdirectories or branches can be created concurrently by including them in the same command line. For example, the following would additionally create a branch called *result/subjects* that begins in *ECE* and *CSE* subdirectories, as follows:

```
$mkdir -p sanrit/ECE/schemes/subjects sanrit/CSE/schemes/subjects
 sanrit/CSE/result/subjects sanrit/ECE/result/subjects
```

Table 10.14, lists the various options of the *mkdir* command.

- **The dot(.) and double dot (..) Directory Entries with mkdir command:**
  When the mkdir command creates a directory, then it automatically places two entries in each directory: a single dot (.) and a double dot (..). The (.) is identical to the pathname of the working directory and can be used in its place; the (..) is equal to the pathname of the parent of the working directory. Normally, you cannot see these entries by using the *ls* command because these are hidden due to their file-names beginning with a period.

- **rmdir: remove an empty directory:**
  The *rmdir* command is used to remove a directory that is empty from the file system in Linux. The rmdir command removes each directory mentioned in the command line only if these directories are empty. If the mentioned directory has some files and subdirectories in it, then such directories would not be removed by the *rmdir* command.

  For example, to remove a directory called mydir, you use the following command:

```
$ rmdir mydir
```

To remove empty directories *mydir1*, *mydir2*, and *mydir3* on a single command line:

```
$rmdir mydir1 mydir2 mydir3
```

If you want to delete the hierarchy of directories from *ECE* to *subjects* (ECE/schemes/subjects) directories that were created in the above section, use the -p

TABLE 10.14

List of Various Options of *mkdir* Command

Option	Notation	Description
–p	--parents	Confirms that each given directory exists and creates any absent parent directories for each input argument.
-m	--mode	Sets the mode of created directories to a given *mode* value. See the chmod command section for details.

option with the *rmdir* command, but all these directories must be empty; otherwise, it will give an error message:

```
$rmdir -p ECE/schemes/subjects
```

This command will first remove the child directory and then remove the parent directory in the mentioned directory path. The major limitation with the *rmdir* command is that the specified directories must be empty. To overcome this limitation, you can use the *rm* command with the -r option to delete nonempty directories recursively in the directory hierarchy.

- **rm: remove files and directories:**
  The rm command is used to remove (delete) files and directories, such as:

```
$rm item...
```

where the item is the name of one or more files or directories.

furthermore, it is easy to remove the entire hierarchy of directories by using the *rm* command with its -r option; for example, the following command would delete the entire ECE directory tree, including the files in it, if any:

```
rm -r ECE   //Directory tree of ECE directory is ECE/schemes/subjects
```

rm -r is a very powerful command therefore you should be very careful while using it. Table 10.15 lists the various options associated with the *rm* command.

10.6.4 Displaying Directory Contents: ls

ls command is used to list the files and directories in the present working directory.

```
$ ls
Desktop  Documents  Music  Pictures Public User Videos
$
```

See Section 10.5.2 for more details of the ls command.

TABLE 10.15

List of *rm* Command Options

Option	Notation	Description
-i	--interactive	Seek the confirmation before deleting an existing file. If this option is not mentioned, straightaway, rm will delete the directory.
-f	--force	Ignore nonexistent files and do not prompt. This overrules the interactive option (-i)
-r	--recursive	Recursively delete directories. It means you can delete a directory along with its subdirectories and files also. To delete a directory, this option must be mentioned, for example: `$rm -r file1 mydir1  //Delete file1 and mydir1 and its contents`
-v	--verbose	Displays informative messages as the deletion is performed.

10.6.5 Moving through Directories cd

In Linux *'cd'* (change directory) command is one of the most important and widely used commands. With this command, you can navigate the directory tree and move to the desired directory from the present place to check files, programs/application/script, and other important tasks. So, with the use of the cd command, you can make another directory as current working directory. The basic syntax of the cd command is:

```
$cd directory_path
```

Type cd followed by the *directory_path* of the desired working directory. The *directory_path* may be supplied as the absolute path or relative path of the destination directory. For example:

- To change from the current working directory to *bin* directory(/usr/bin).

```
$cd /usr/bin
$ pwd
/usr/bin
$
```

- To move from the current working directory to its parent directory:

```
$cd ..
```

- To return to your login home directory from anywhere in the directory tree,

```
$cd
```

Some useful shortcuts of cd command are listed in Table 10.16 that change the current working directory so quickly.

10.6.6 Locate Directory: pwd

After successfully logging into the Linux system, certainly, you will see yourself at the terminal or shell prompt, and you may not be aware of where you are in the Linux file system hierarchy or the directory tree. Therefore, to know the exact information on your location in the directory tree structure, you need to use the *pwd* command. *pwd* stands for **print working directory**. It prints the absolute path of the current working

TABLE 10.16

Command cd Shortcuts

Shortcut	Description
cd	Switches the current working directory to the user home directory.
cd ..	Switches the current working directory to its parent directory.
cd -	Switches the current working directory to the previous working directory.
cd ~username	Changes the working directory to the home directory of the *username*. For example, cd ~sanrit the directory to the home directory of user *sanrit*.

directory, i.e., starting from the root (/). To display your current working directory, use this command:

```
$ pwd
/home/sanrit
```

At any given point in time, the *current directory* is the *directory* in which a user is working at a given time.

Every user is always working within a directory. The *pwd* command is a shell built-in command, which resides in the *bin* directory(/bin/pwd) and is available with most of the shell – bash, Bourne, ksh,zsh, and so on. $PWD is an environmental variable that stores the path of the current directory.

```
$ echo $PWD
/home/sanrit
```

This command has two flags or options, which are as follows:

```
pwd -L: Prints the symbolic path.
pwd -P: Prints the actual path.
```

10.6.7 Scanning Directories: opendir, readdir, telldir, seekdir, and closedir

As you are aware, Linux is written in c languages. A system call is a routine that can be invoked from a C program to access the system resources to execute the command. This section describes the directory scanning system, *opendir, readdir, telldir, seekdir,* and *closedir,* written in C language function [4]. Directories are also files, so they must be opened, read, and written similarly to the regular files. The format of a directory is not consistent across the various Linux distributions. Linux offers several library functions to handle a directory operation. It is worth mentioning here that you cannot directly write a directory. Only Linux kernel may do that by using an open, read, and close system call such as ordinary files.

- **opendir():opens a directory**:
  The opendir() function opens a directory stream corresponding to the directory *dirname* and returns a pointer to the directory stream. The stream is positioned as the first entry in the directory. The basic syntax of the opendir() system call is given as:

  ```
  #include <sys/types.h>
  #include <dirent.h>
  DIR *opendir(const char *dirname);
  ```

  The opendir() function returns a pointer to the directory stream or NULL if an error occurred.

- **Readdir: reads a directory:**
  The readdir() function returns a pointer to a *dirent* structure representing the next directory entry in the directory stream pointed to by *dirp*. It returns NULL on reaching the end of the directory stream or if an error occurred. The basic syntax of a readdir() system call is given as:

```
#include <sys/types.h>
#include <dirent.h>
struct dirent *readdir(DIR *dirp);
```

A directory maintains the inode number and filename for every file in its folder. As expected, these two parameters are members of the *dirent* structure that is returned by readdir. Every invocation of readdir fill up this structure with information related to the next directory entry (i.e., the next filename).

In the glibc(GNU C Library) implementation, the *dirent* structure is given below:

```
struct dirent {
ino_t          d_ino;         /* Inode number */
off_t          d_off;         /* Not an offset */
unsigned short d_reclen;      /* Length of this record */
unsigned char  d_type;        /* Type of file */
char           d_name[256];   /* filename */
};
```

The only fields that are mandated by POSIX.1 (The Portable Operating System Interface) in the dirent structure are directory name, *d_name*, and directory inode *d_ino*. The other fields are not present on all the systems as they are unstandardized.

The fields of the dirent structure are given below:

- *_ino*: This is the inode number of the file.
- *d_off*: The value returned in d_off is the same as would be returned by calling telldir() at the current position in the directory stream.
- *d_reclen*: This is the size (in bytes) of the returned record.
- *d_type*: This field contains a value indicating the file type, making it possible to avoid the expense of calling lstat(), if further actions depend on the type of the file. In glibc, it uses the following macro constants for the value returned in *d_type*:
  - DT_BLK   This is a block device.
  - DT_CHR   This is a character device.
  - DT_DIR   This is a directory.
  - DT_FIFO   This is a named pipe (First In, First Out, FIFO).
  - DT_LNK   This is a symbolic link.
  - DT_REG   This is a regular file.
  - DT_SOCK   This is a Unix domain socket.
  - DT_UNKNOWN The file type could not be determined.
- *d_name*: This field contains the null-terminated filename

The readdir() function returns a pointer to a *dirent* structure or NULL if an error occurs or the end-of-file is reached.

- **telldir: returns current location in directory stream:**
  The telldir() function returns the current location associated with the directory stream *dirp*. The basic syntax of telldir() system call is given as

```
#include <dirent.h>
 long telldir(DIR *dirp);
```

The telldir() returns the current location of the DIRpointer. This location can be used as an argument to seekdir to set the pointer to a specific location. The telldir()function returns the current location in the directory stream or –1 if an error occurs.

- **seekdir: sets the position of the next readdir()call in the directory stream:**
  The seekdir() function sets the location in the directory stream from which the next readdir()call will start. seekdir() should be used with an offset returned by telldir(). The basic syntax of seekdir() system call is given as:

```
#include <dirent.h>
void seekdir(DIR *dir,off_t offset);
```

The seekdir() function returns no value.

- **closedir():close a directory:**
  The closedir() function closes the directory stream associated with dirp. A successful call to closedir() also closes the underlying file descriptor associated with dirp. The directory stream descriptor dirp is not available after this call. The basic syntax of closedir() system call is given as:

```
#include <sys/types.h>
#include <dirent.h>
int closedir(DIR *dirp);
```

The *closedir()* function returns 0 on success. On error, -1 is returned, and errno is set appropriately

10.7 Archiving and Compressing Files

One of the key tasks of a computer system's users and administrator is to keep and secure the system's data. There are many ways to do this, but one of the widely used practices is to keep a backup of the system's files at regular intervals. For easy backup, it is useful to store a group of files in one file and transfer them to another directory, or even other computer systems. It is also useful to compress huge files because compressed files take up less memory space [6]. It is important to recognize the difference between an *archive file* and a *compressed file.*

- An archive file is a collection of files and directories stored in one file. The archive file is not compressed, and it uses the same amount of memory space as all the individual files and directories would take combined.
- A compressed file is a collection of files and directories that are stored in one file *and* stored in a way that uses less disk space than all the individual files and directories would take combined.

An archive file is not a compressed file, but a compressed file can be an archive file. In the subsequent sections, you can see the description of various commands used for file compression and archive files.

10.7.1 Archiving and Compressing Files with File Roller

Gnome provides a File Roller tool that operates as a GUI front-end to compress, decompress, and archive files in common Unix and Linux format. File Roller is an archive manager and supports tar, bzip2, gzip, zip, jar, compress, lzop, and many other archive file formats. You can inspect the contents of archives and extract and create new compressed archives when required. When you create an archive, you determine its compression technique by specifying its filename extension, such as .gz for gzip or .bz2 for bzip2. You can also select the different extensions from the File Type menu or enter the extension yourself. To compress and archive files, you can select a combined extension similar to *.tar.bz2*, which both archives with tar and compresses with bzip2.

To start File Roller, select Archive Manager from the Applications (the main menu on the GNOME panel) => System Tools submenu. File Roller is also integrated into the desktop environment and Nautilus (file manager), which allows you to create archives of your files and directories.

10.7.2 Archive Files and Devices: tar

The *tar* command creates archives for files and directories. The term *"tar"* stands for tape archive that combines multiple files into a single large file. With the tar command, you can archive exact files, update, and add new files to an archive. You may also archive entire directories, including their files and subdirectories, and the same can be reinstated from the archive. The *tar* function was originally intended to generate archives on tapes. A *tar* function is a perfect tool for creating backups of your files or combining several files into a single file for transmission across a network (File Roller is a GUI for tar). Automatically, GNU Network Object Model Environment (GNOME) and K Desktop Environment (KDE) desktops can display the contents of a tar archive file. You often see filenames that end with the extension *.tar* or *.tgz*, which indicate a "plain" *tar* archive and a gzipped archive, respectively.

The syntax for the tar command:

```
$tar option filename
```

or

```
$tar option pathname
```

By using the tar utility with -f option, you may archive files to a specific device or a file with the name of the device or file. You can extract files and directories from the archive using the *x* option. Similarly, The *xf* option extracts files from an archive file or device. Furthermore, you use the *r* option to append or add files to the existing archive. Some key options of the *tar* command are listed in Table 10.17 that help to perform various operations on the archive files such as create, extract, append, list, and many more. In Linux, most commands use their options starting with the hyphen (–) symbol, but the hyphen (–) in the tar command options are optional.

TABLE 10.17

tar Command Options

Option	Description
c	Creates a new archive.
x	Extracts an archive.
r	Append specified files or directories to the existing archive file.
t	List the contents of an archive.
U	Updates an archive with new and altered files.
w	Enable user confirmation before archiving every file.
M	Creates a multiple-volume archive that may be stored on several floppy disks.
f *archive_name*	Permits you to specify the filename of the archive.
f *device-name*	Saves a *tar* archive to a specific device names such as the floppy disk (/dev/fd0) or tape.
v	Displays the progress of archive files.
z	Compresses or decompresses archived files using gzip.
j	Compresses or decompresses archived files using bzip2.

- **Create a tar archive file**:
  Let's create a tar file of */usr/data* directory and */root/hello.cpp* file, by using the following command:

  ```
  # tar -cvf myarchive.tar /usr/data /root/hello.cpp
  ```

  The above command will create a tar file with the name *"myarchive.tar"* in the current folder. This archive file comprises all the files and directories of /usr/data folder and `hello.cpp` file.

- **Append or add files to the end of the archive or tar file**:
  By using *-r* option with the tar command, you can add or append a file to the existing tar file. For example, add */etc/fstab* file in *"myarchive.tar"*

  ```
  # tar -rvf myarchive.tar /etc/fstab
  ```

- **Extracting files and directories from the tar file**:
  By using the -x option, you can extract the files and directories from the existing *tar* file. For example, extract the content of the above-created tar file, namely *myarchive.tar*,

  ```
  # tar -xvf myarchive.tar
  ```

  The above command will extract all the files and directories of the *myarchive.tar* file in the current working directory.

10.7.3 File Compression: gzip, bzip2, and zip

If you are familiar with the Microsoft Windows OS, then you are very much familiar with zip files. The zip program permits you to easily compress large files into smaller files to reduce the memory size of files. Similarly, the Linux OS offers several file compressions programs, which are mentioned in Table 10.18.

TABLE 10.18

File Comparison Utility

Utility	File Extension	Description
bzip2	.bz2	Uses the Burrows – Wheeler block-sorting text compression algorithm and Huffman coding.
gzip	.gz	The GNU Project's compression utility; uses Lempel – Ziv coding.
zip	.zip	The Unix version of the PKZIP program for Windows

- **gzip—compress files:**

    In the Linux system, mostly software uses the GNU's Not Unix (GNU) gzip and gunzip utilities. The gzip program is used to compress one or more files and replace the original file with the compressed version of the original file with .gz extension. The gunzip program is used to uncompress or restore compressed files to their original version. For example:

```
$ gzip myfile1
$ ls
 myfile1.gz
$
```

In the above example, you can compress myfile1 by using gzip command. This replaces the file with a compressed version of the input file with the extension .gz, namely myfile1.gz.

```
$ gzip myfile1 myfile2 myfile3
$ ls
 myfile1.gz myfile2.gz myfile3.gz
$
```

In a similar way, you can also compress archived tar files and the output files stored with the extension (.tar.gz.). Such compressed archived files are normally used for transmitting very huge-sized files over the network.

```
$ gzip myfile.tar
$ ls
myfile.tar.gz
$
```

To decompress a gzip file, use gzip with the -d option or the command gunzip. These commands decompress a file having the .gz extension and replace it with a filename without .gz extension. It is optional to type the .gz extension on the command line along with the filename. For example:

```
$gunzip myfile1.gz
$ls
myfile1
$
```

or

```
$gzip -d myfile1.gz
$ls
myfile1
$
```

Table 10.19 lists the different *gzip* options.

- **bzip2: compression with a higher rate**:

    Another widespread compression program is bzip2. The bzip2 program uses Burrows – Wheeler block-sorting text compression algorithm and Huffman coding algorithm for file compression, which attain higher levels of compression ratio. Mostly, it works in a similar way as gzip except with the -r option. You compress files using the bzip2 command and decompress files with bunzip2. A file compressed with the bzip2 command is represented with the extension .bz2.The bzip2 command creates files with the extension .bz2. You can use the *bzcat* command to redirect the compressed data to the standard output. For example:

```
$bzip2 myfile1
$ls
myfile1.bz2
$
```

To decompress, you can use *bunzip2* command to decompress a bzip file, as given below:

```
$bunzip2 myfile1.bz2
$ls
myfile1.bz2
  $
```

TABLE 10.19

List of gzip Options

Option	Description
-c	Sends a compressed version of files to the standard output; each file listed is separately compressed: `gzip -c mydata mydata2 mydata3 > myfiles.gz`
-d	Decompress the file similar to gunzip.
-f	Force compress even if a compressed version of the original file already exists.
-h	Display usage information.
-l *file-list*	Displays compression statistics for each listed compressed and uncompressed size of files. `gzip -l myfiles.gz`
-r *directory-name*	Recursively searches for the mentioned directories and compresses all the files within them.
-t	Tests the integrity of a compressed file.
-v *file-list*	Displays each compressed or decompressed file with its name and the percentage(%) of its reduction in size. It may be specified with—verbose.
-number	Set limits for speed and size of the compression; the range is from—1(fastest, least compression) to—9 (slowest, most compression). A lower number gives greater speed but less compression. The values 1 and 9 may also be expressed as --fast and --best, respectively. The default value is—6.

TABLE 10.20

Zip Command Option

Mode	Description
-r	Archives and compresses the directory.
-f	Updates a particular file in the Zip archive with an updated version.
-u	Replaces or adds files.
-d	Deletes files from the Zip archive.

The *bzip2* program also comes with the *bzip2recover* program, which helps to recover the damaged .bz2 files.

- **Zip: package and compress files**

  A zip program is a compression tool as well as an archiver. You compress a file using the zip command and create a Zip file with the .zip extension. The Zip file extension is very much familiar to Windows users, as it reads and writes .zip files. Preferably, Linux users mainly use the zip command for exchanging files with Windows systems. You can also use the - argument to have a zip read from the standard input. To compress a directory, use the -r option. For example:

```
$ zip datafile1
$ ls
datafile1.zip
$
```

The various options of zip command are listed in Table 10.20.

To decompress and extract the Zip file, use the unzip command.

```
$ unzip datafile1.zip
```

10.8 File and Directory Attributes: ls –l, ls –d

- **File attributes: ls –l:**

  The ls command with -l option is used to list out the major file attributes of various files and subdirectories of a directory. The -l option is meant to display the content of a directory in a long format, which comprises useful information, referred to as file attributes, related to each file as follows:

```
$ ls -l
Total 5
-rw-r--r-- 1 root root 45326 Jul 12 11:06 computer1.png
-rw-r--r-- 1 root root 45326 Sep 23 08:05 file1.txt
drw-r--r-- 2 root root 75234 Aug 18 12:04 employees
-rw-r--r-- 1 root root 85453 Jul 22 06:07 computer1.png
drw-r--r-- 2 root root 45326 Aug 11 16:15 file2
$
```

The above list shows seven marked fields in nine columns with the path of each filename ordered in American Standard Code for Information Interchange

TABLE 10.21

Description of Various Fields of ls -l Output

Field		Description
-rw-r--r--	Type and permissions	Ten characters of the first column show the file type and access permission of the file. The first character indicates the type of file. Among the different types, a leading dash means a regular file, whereas a d indicates a directory. The next three characters are the access rights for the file's owner; the next three are for members of the file's group, and the final three are for everyone else.
1	Links	The second field indicates the number of links associated with the file.
root	Ownership	The username of the file's owner. Every file has an owner. The third field shows the owner's name of the files.
root	Group ownership	A user also belongs to a group, and the fourth field shows the group owner name of the files.
45326	Size	The fifth field shows the file size in bytes.
Jul 12 11:06	Last modification time	The sixth field displays the date and time of the last modification of the file in three columns. The year is displayed only if more than a year has passed.
computer1.png	Filename	The last field displays the name of the file, which can be up to 255 characters long.

(ASCII collating sequence. Here, each field represents a file attribute, and all the mentioned attributes are stored in the inode (see Section 10.4) except the filename. Table 10.21 describes the meaning of each file attribute.

- **Directory attributes: ls -ld:**

    For a selective listing, you can use ls -l with filenames as arguments to display file attributes. But if you want to display only directory details of a directory, then you need to combine the -l and -d options with the ls command to display only directory details:

```
$ ls -ld
drw-r--r-- 2 root root 75234 Aug 18 12:04 employees
drw-r--r-- 2 root root 45326 Aug 11 16:15 file2
$
```

A directory is also a type of file; therefore, its attributes such as permissions, link count, ownership, group name, size, and modification time are the same as ordinary files.

10.9 File Permissions: chmod, chown, chdir, getcwd, unlink, link, symlink

Due to the multiuser feature of the Unix/Linux OS, every file and directory created in Linux comprises some set of permissions that regulate accessing permission among various users. In this section, we describe the various system calls that perform the following operations such as file permission, change ownership, change directory, get a current working directory, linking file, unlinking file, and symbolic link of a file [7].

- **chmod(): change the file permissions**:

    The chmod command allows you to change the file permission for files and directories. The syntax of the *chmod* command is:

    ```
    #chmod options mode filename
    ```

    The mode parameter permits you to set the file permission settings using either the octal number or symbolic notation. As discussed in earlier sections, you can view the file permission by using the `ls -1` command. In the output of `ls -1`, the 10 characters of the first column show the file type and access permission of a file or directory, see Table 10.21.

    ```
    $ ls -1
    -rw-r--r-- 1 root root 45326 Jul 12 11:06 computer1.png
    ```

    The file permission of a file or directory may comprise read, write, and execute permission. This file permission is applied to three different user categories, namely the owner user, the group user, and all other users (not belonging to any groups) [8]. The structure of a file's permission string is given as in Figure 10.3.

    The read permission is represented by r, write by w, and execute by x. But, an empty place is denoted as no permission and represented by a dash - ,(except the first dash -,which represents the file type in a 10-character file permission format). For example, the owner has only read and execute permission, the owner's permission for file access is *r-x*. When a file is created, it is automatically given some read, write, and execute permission, referred to as default permission, which can be set using the *umask* command. You may change these permissions to any combination as required as mentioned in Figure 10.3. To avoid any modification in the file, you can set only read permission for all users; for example, a regular file has only read permission for all user categories, then the file permission string is given as in Figure 10.4.

- **Octal representation**

    The mode parameter permits you to set the file permission settings using either the octal number or symbolic notation. In the octal representation, you just use a three-digit octal number *(d1d2d3)* to assign the file permission. Each digit in an octal number represents three binary digits *(b1b2b3)*, which are easily set with read (r), write(w), and execute (x) permission to the three user categories respectively. The three octal digit maps with the users' categories are shown in Figure 10.5 and Table 10.22 shows the binary number mapping with the file permission(r,w,x)

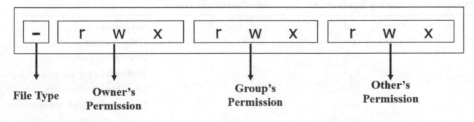

FIGURE 10.3
File permission string layout.

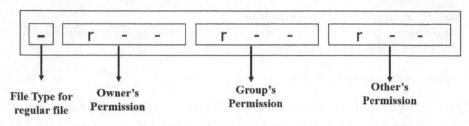

FIGURE 10.4
A file with only read permission for all.

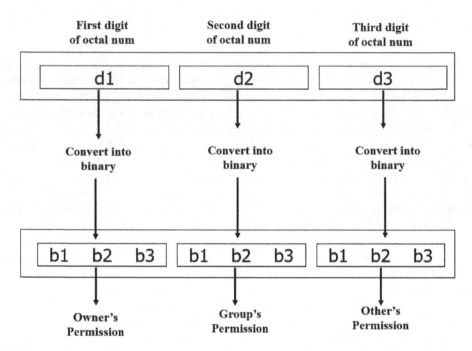

FIGURE 10.5
Octal number mapping with file permission.

TABLE 10.22

File Permission Mapping with Respect to Binary and Octal

Octal Digit	Binary Equivalent	File Permission	Description
0	000	---	No permission.
1	001	--x	Only execute permission.
2	010	-w-	Only write permission.
3	011	-wx	Write and execute permission.
4	100	r--	Only read permission.
5	101	r-x	Read and execute permission.
6	110	rw-	Read and write permission.
7	111	rwx	Read, write, and execute permission.

By using three octal digits, you can set the file permission for the owner, group owner, and others. For example:

```
#ls -l
-rw-r--r-- 1 root root 5224 Jul 16 10:06 hello.c
#chmod 757 hello.c
#ls -l
-rwxr-xrwx 1 root root 5224 Jul 16 10:06 hello.c
```

* **Symbolic representation**

    The chmod command also supports a symbolic notation for specifying file modes that comprise three characters. The symbolic representation format for specifying file permission is given as follows:

```
chmod [ugoa] [+-=] [rwxstugo]
```

The above syntax comprises three sets of characters: the first group of characters states to whom the new permissions apply; the second group of symbols is used to specify whether you want to add permission or remove permission or set permission to the value, and the third group of characters defines permission used for the setting. Table 10.23 describes these groups of characters in detail. Table 10.24 lists some examples of symbolic notation to set permission.

The chmod system call in C language is defining as

```
#include <sys/stat.h>
#include <stdio.h>
int chmod(const char *pathname, mode_t mode);
```

TABLE 10.23

chmod Symbolic Notation

Option (Symbolic Notation)	Description
The first set of characters, [ugoa]	
u	For the owner user.
g	For the group user.
o	For the other users.
a	For all the user's categories (For everyone, i.e., the combination of u, g, and o).
The second set of characters, [+-=]	
+	Adds permission.
−	Removes permission.
=	Assigns an entire set of permissions.
The third set of characters, [rwxstugo]	
r	Sets read permission.
w	Sets write permission.
x	Sets execute permission.
s	Sets user ID and group ID permission.
t	Saves the program text.
u	Sets the permissions to the owner's permissions.
g	Sets the permissions to the group's permissions.
o	Sets the permissions to the other's permissions.

TABLE 10.24

List of Examples of Symbolic Notation

Symbolic Notation	Description
u+x	Add execute permission for the owner
u-w	Remove write permission from the owner.
+x OR a+x OR ugo+x	Add execute permission to everyone
g-rw	Remove the read and write permissions from the group user.
ug=rx	Set read and execute permission to the owner and the group.
u+x, go=rx u+w, go=rw	Add write permission for the owner and set read and write permissions for the group and others. Numerous symbolic notations may be specified separated by commas.

- **chown(): change file ownership**

  Sometimes, you are required to change the ownership of a file. Linux offers the *chmod* command to change the owner of a file. So, the *chown* command changes the owner of the file from the owner to the other user. The syntax of the *chown* command is:

  ```
  #chown options owner[:group] file(s)
  ```

  The chown command transfers ownership of a file to a user, and the syntax shows that it can also change the file group owner depending on the first argument of the command. Unix also offers another command to change the file's group owner. It is the *chgrp* (change group) command. The chgrp command shares a similar syntax with chown. But with the added feature, the chown command changes the owner (user ID) and group (group ID) of a file. Changing ownership of a file requires superuser permission.

  For example, you can specify either the login name or the numeric UID for the new owner of the file:

  ```
  #ls -l
  -rwxr-xrwx 1 root root 5224 Jul 16 10:06 hello.c
  # chown sanrit hello.c
  # ls -l hello.c
  -rwxr-xrwx 1 sanrit root 5224 Jul 16 10:06 hello.c
  #
  ```

  Similarly, the user changes the group name for *hello.c* file and *employees* directory to the *book*1 group. The ls -l command then replicates the group change in the following example:

  ```
  #ls -l
  drw-r--r-- 2 root root 75234 Aug 18 12:04 employees
  -rwxr-xrwx 1 sanrit root 5224 Jul 16 10:06 hello.c
  #
  # chgrp book1 hello.c employees
  #ls -l

  drw-r--r-- 2 root book1 75234 Aug 18 12:04 employees
  -rwxr-xrwx 1 sanrit book1 5224 Jul 16 10:06 hello.c
  ```

The *chown* command can combine the *chgrp* operation along with *chown* operation by attaching a group to the new owner with a colon as follows:

```
#chown user1:book2 hello.c
-rwxr-xrwx 1 user1 book2 5224 Jul 16 10:06 hello.c
```

The chown() system call in C language is defining as

```
#include <stdio.h>
#include <unistd.h>
int chown(const char *pathname, uid_t owner, gid_t group);
```

- **chdir():changes the current working directory**:
   Now, you move on to the directory; first, to navigate to it and then to read its entries. A program can navigate directories in almost a similar way as a user moves around the Linux file system by using the change directory "cd" command. This program can use the *chdir()* system call to navigate directories.
   The chdir() system call in C language is defining as

```
#include <unistd.h>
int chdir(const char *pathname);
```

The chdir()system call changes the calling process's current working directory to the mentioned relative or absolute pathname in the pathname argument. The *chdir* system will return 0 for successful execution, otherwise -1 on error.

- **getcwd: get pathname of the current working directory**:
   In Linux, the pwd command is used to know the pathname of the current working directory. Similarly, the pwd command can use *getcwd()* system call to get the pathname of the current working directory. The chdir() system call in C language is defined as

```
#include <unistd.h>
char *getcwd(char *buffer, size_t size);
```

The **getcwd()** function gets an absolute pathname of the current working directory and writes into the array pointed to *buffer,* which is of length *size.* If the current absolute pathname of the current working directory would *exceed the size of the buffer* than *size* elements, **-1** is returned, and *errno* is set to **ERANGE**; a program should check for this error and allocate a larger *buffer* if required. If successful, getcwd() returns a pointer to the buffer. If the *buffer* is NULL, the behavior of **getcwd()** is undefined.

- **unlink(): removing a hard link**:
   The unlink() system call removes the directory entry for the file, provided as the argument **pathname*. In other words, the unlink() system call removes hard links of a file. If there is no other hard link to the file data, the file data itself is removed from the system. The unlink() system call, in C language, is defined as

```
#include <unistd.h>
int unlink(const char *pathname);
```

It returns 0 if the unlinking is successful, –1 on an error.

- **link() : creating a hard link:**
  The ln command is used to create both hard links and symbolic links of a file. In a similar way, the link () system call creates a new hard link to an existing file. The *link()* system call, in C language, is defined as

```
#include <unistd.h>
int link(const char *old_path, const char *new_path);
```

  In the above function, the *link()* system call generates a new link using the pathname mentioned in *new_path*. If *new_path* is already existing, then it is not over-written; rather, it gives an error (EEXIST) result. It returns 0 if the linking is successful, –1 on an error.

- **symlink: creating a symbolic link:**
  The *symlink()* system call creates a new symbolic link, *path2*, to the pathname specified in *path1*. To remove a symbolic link, you can use the system call *unlink()* function. The *symlink()* system call, in C language, is defined as

```
#include <unistd.h>
int symlink(const char *path1, const char *path2);
```

  If the pathname is given in *path2* already exists, then the call fails and gives an error (EEXIST). The pathname mentioned in *path1* may be an absolute or a relative path. It returns 0 if the symbolic linking is successful, –1 on an error

10.10 Summary

Most things in Linux are based on the concept of files and file management. A filename is a collection of information. If it starts with a dot (.), then such a file refers to a hidden file-name. Files may be of two types: normal files and directories. The first directory in the file system is called the root directory, which contains files and subdirectories. A home directory is also called a login directory, which is a subdirectory of home that serves as a repository for a user's personal files, directories, and programs. There are various operations that can be performed on individual files such as creating a file, printing a file, weaving, listing, and so on. File system paths are the paths to directories within the system. In a file system, a file is represented by an inode, a kind of serial number containing information about the actual data. We have certain utility commands, such as the search command and linking. We can move through, scan, and locate the directories. A zip program is a compression tool as well as an archiver. By using the -r option, you can compress the contents of the directories. A directory is also a type of file, and therefore, its attributes such as permissions, link count, ownership, group name, size, and modification time are the same as ordinary files. Every file and directory created in Linux comprises some set of permissions that regulate accessing permission among various users, such as file permission, change ownership, change directory, get a current working directory, linking file, unlinking file, and symbolic link of a file.

10.11 Review Exercises

1. Define, how to see and recognize the system file, and how to print the absolute path of your present place of working in the Linux file system?

2. Can we create multiple directories by a single invocation of the *mkdir* command? If yes, write the command steps.

3. In which directory are all the libraries stored in the file hierarchy structure?

4. Which directory is used to store the Kernel information needed to start the OS on startup?

5. Which directory contains read-only user data and command to view all the kernel parameters?

6. How to see the user id of all logged users? List the key differences between Set-user Identification (SUID) and Set-group identification (SGID)?

7. How do you make permanent changes to any file, residing in the */proc* directory?

8. How can you restrict a normal as well as the root user from making any changes as well as deleting any file?

9. Write the various key parameters of the kernel, and how can you see all such parameters?

10. Which command helps in compressing and archiving a file?

11. Write down the commands for the following queries.

    a. How to *untar* multiple files from tar, tar.gz, and tar.bz2 file?

    b. How to extract a group of files using wildcard?

    c. What are home directories and system directories?

    d. Give a command to create, display, and delete Directories.

    e. Write commands to scan the directories.

12. Can we provide multiple filenames as arguments to gunzip. True or false?

13. Write a shell script to get the current date, time, username, and current working directory.

14. Write a short note on inodes. Draw and elaborate the Linux file system architecture?

15. Write down the output of the following command.
    (i) mkdir file1/file2/file3, (ii) mv mango orange (iii) rmdir -r /usr/a/b (iv) mv mango /orange.

16. Elaborate the file permissions *chmod* command syntax and how can you set various security permissions to a file from authorized access or execution.

    a. Write a command to remove the write permission of a file.

    b. You execute a command cat *file1* and the output is "--command not found", Justify whether *cat* command and file1 both are present in the system.

    c. Write a command to add and write permission to all group users.

17. Describe *seekdir* functionality, and write the command to search and link a file.

18. Extract the contents of lines between two specific lines (say 5 to 10) from one file to another.

19. How would you know the available number of printers over the local network? How will you distribute the printing request among the available printers? Write a command for each.

20. Write a command to move all the **.mp3** extension files in the home directory to the **/usr/music** directory in the home directory.

21. How do device driver files help in accessing the device? Write the command to print the list of all device drivers with its details.

22. How to hide files and directories of a folder, so that they are not visible using the *ls* command without any special argument?

23. Write a bash script to search for a file in the whole file system that starts with the character "s" and stores their hard disk usage information in a separate file in the home directory with the name "out.txt".

24. Answer the following:

    a. What do "." and ".." symbols represent in any directory?

    b. What is the function of the "**cd –**" command?

25. Write a script to obtain a list of all the files that occupy a hard disk space greater than 12 KB, and then store them in a file and mail it to your friend.

26. Write the command that displays the various permissions that files have, and then using the *awk* command, print all the files that have only write permissions.

References

1. Wale Soyinka. 2012. *Linux Administration: A Beginner's Guide,* The McGraw-Hill.
2. Mark G. Sobell. 2012. *A Practical Guide to Fedora and Red Hat Enterprise Linux.* Prentice Hall.
3. Richard Blum and Christine Bresnahan. 2015. *Bible- Linux Command Line & Shell Scripting.* John Wiley & Sons.
4. J. Purcell. 1997. *Linux Complete: Command Reference.* Red Hat Software, Inc.
5. Richard Petersen. 2008. *Linux: The Complete Reference.* The McGraw-Hill.
6. Michael Stutz. 2001. *The Linux Cookbook: Tips and Techniques for Everyday Use.* Free book under GNU Free Documentation License.
7. Neil Matthew and Richard Stones. 2001. *Beginning Linux Programming.* Wrox Press Ltd.
8. William Stallings. 2005. *The Linux Operating System.* Prentice Hall.

11

Linux File Systems

Chapter 10 looked at Linux file system management including file operations, directories operations, file permission, and other related operations with files. This chapter describes the various partitioning methods and file system layout including mount and unmounting file systems. This chapter also highlights techniques to check and repair file system and virtual file system. This chapter concludes with describing the procedure for conversion of the file system from ext2 to ext3 and then ext3 to ext4 file system.

11.1 Introduction

A file system is a planned arrangement of regular files and directories. By using the *mkfs* command, you can create a file system. Linux arranges files into a hierarchically attached set of directories. Each directory may comprise files or directories or both. In this regard, directories may perform two key functions, either directory keeps files or directory connects to other directories, much similar as tree, where branches are connected to other branches. With this similarity, Linux file system structure is also regularly referred as a tree structure.

There are various kinds of file systems are available. Each file system has its own structure and properties like accessing speed, flexibility, size, security, and many more. As you know that some file systems have been designed explicitly for optical disks, like as ISO 9660 file system for CD ROM, Universal Disk Format (UDF) file system for DVD (Digital Versatile Disc or Digital Video Disc) etc. which are supported by all operating systems. The greatest strength of Linux operating system is that it supports a wide variety of file systems, and some are listed below:

- the traditional Linux file system, like ext, ext 2, ext3 ext4 and others; ;
- Unix file systems like Minix, System V, BSD file system, and others;
- Microsoft's file systems, like FAT, FAT16, FAT32, and NTFS;
- Optical media file system, like ISO 9660 CD-ROM file system, UDF DVD;
- Apple Macintosh's HFS file system;
- Network file systems like Sun's (NFS file system), IBM (SMB file system), Microsoft's (CIFS file system), Novell's (NCP system), Coda file system developed at Carnegie Mellon University, and many more;
- Various journaling file systems like ext3, ext4, ReiserFS, JFS, XFS, Btrfs, and others.

The details of various currently known file systems by kernel can be viewed from **/proc/ filesystems** file (see Section 11.7).

As described earlier that, in Linux, everything is in the form of a file. So, Linux treats that Shell is file and directories and devices are also a file. Here, devices are refereed as hard disk, mouse, DVD-ROM, USB disk, printer etc. There are several types of files, but primarily, there are three basic types of files:

- **Ordinary file**: Refer as a regular file comprises with only data as a stream of characters.
- **Directory file**: Refer as a folder comprises with files and subdirectories.
- **Device file**: Refer to a device or peripheral comprises with its device drive.

In the subsequent sections, you may see disk device details. Each device on the system has its own corresponding device file. Device files reside within the file system, under the */dev* directory as like the other files. Each device type has its own corresponding device driver installed in kernel, which handles all I/O operation for the device. Devices may be divided into two parts:

- **Character devices**: Handle stream of data (character-by-character), for example, keyboards.
- **Block devices**: Handle block of data (Group of Characters at a time). The size of data block depends on the device type of device like 256 or 512 bytes or others. Hard disks and tape drives are the example of block devices.

Each device file is identified by device ID (a major ID number and a minor ID number). The major and minor IDs of a device file are shown by using ls –l command. These ID entries are noted in the inode for the device file (see Section 10.4).

- **Disk devices (or Disk drives)**
  A disk drives is a physical device for storage of data that comprise one or more disks (or platters) rotating at a high speed. Data are stored on the disk surface that read/write with help of heads. The heads are moved outwardly across the disk to locate the data. Each disk further divided into a set of concentric circles, referred as tracks, which further divided into several sectors, referred physical blocks. The size of each physical block is generally 512 bytes or multiple of 512 bytes. It represents the smallest unit of information that the head can read or write. The modern hard drives have come with huge store space (in TB—Terabytes) along with high drive speed (around 7,000 rpm) and read speed (around 100 MBps—Megabytes per second).

11.2 Disk Partitioning

Regular files and directories are typically residing on hard disk devices. Storage devices which include but are not limited to hard drives or USB drives need to be structured in some way before they can be used to store data. Partitions are the small separate sections in which the large storage devices are divided. Another advantage of partitioning is to have isolated sections, which allows the capability to each subsection to perform as they are the own master, or we can say as a complete new hard drive. It usually comes handy in the need of having to run multiple operating systems.

There is a plethora of tools available for the purpose of creating, removing, and altering these partitions in Linux. The popular command in Linux for disk partition is *parted command* [1]. This command is very helpful while dealing with large disk devices and several disk partitions. There are other commands as well which are more commonly used like the *fdisk* or the *cfdisk*, and the difference between these is given below:

1. **GPT format**: One of the advantages of using parted command is that it is capable of creating Globally Unique Identifiers Partition Table, more commonly known as the GPTs, where the other two commands lack this ability and they are limited to using DOS partition table.
2. **Larger disks**: Another point which makes the use of parted command superior is its ability to handle large file sizes. Only 2 TB and in some cases 16 TB disk space could be handled by using DOS partition table, but this capability extends to 8ZiB of space when using the parted command.
3. **More partitions**: Including both the primary and the extended partitions, DOS partition tables only allow 16 partitions, whereas by default we can have 128 partitions in GPT and can go for many more.
4. **Reliability**: Another important aspect of handling data is that of reliability. Where the DOS partition only saves one copy, the other method of GPT is more reliable which keeps two copies one in the beginning and the other one in the end also making an efficient use of CRC checksum to check the integrity of the partition table, which was the missing component in the DOS partition.

Taking into the consideration the call of the hour where we have to work with larger disks and require enhanced flexibility, the use of parted is recommended over the other commands. When we install the operating system, it takes care of the disk partitioning. The use of parted command is more in play when we have to add a storage device to an existing system.

Mostly, there are two main tools that may be used to do most partitioning tasks and are the *parted* and *fdisk* utilities. Both the commands are almost similar. But *fdisk* command is small and command-line partitioning tool. On the other side, *parted* tool is much more user friendly and has many more built-in features than other tools. Therefore, following section explained the step-by-step method to partition a storage device by using *parted command*. A sentence of caution before use, try this on a new device or on the one where wiping out of data is not a big concern.

Step 1: Partition listing: The use of *parted -l* is to identify the storage device intended for the partition. Usually, the first hard disk contains the operating system, i.e., /dev/sda or /dev/vda; therefore, other place of storage device needs to be looked for this purpose.

Step 2: Opening the storage device: After the device is selected, use of *parted* comes into play. It is imperative to indicate the specific device of use. Random selection of space occurs if nothing is mentioned and only the parted command is executed.

Step 3: Partition table set-up: The partition table type needs to be set to GPT and then press "Yes" to accept it. For this purpose, the *mklabel* and the *mktable* commands are used. The partition tables supported are aix, amiga, bsd, dvh, gpt, mac, ms-dos, pc98, sun, and loop. It is important to note that *mklabel* is not for making partitions rather it is used for making partition tables.

Step 4: **Review of the partition table**: To show the information about the storage device.

Step 5: **Get help**: The following command could be used to find out how a new partition could be made.

```
#(parted) help mkpart
```

Step 6: **Making the partition**: To make a partition of 2048 MB on partition 0, we need to type the following command:

```
#(parted) mkpart primary 0 2048MB
```

In a DOS partition table, the partition types are primary, logical, and extended. Whereas, in GPT partition table, the partition name is used as the partition type as well which makes it necessary to provide partition name in GPT. In above command statement, primary is a partition name and not the partition type.

Step 7: **Save and quit**: Whenever the parted is quitted, the changes are automatically saved. Below given is the command to quit the parted:

```
#(parted) quit
```

Whenever adding a new storage device, it is imperative to identify the correct and intended disk before initiating the above procedure. If by chance the change is made to the disk containing the operating system, it would make the system vulnerable and prone to a problem where it might become unbootable.

NOTE: You can also use *fdisk* command to verify the number, type, and size of partitions on a disk. The command *fdisk –l* shows all partitions on a disk. The Linux-specific */proc/partitions* file displays the major and minor device numbers, size, and the name of each disk partition on the system.

11.3 Disk Partition into File System

As discussed earlier, that each disk drive is divided into one or more (nonoverlapping) partitions. It means once a partition is fully occupied with data, then the data cannot automatically overflow or store onto alternative partition. Therefore, each partition is treated as a separate device by the kernel that residing in file system under the */dev* directory. If you are adding a new type of disk (like SCSI disk or IDE disk) on a system, you must be ensuring that your kernel recognizes and supports the newly added disk.

Many things can be done with a partitioned disk like installing an operating system into a single partition. In the same way, you can install many different OSs into other separate partitions. Such situation is commonly mentioned as "dual booting" configuration, and the different partitions may also use to install other user application programs or system functions.

In Linux, each hard disk is assigned with a unique device name which is kept under the /dev directory. Each hard disks start with the name sdX (for SCSI) or hdX (for IDE), where

X can range from a to z character. With each character letter, it represents an individual physical hard disk. For example, if there are two SCSI hard disks, then the first hard disk would be /dev/sda and the second hard disk would be /dev/sdb. In the similar way, if the system has three hard drive (One disk is SCSI and other two are IDE disks, then disk names are /dev/sda, /dev/hda, /dev/hdb respectively). Further, if you are creating partition on a hard disk, say on /dev/sda then Linux treats each partition as a separate device and allocates a corresponding device file name as /dev/sdXY, where X is the device letter (as described above) and Y is the partition number (in numeric). Hence, if the hard disk /dev/sda has three partitions, then the first partition on the /dev/sda disk is /dev/sda1, the second partition would be /dev/sda2, and the third would be /dev/sda3. The similar nomenclature can be followed in IDE disk drive also. Presently, the majority of computers are using SCSI disk drives.

A disk partition may hold any type of information, but to install the Linux operating, hard drive may have three major kinds of partitions:

- **root partition**: The standard root partition (indicated with a single forward slash, /) is comprised of system configuration files, most basic commands and server programs, system libraries, some temporary space, and the home directory of the administrative user.
- **data partition**: Linux regular data and other user data or application.
- **swap partition**: Expansion of the computer's physical memory for memory management by kernel. Swap space (specified with *swap*) is only available for the system itself and is hidden from view during normal operation.

Most systems contain a root partition, one or more data partitions, and one or more swap partitions.

11.4 File System Layout

Hard drive may divide into various partitions for better utilization of memory resource. Each partition has only one file system, and it means you cannot install more than one file system in one partition. But more than one partition has same file systems on a disk. When we talk about Linux file system, it is divided mainly into four blocks [2] which are given as follows:

1. Boot Block
2. Super Block
3. Inode Block
4. Data Block

Partition is divided into fixed size data blocks which are used to store the contents of the file. But now the question arises that what if the file is larger than that of the block size, the simple answer is that multiple blocks are to be combined, and this file is then stored at these multiple blocks. The relationship between disk partition and Linux file system is shown in Figure 11.1.

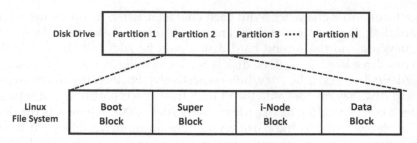

FIGURE 11.1
File system layout.

- **Boot block**: The first block of file system is boot block, and then question arise what does it actually store? The name of this block is quite intuitive to answer this question, that the initial bootstrap program which is employed to load the operating system, is stored here. Generally, the first sector consists of the initial bootstrap program which reads further a larger bootstrap program from the next couple of sectors, and onward to load the installed OS. Although only one boot block is required for the operating system, but all file systems have a boot block which are mostly unused.

- **Super block**: Every file system has two components: One is the super block, and the other is the duplicate super block, and it consists of information about file system, including

  1. File system type (ext2, ext3,...)
  2. The size of logical block in file system
  3. Pointers to a list of free blocks.
  4. The inode number of the root directory and size of inode table
  5. Size of file system in logical block

  Distinct file systems residing on the same physical device (but in different partitions) can be of different types and sizes and have various parameter settings (e.g., block size). This is main cause to divide a disk into numerous partitions.

- **inode block**: Each file or directory in the file system has assigned with a unique entry, referred as inode number in the inode table. inode number is an integer numerical value used to recognize every file or directory in the Linux systems. About the file info structure is as given below:

  1. Indication of the file ownership.
  2. Type of file (e.g., regular, directory, special devices, and pipe).
  3. Permissions of file access. It may consist of setuid (sticky) bit set.
  4. Last access and modification time.
  5. File links and the aliases.
  6. Pointers to the data blocks for the file.
  7. File size in bytes for regular files, and for special devices, the major and minor device numbers.

The inode table is sometimes also referred as the i-list.

- **Data blocks** are pointed by pointers included in inode. There are 15 pointers for each node.
  1. Data blocks are directly pointed by the first 12 pointers.
  2. There is an indirect block which contains pointers to data blocks, and this block is pointed by the 13th pointer.
  3. 128 addresses of singly indirect block, is saved in the doubly indirect block, which is pointed by the 14th pointer
  4. Further, the pointers to the doubly indirect blocks are saved in triply indirect box which is pointed by the 15th pointer.

The excessive majority of memory space in a file system is used for data blocks that store the files and directories residing in the file system.

11.5 Managing File System

As mentioned in Chapter 10, everything in Linux is a file, which makes the management of file systems in Linux crucial chores to do, Linux supports a wide variety of file systems for various operating system and other applications. There are many file system types are available, including Linux file system, like iso9660, FAT, FAT16, FAT32, ext, ext2, ext3, ext4, msdos UDF, hpfs, JFS, minix, ntfs, ncpfs nfs, proc, vfat, ReiserFS, smb, sysv, umsdos, XFS, xiafs, and many mores. Generally, Linux system recognizes only few essential file systems. If you want to access the data of any external hard drive, having the different file system (not available under Linux), then you need to mount (see subsequent section) that hard drive by mentioning the file system type which is already present in system. Hence, the subsequent section describes Linux file systems and the various way to manage external file system with Linux system.

11.5.1 File system Types

Given below are the file formats along which the standard Linux distribution provides the option of partitioning disk. Each of the given file types has a special meaning associated with it, and some are ext2, ext3, ext4, jfs, ReiserFS, XFS, and Btrfs file systems.

- **ext2, ext3, ext4**: Extended File system (ext) was developed by MINIX, and all of these are the successive versions of it. The ext2, which is the second extended version, was an improvement. Improvement in performance was added by ext3 and ext4 along with performance improvement provided additional features too (see Section 11.11).
- **JFS**: IBM for AIX Unix developed the Journaled File System (JFS) in 1990, and this was used for its AIX flavor of Unix as an alternative option of ext file system. Whenever stability is required with the use of minimal resources, JFS is used. Outdoor of the IBM Linux offerings, the JFS file system is not generally used, but it is used as an alternative to ext4 system. But you may use it for your Linux system, and CPU power limitation is very well tackled by the JFS.

- **ReiserFS**: Hans Reiser developed the first journaling file system for Linux in 2001, known as ReiserFS. The ReiserFS file system supports only writeback journaling mode, means writing only the inode table data to the journal file. Due to this feature, the ReiserFS file system is one of the faster Linux journaling file systems. Some improved performance and advanced features were also introduced as an alternative to ext3. Support for the dynamic system extension was the highlight of this file system but it went out of business as it lacked certain performance aspects.

- **XFS**: It is a faster version of JFS which aimed at parallel I/O processing. This file system is still being used by NASA on their 300+ terabyte storage server. Silicon Graphics Incorporated (SGI) initially developed XFS in 1994 for its commercial IRIX Unix system. But in 2001, it was offered to the Linux environment for common use. The XFS file system is used as the default file system in mainstream Linux distributions like RHEL. The XFS file system uses the writeback mode of journaling and allows online resizing of the file system, like the Reiser4 file system. But XFS file systems can only be expanded and not contracted.

- **Btrfs**: Oracle started development on Btrfs fie system in 2007, also known as B-Tree File System. Some of the features of the Btrfs file system include fun administration, fault tolerance, large storage configuration, repair system, reliability, and the ability to dynamically resize a mounted file system. The Btrfs is recognized as default file system for openSUSE Linux distribution. It is also offered in other Linux distributions, like RHEL but not as the default file system.

Among the above-mentioned Linux file system, Ext4 is the favored and most commonly used file system in Linux Distributions. But in some situation, XFS and ReiserFS are used. Further, many File formats are not presented under Linux system but are offered by other OSs like FTA32, NTFS by Microsoft, and HFS by Apple/Mac os. Many of them may be used by Linux system by using some certain procedures like mounting technique (see Section 11.11 for more details).

11.6 Mounting File Systems

The process of accessing a file system in Linux is known as mounting because Linux preserves a single file system tree with devices attached at numerous places (in file system tree). The file system can be mounted to any directory, and by entering the directory, one can access the contents of it. When we talk about the place at which mounting is occurred, that place is referred as mount point. The *mount* and *umount* commands are used for mounting and unmounting file systems [3]. This is usually an empty directory. Mount command is present in almost each Linux operating system, which can be used to mount file system to any directory. The syntax of mount command is given below:

```
# mount -t <filesystem _type> <device_name> <directory_path>
```

The superuser may execute the mount command. This command attaches the file system of the mentioned device name into the main Linux file system hierarchy at the specified directory location. You provide the absolute path of that directory location, referred as mount point. Linux maintains a single file system tree after the devices attached to main file system.

To mount an already added disk /dev/sdb on the system to /data directory, the following mount command is executed:

```
# mount /dev/sdb /data
```

File system on the disk is automatically detected by the mount command, but sometimes it asks to specify the file system type along with the command, as given in example:

```
# mount -t ext4 /dev/sdb /data
```

To list the currently mounted file systems, you can use the *mount* command with no arguments, for example,

```
$ mount
/dev/sda4 on / type ext4 (rw)
proc on /proc type proc (rw)
sysfs on /sys type sysfs (rw)
/dev/sda8 on /home type ext3 (rw,acl,user_xattr)
/dev/sda1 on /windows/C type vfat(rw,noexec,nosuid,nodev)
```

Basically, the mount offers four types of information as follows:

- The device filename of the media,
- The mount point in the main tree directory where the device is mounted,
- The type of file system,
- The access permission of mounted device.

The type of file system of device refers file system category under which the disk is formatted. Linux

Supports and recognizes numerous file system types. Let's suppose, if you are connecting removable media devices with your Windows operating system installed computers, then you are mostly use the following types:

- **vfat:** Windows long file system
- **ntfs:** Windows advanced file system (Used in Windows XP, Vista, Window7)
- **ReFS:** Microsoft Windows 8 Server file system
- **iso9660:** The standard CD-ROM file system
- **UDF:** The standard DVD file system

If you want to access the data from USB media device, then you have to mount the USB device (/dev/sdb1) at location /mnt/USB, by using the following command:

```
#mount -t vfat /dev/sdb1 /mnt/USB
```

After attaching the USB media device in the /mnt/USB directory of main tree, then the superuser has full right to access the device data. Table 11.1 listed some key option of mount command.

TABLE 11.1

Mount Command Options

Option	Description
-r	Mounts the device in read-only mode
-a	Mounts all file systems mentioned in the */etc/fstab* file
-F	Mounts all file systems at the same time when used with the -a option
-v	Verbose mode
-n	Mounts the device without registering it in the */etc/mstab* mounted device
-l	Adds the file system labels repeatedly for ext2, ext3, or XFS file systems file
-s	Overlooks mount options which are not supported by the file system
-t fstype	Mention the type of file system being mounted. Linux supported many other file systems other than the ext2/ext3/ext4/Btrfs standard. For example, FAT, VFAT, NTFS, ReiserFS, UDF, iso9660, and so on.
-p *num*	For encoded mounting, reads *num* form the file descriptor
-U *uuid*	Mounts the device with the mentioned *uuid*
-L label	Mounts the device with the mentioned *label*
-w	Mounts the device as read-write by default.

- **Unmount filesystem**

  Any mounted file system on the system can be unmounted using the *umount* command. To unmount the currently mounted disk, run the *umount* command with disk name or mount point. The syntax for the *umount* command is very simple as follows:

```
#umount [directory_path | device_path ]
```

The *umount* command offers you the option of mentioning the media device by either its device location path or its mounted directory absolute path, as following example:

```
# umount /dev/sdb1          //Unmount the device
# umount /mnt/data          //Unmount the directory mount point

# umount /mnt/data
umount: /mnt/data: device is busy
```

// means you are in the mount point directory, so you move out form the path, either move to home directory or any other directory//

```
# cd                        // return to home directory
# umount /mnt/data
# ls -l /mnt/data
total 0
#
```

11.6.1 Automounting Devices

Some time, you required to mount disk or disk partition during system boot. For this, system uses */etc/fstab file* which comprise a list of all partitions available for automatic mounting to the system during booting. Disks are mounted using */etc/fstab* file, are referred as automounting devices. The file */etc/fstab* maintained manually by the system administrator and needs to be edited and new entry to be made to mount the partition automatically during booting.

During the boot process, this list is read, and the items mentioned in it are automatically mounted with the options mentioned in that file. The format of entries in the */etc/fstab* file is as follows:

```
<mount_device_name> <mount_directory_point> <filesystem type> <mount
flags(option)> <filesystem_backup_freq> < Filesystem_check_num     >
```

In the above syntax, each line comprises six fields that are describe in Table 11.2

The below mentioned line needs to be appended at the end of file while editing /etc/fstab. Change /dev/sdb with intended disk name.

```
/dev/sdb /mnt/data  ext4   defaults    0 0
```

Now to immediately mount all disks defined in /etc/fstab file, run the mount -a command

```
# mount -a
```

TABLE 11.2

Description of Various Fields of Each Line of */etc/fstab* file

Field	Field Names	Descriptions
1	mount_device_name	This field represent the actual name of a device file related with the physical device, like as /dev/hda1 (the first partition of the master device on the first IDE device). Similarly, to mount the USB device then its device name is /dev/sdb1. You can use *fdisk-l* command to see the list of currently attached device name.
2	mount_directory_point	The directory where the device is to mount or attached to the main file system tree.
3	filesystem type	Linux supports many file system types to be mounted. Most popular Linux file systems are ext3 and ext4, but numerous other file systems are supported, like FAT16 (msdos), FAT32 (vfat), NTFS (ntfs), CD-ROM (iso9660), and UDF.
4	mount flags(option)	File systems can be mounted with various mount options like ro (read only), rw (read write), and default.
5	filesystem_backup_freq	A single number that used to control the operation of file system backups by *dump* command. This field is used only in the /etc/fstab file, and this field is set to be 0 for /proc/mounts and /etc/mtab,
6	filesystem_check_num	A single number that used to control the sequence in which *fsck* command checks file systems at system boot time.

11.6.2 Creating File System

Writing information to the device and creating order of the empty spaces is what we call as the creating a file system. A small share of the space is taken by the file system–related data. The remaining chunk of space on the disk drive is divided into small and equally sized segments called the blocks. The *mkfs* is the command that one uses to build a Linux file system on a device or the hard disk partition. In other words, before you can start storing the data on the partition, you must format or install a particular file system on that partition, so that Linux can use it for storing the data.

Given below is the syntax to write the *mkfs* command-

```
# mkfs [options] device
```

Different file system builder utilities, which include mkfs.ext2 and mkfs.ext4, are provided with a front end using the mkfs command. We can directly execute these utilities from the command line. To specify the type of file system to be built, -t fstype option is included with the mkfs wrapper. The default file system type, ext2, is made if nothing is clearly specified. The Linux system does not install all file system utilities by default, so the below command will show the available file systems builders (mkfs* commands) in Linux system:

```
$ls /sbin/mkfs*
```

Each file system type uses its individual command line program to format partitions, and Table 11.3 lists some command line program for the different file systems.

Each file system command has many command line options that permit you to modify the various parameters which define the way of file system creation in the partition. These parameters include the block size, fragment size, blocks per group, journal options, number of inodes, and other available parameters. Use *man* command to see all the command line options available with respective file system.

The default mentioned in the */etc/mke2fs.conf* configuration file is used, when none of the option is included.

TABLE 11.3

List of Various Command Line Programs to Create File systems

Command Line Program	Description
mkefs	Creates an ext file system
mke2fs	Creates an ext2 file system
mkfs.ext3	Creates an ext3 file system
mkfs.ext4	Creates an ext4 file system
mkdosfs	Creates an MS-DOS file system
mkfs.fat	Creates a FAT16, FAT 32 file system
mkntfs	Creates an NTFS file system
mkfs.xfs	Creates an XFS file system
jfs_mkfs	Creates a JFS file system
mkreiserfs	Creates a ReiserFS file system
mkfs.zfs	Creates a ZFS file system
mkfs.btrfs	Creates a Btrfs file system

You can use the **fdisk**- l command to see the list of all devices identified by the system after recognizing the devices or partitions. Let's, if you want to create Linux ext4 file system on */dev/sdb1* device, then you must use the following commands:

```
# mkfs.ext3 /dev/sdb1
< program will display lots of information during
creating ext4 file system>
............
............
#
```

This would format the device at /dev/sdb1 with an ext4 file system. After you create the file system, then subsequent step is to mount it on the main file system tree at any directory, i.e., mount point (preferably, mount in /mnt directory, but you can mount at any point on the main file system tree). So that you can access and store data in the new file system. The following steps are required to perform the mounting process:

```
Step 1: Move from current working directory to /mnt directory
# cd /mnt

Step 2: Create a subdirectory in /mnt directory
# mkdir /mnt/my_partition

Step 3: Mount the device /dev/sdb1 into /mnt/my_partition directory
#mount -t ext4 /dev/sdb1 /mnt/my_partition
#
```

the mount command adds the new device (*/dev/sdb1*) to the mount point */mnt/my_partition*. The -t option on the mount command indicates file system type, ext4, which you are mounting. Now you can save and access the files and folders of new device (*/dev/sdb1*) through */mnt/my_partition* directory.

The fsck (file system consistency check) command is used to check and repair a damaged file system.

11.6.3 Checking and Repairing File systems: fsck

In Linux, file systems are responsible for organization of storing and recovering of the data. With time, there might arise some issues with the file system, it might get corrupted, or there is possibility some parts become inaccessible. Therefore, the integrity of the file system needs to be checked in case of such inconsistencies. The fsck (file system consistency check) command is a utility which is used to check and repair a damaged file system. This can either be done manually or automatically during the boot time.

As mentioned earlier that */etc/fstab* file have entries of all partitions details for automatic mounting to the system during booting. In the format of each entry in */etc/fstab* file, you saw some numeric digits at the end of each line. At each time when the system boots, *fsck* command regularly checks the correctness of the file systems before mounting them. The last number in each *fstab* entry mentions the sequence in which the devices are to be checked. There are various different conditions when one might want to run this fsck command, given below are some of it:

- Failure in the booting of system,
- Encounter of input or output error mainly when file system becomes corrupt,
- Unexpected working of attached drives (including flash drives/ SD cards).

- **Available options for fsck**
  Superuser privileges or root is required to run the fsck command. This command can be used with various arguments which depends on the usage of specific case. Some of the most widely used and important options are given as follows:
  - **-A**: Used for checking all file systems. The list is taken from /etc/fstab.
  - **-C**: Show progress bar.
  - **-l**: Locks the device to guarantee no other program will try to use the partition during the check.
  - **-M**: Do not check mounted file systems.
  - **-N**: Only show what would be done—no actual changes are made.
  - **-P**: If you want to check file systems in parallel, including root.
  - **-R**: Do not check root file system. This is useful only with "-A".
  - **-r**: Provide statistics for each device that is being checked.
  - **-T**: Does not show the title.
  - **-t**: Exclusively specify the file system types to be checked. Types can be comma separated list.
  - **-V**: Provide description what is being done.

  It is imperative to check that the partition on which fsck will be run is not mounted beforehand. If the partition is mounted, use the below mentioned command to unmount it; otherwise, fsck command will not work:

```
# umount /dev/sdb
```

After which *fsck* can be safely ran with

```
# fsck /dev/sdb
```

- **Understanding fsck exit codes**
  This command will return an exit code after the execution, which can be seen in fsck's manual by running the following command:

```
# man fsck

    0    No errors
    1    Filesystem errors corrected
    2    System should be rebooted
    4    Filesystem errors left uncorrected
    8    Operational Error
    16   Usage or Syntax error
    32   Checking canceled by user request
    128  Shared-library error
```

- **Repair Linux file system errors**

    It may happen sometimes that there is more than one error on the file system. In such cases, it is usually required that *fsck* automatically attempts the error correction which can be done with the following commands:

    ```
    # fsck -y /dev/sdb
    ```

    The -y flag is used for automatic yes to any prompt by the *fsck* command to correct multiple errors. In the similar fashion, the below mentioned command can be executed on all file systems without root.

    ```
    $ fsck -AR -y
    ```

    On Unix-like file systems, recovered portions of files are positioned in the *lost+found* directory which is situated in the root of each individual file system.

11.7 Errors: strerror, perror

Errors are very common whenever we are dealing with file systems. Errors can occur for numerous causes, like nonavailability of resource, unacceptable call arguments, I/O operational failures, and many other reasons. If such error happened, then a system call returns -1 and kernel sets the static (global) variable, represented as *errno*, which is a positive integer number. Further, this positive number, denoted by a symbolic constant, is linked with an error message, which is to be printed in a form that makes sense to the user.

The perror ()and strerror() library functions are used to print an error message based on the errno value [4]. In total, what exactly perror() does is that it takes errno value and prints it in a nicer form. The function strerror is almost same as that of perror() but is unique in its term that it returns a pointer to the error message string for a given value. Including the <errno.h> header file provides a declaration of errno, as well as a set of constants for the various error numbers. Hence, there are two main things, and you can take care when an error occurs:

- Use perror() library function to print the error message connected with errno.
- By checking errno, you may know the reason of the error.

The strerror()function returns the error string corresponding to the error number, as given in its *errnum* argument in below mentioned system function:

```
#include <string.h>
char *strerror(int errnum);
```

The perror()function is used to print the string pointed to by its *msg* argument, along with a message corresponding to the current value of errno. The function follows below syntax:

```
void perror(const char *msg);
```

TABLE 11.4

Linux System Error Code Table

Error Number	Symbolic Constant (Error Code)	Error Description
1	EPERM	Operation not permitted
2	ENOENT	No such file or directory
5	EIO	I/O error
6	ENXIO	No such device or address
8	ENOEXEC	Exec format error
12	ENOMEM	Out of memory
13	EACCES	Permission denied
16	EBUSY	Device or resource busy
19	ENODEV	No such device
20	ENOTDIR	Not a directory
22	EINVAL	Invalid argument
24	EMFILE	Too many open files
26	ETXTBSY	Text file busy
27	EFBIG	File too large
28	ENOSPC	No space left on device
39	ENOTEMPTY	Directory not empty
100	ENETDOWN	Network is down
115	EINPROGRESS	Operation now in progress

For example, EIO has the value 5, which indicates I/O error. In Linux operation system, the Table 11.4 shows the list of some error numbers along with its symbolic constant and its descriptions.

11.8　The /proc File System

There is a virtual file system (VFS) which is created on fly at the time of system boot, and immediately gets dissolved at the time of system shut down. This virtual file system is known as */proc* file system (procfs) [4]. The /proc file system is considered as the control and information hub for kernel as it consists of useful information related to the processes currently in execution. Like other file system, the */proc* directory is the mount point for the *proc* file system. The /etc/fstab file has a distinct entry for /proc with a file system type of proc and no device mentioned as follows:

```
none /proc proc defaults 0 0
```

The Linux kernel is core part of system for various user and system operation, and it is very important that there be a technique for exchanging information with the kernel. Generally, this is done through some special functions or system calls. Hence, the *proc* file system was created for better communication between users and the kernel. The entire file system is particularly thought-provoking because it does not truly exist on disk anywhere. It is completely a concept of kernel information.

All files in the /proc directory correspond either to a kernel function or kernel variables set in the kernel, respectively. For example, to see the details of memory status information of a system, then we must see *meminfo* file under the /proc directory, i.e., */proc/meminfo*, as follows:

```
#cat /proc/meminfo
```

The kernel will dynamically create the report, showing memory information, and give to cat command to display on screen. Table 11.5 lists some subdirectories and files under /proc directory that provide some useful information about current hardware and process to manage Linux system [5].

- **Obtaining information about a process: /proc/PID**

    The /proc file system acts as a medium of communication within the kernel space and user. The /proc file system discloses a range information regarding each executing program (or processes). Each /proc/PID subdirectory comprises files and subdirectories that give information about the process whose ID matches PID. For each PID of a process, there exists a dedicated directory, and this could be seen when one lists the directories. Below given command can be used to check the directories:

```
#ls -l /proc | grep '^d'
```

The PID of any process in execution can be obtained using *ps* command.

```
#ps -aux
```

TABLE 11.5

List of Some /proc Subdirectories and Files

File Name	Description
/proc/*num*	Every process has a directory known by its number. For example, /proc/1 is the directory for process 1.
/proc/cpuinfo	Information about the CPU like its make, model, type, performance, and others of system.
/proc/dma	Information about currently used DMA channels.
/proc/diskstats	Information about each logical disk devices including its device number.
/proc/filesystems	Lists of the file systems supported and configured into the kernel.
/proc/devices	Information about all the currently configured device drivers in running kernel.
/proc/ modules	Lists the kernel modules currently loaded.
/proc/meminfo	Display the summary memory usage.
/proc/scsi	Information about all the connected SCSI devices.
/proc/ioports	Shows the list of I/O ports currently in use.
/proc/tty	Displays the information about the current running terminals.
/proc/version	Displays the information about the Linux kernel version, distribution number, the system on which it was compiled along with date and time of compilation, and any other pertinent information relating to the version of the currently running kernel.
/proc/swaps	Details status information of swap partitions, volume, and/or files
/proc/interrupts	Displays the information about currently in-use interrupts.
/proc/net	Lists status information about network protocols.
/proc/uptime	Displays system uptime.

TABLE 11.6

List of Directory Under /proc/PID

Directory	Description
/proc/PID/cmdline	Command line arguments.
/proc/PID/cpu	Current and last CPU in which it was executed.
/proc/PID/cwd	Link to the current working directory.
/proc/PID/environ	Values of environment variables.
/proc/PID/exe	Link to the executable of this process.
/proc/PID/fd	Directory, which contains all file descriptors.
/proc/PID/maps	Memory maps to executables and library files.
/proc/PID/mem	Memory held by this process.
/proc/PID/root	Link to the root directory of this process.
/proc/PID/stat	Process status.
/proc/PID/statm	Process memory status information.
/proc/PID/status	Process status in human readable form.

The /proc/PID directories are volatile in nature, because such directories come into existence when a process is created along with its process ID and vanishes when that process terminates from the system. It means, for each executing process, a subdirectory is included in /proc directory including the kernel processes. In directories named /proc/PID, following are the subdirectories present as mentioned in Table 11.6.

11.9 The Linux File System: ext, ext2, ext3, ext4, Journaling

A *file system* is the procedures and data structures that an operating system uses to store, organize, and access files and/or directories on a disk or partition. The original file system, known as **extended file system, or ext,** was introduced in 1992 as the first file system for Linux operating system, i.e., file system for Linux kernel. The extended file system (ext) has metadata structure inherited by the Unix File System (UFS) to overcome certain restrictions of the MINIX file system. MINIX was a very small Unix-like operating system for IBM PC microcomputers that was developed by Andrew Tannenbaum for teaching purposes and released its source code in 1987.

The **ext** was the first in the series of **extended file systems**, but other upgraded versions in the extended file system series are given as follows:

- **ext2, the second extended file system.**
- **ext3, the third extended file system.**
- **ext4, the fourth extended file system.**

The majority of recent Linux distributions use *ext4* file system as default Linux file system. But in earlier Linux distributions, they used *ext3 or ext2* as default file system, and if you go back in early 1990s, they used *ext* as default file system. The following section briefed the key features of Linux file systems, namely ext2, ext3, and ext4.

- **ext2: the second extended file system**
  - Ext2 means the second extended file system for Linux kernel.
  - It was developed by Remy Card in 1993.
  - Limitations with the original ext file system motivated the development of this file system.
  - There is no journaling feature available in this file system.
  - As there is no requirement of the overhead journaling, therefore this file system is recommended for the flash drives and USB drives.
  - Maximum filename length: 255 bytes (255 characters).
  - 16 GB–2 TB is the maximum individual file size.
  - 2 TB–32 TB is the overall ext2 file system size.
  - Maximum number of files: 10^18
  - Date range: December 14, 1901–January 18, 2038.
- **ext3: the third extended file system**
  - Ext3 means the third extended file system.
  - It was developed by Stephen Tweedie, and he introduced this in the year 2001.
  - This file system is available from Linux Kernel 2.4.15.
  - It supports journaling file system feature.
  - There is a dedicated area available in the file system for the journaling, which is used for tracking the changes. Journaling ensures that there is less chance of system corruption in case of the system crashes.
  - Maximum filename length: 255 bytes (255 characters).
  - 16 GB–2 TB is the maximum individual file size.
  - 2 TB–32 TB is the overall ext3 file system.
  - Ext2 file system can be converted to the ext3 file system without extra backup or restore.
  - Date range: December 14, 1901–January 18, 2038.
  - Maximum number of files is variable.
  - Following are the three types of journaling available in the ext3 file system:
    - Journal—It is used to store the metadata and content.
    - Ordered—Journal only saves the metadata. It is only after writing the content to disk that metadata are journaled, which is the default.
    - Writeback—Metadata can either be journaled before or after the content written to the disk.
- **ext4: the fourth extended file system**
  - Ext4 is the fourth extended file system.
  - It was introduced in 2008.
  - Linux kernel 2.6.19 and later supports the ext4.
  - It supports journaling file system feature.
  - Main advantage is that it supports large individual file size as well as overall file system size.

- The limit of individual file size ranges anywhere between 16 GB and 16 TB.
- 1 EB is the maximum ext4 file system size. EB stands for exabyte, 1 EB = 1024 PB (petabyte) and 1 PB = 1024 TB (Terabyte).
- In contrast to the ext3 limit of 32,000, it can store a maximum of 64,000 subdirectories.
- Date range: December 14, 1901–April 25, 2514.
- Without having the need to upgrade the ext3, fs can be mounted to ext4.
- Maximum number of files is four billion.
- Various other features that were introduced in ext4 are allocation of multiblock, allocation with delay, journal checksum, fast fsck, etc. These features have improved the overall performance and reliability of the fs when it is compared to ext3.

- **Journaling file system**
  - Journaling file systems offer numerous advantages over static file systems and provide more safety to the Linux system. The ext3, ext4, and ReiserFS file systems introduced journaling features to Linux kernel. It offers a new level of data safety to the Linux system. In journaling, instead of writing data straight to the storage device and update the index table (inode), journaling file systems first write data (file changing data) into a temporary file, referred as *journal file*. After data are successfully written to the storage device and subsequently to the inode table, the journal entry is removed or deleted.
  - This feature is very useful in the case of system crashed or power failure before the actual data are written to the storage device. In such case, the journaling file system reads left over data from journal file and writes back to the storage device. Normally, Linux uses three different methods of journaling with its own level of protection, which are described here:
    - **Data or journal mode**: It is referred as lowest risk mode in which both data and inode (metadata) are journaled. In this mode, data losing risk is lower but lacking in performance.
    - **Ordered mode**: In most Linux distributions, it is the default mode in which only inode (metadata) is written to the journal, but not removed until file data are successfully written. In case of system crash, this mode ensures that the inode (metadata) associated with incomplete writes is still in the journal, and the file system can sanitize those unfinished writes while rolling back the journal. It offers good balance between performance and safety.
    - **Writeback mode**: It is referred as least safe mode in which only inode (metadata) is written to the journal, but no control over when the file data are written. It gives higher risk of losing data, but still it is a good option as compared to not using journaling.

11.10 Logical Volume Management (LVM)

There is a storage device management technology known as the LVM (Logical Volume Management) which enables the users with the power to pool and abstract physical layout of storage component, which subsequently leads to hassle-free and flexible administration. Existing storage devices can be clubbed into groups and as per requirement, the allocation, the of the logical units is performed from this combined space. This is possible because of the present iteration, i.e., LVM2 which utilizes the device mapper Linux kernel framework.

Some of the major advantages of using the LVM include the increased level of abstraction, control, and flexibility. There are meaningful names given to the logical volumes like the "databases" or "root-backup".

- **LVM storage management structures**
    - The principle of working of the LVM is layering abstraction on top of physical storage devices. Given below are the basic layers that LVM uses:

- **Physical Volumes:**
    - LVM utility prefix: pv…
    - LVM uses physical block devices or other disk-like devices as raw building material for greater levels of abstraction.
- **Volume Groups:**
    - LVM utility prefix: vg…
    - Volume groups are storage pools which are created by combination of physical volumes by the LVM.
- **Logical Volumes:**
    - LVM utility prefix: lv… (generic LVM utilities might begin with lvm…)
    - Any number of logical volumes can be sliced from a volume group. These are same as that of partitions on a physical disk but with greater level of flexibility. These are the basic part that users and applications interact with.

The storage space available on a system is unified by combination of physical volumes in the form of volume groups.

11.11 Virtual File System (VFS)

In Linux, every available file system differs in terms of its implementation and operations such key differences include allocation method of blocks to a file, way to organized directories, size of blocks to store the data etc. Therefore, if a program worked with files, must understand and aware the specific details of each file system. So, creating and writing such programs that worked with different file systems features would be very difficult and tough task. The virtual file system (VFS, sometimes also known as the virtual file switch) is the software layer in the kernel that resolves this problem by providing an interface layer for file system operations to the user program. This interface layer within the kernel allows different file system implementations to coexist [2].

In another word, VFS is an interface within the kernel and the file system, which sits above file system and is kind of abstraction layer. Different file systems can be accessed by client through this VFS. It is a kind of a manageable container which gives the functionality of file system. The file system registers itself with the VFS at the time of its initialization. This takes place when the operating system initializes itself at start-up. The

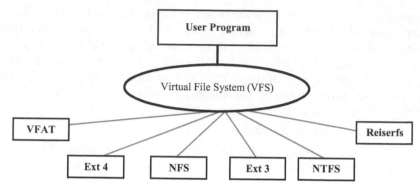

FIGURE 11.2
Virtual file system layout.

real file systems are developed as loadable modules or otherwise built directly into the kernels. Cache of directory lookups is also maintained by VFS in the form of records, and main motive of this is easy and fast retrieval of inodes of frequently accessed directories. Without client application, i.e., the original file system, VFS is capable of transparent access of local as well as network devices. Another important feature of VFS is that it can bridge the disparity gap between different operating systems like windows, Mac OS, or the Linux file systems as shown in Figure 11.2.

The key ideas of VFS are mentioned as follows:

- The VFS provide a common interface for file system operations. All programs that work with files mention their operations with respect to this common interface.
- Each file system offers an implementation for the VFS interface.

The VFS interface allow operations matching to all of the traditional system calls for working with file systems and directories like open(), read(), write(),truncate(), lseek(), close(), stat(), mmap(), mkdir(), mount(), umount(), link(), unlink(), symlink(), and rename().

11.12 File System Conversion

The Linux file systems Ext2 and Ext3 have been very old and almost outdated. Therefore, it is necessary to convert the old file systems to the latest one, Ext4 file system. The Ext4 file system is more reliable and faster with respect to the previous versions of file system (Ext2 and Ext3). If you are worried that you must reinstall the system for such conversion, then you do not worry. You can simply convert your existing file system to Ext4 by keeping your present data and application in Linux system, as mentioned below. The subsequent section summarizes the steps to conversion of Ext2, Ext3, and Ext4 file systems.

Once you create your partition with the *fdisk* command, you can create a file system on it by using *mkfs* command. The *mkfs* command takes the name of the hard disk partition as a parameter (full pathname) and builds the Linux file system on it. The *mkfs* command calls *mke2fs* utility when it is called for creating an ext2, or ext3, or ext4 file system. For example, install ext3 file system on the partition (device name or full path) */dev/hdc1* and then use the following command:

```
# mkfs -t ext3 /dev/hdc1
```

Here, we use the mke2fs command to create ext2, or ext3, or ext4 file system as follows:

- Ext2 file system can be created with the below mentioned command:

```
#mke2fs /dev/sda1   // create ext2 file system
```

- Ext3 file system can be created with the below mentioned command:

```
#mkfs.ext3 /dev/sda1
Or
#mke2fs -j /dev/sda1
```

- Ext4 file system can be created with the below mentioned command:

```
#mkfs.ext4 /dev/sda1
Or
#mke2fs -t ext4 /dev/sda1
```

11.12.1 Converting ext2 to ext3

You can convert or upgrade the Linux file system in a very simple procedure. Let us assume /dev/sda2, presently have *ext2* file system and mounted on */home* directory. Then, its entries in *fstb* table are as follows:

```
/dev/sda2   /home   ext2   defaults   0 0
```

Then, you can use the following commands and procedures to convert/upgrade ext2 to ext3 file system:

```
#umount /dev/sda2        //Unmount the partition
#tune2fs -j /dev/sda2    //converting to ext3
#mount /dev/sda2 /home   //mount the partition on /home
```

The mounting and unmounting step is optional as ext2 to ext3 fs conversion supports the live file system conversion also.

11.12.2 Converting ext2 to ext4

In the similar fashion, you can use the following commands and procedures to convert/upgrade ext2 to ext4 file system:

```
#umount /dev/sda2                   //Unmount the partition

#tune2fs -O dir_index,has_journal,uninit_bg /dev/sda2
                        // converting to ext4 from ext2
```

After running tune2fs command, you must execute *fsck* to complete file system check and repair that tune2fs has modified as follows:

```
#e2fsck -pf /dev/sda2

#mount /dev/sda2 /home    //mount the partition on /home
```

11.12.3 Converting ext3 to ext4

In the similar fashion, you can use the following commands and procedures to convert/upgrade ext3 to ext4 file system:

```
#umount /dev/sda2                      //Unmount the partition

#tune2fs -O extents,uninit_bg,dir_index /dev/sda2
                                // converting to ext4
```

After running tune2fs command, you must execute *fsck* to fix up some on-disk structures that tune2fs has modified as follows:

```
#e2fsck -pf /dev/sda2

#mount /dev/sda2 /home   //mount the partition on /home

# tune2fs -O dir_index,has_journal,uninit_bg
```

Here,

```
-p option : Automatically repairs the file system.
-f option:  Forcefully checking file system even it appears clean.
```

It is being noted here that the above commands should be executed on a test system where one can afford to lose the data, if error happened.

11.13 Summary

A file system is a planned arrangement of regular files and directories. There are several types of files, but 3 being the basic primary types are ordinary, directory, and device files. Each device in the system has its own device file. Devices may be divided into two parts: character devices and block devices. A disk driven is a physical device for storage of data that comprises one or more disks. Partitions are the small separate sections in which large storage devices are divided. Certain commands are used while dealing with large disk devices which are GPT format, large disks, more partitions, and reliability. There are three basic kinds of partition: root partition, data partition, and swap partition. The file system in Linux is divided into 4 blocks: Boot Block, Super Block, Inode Block, and Data block. Linux distribution provides the option of partitioning disks in the following formats: ext2, ext3, ext4, jfs, ReiserFS, Xfs, and Btrfs. One can access a file system in Linux by mounting.

It preserves a single file system tree with devices attached at numerous places. The mount offers four different types of information: the device filename, mount point, type of file system, and access permission. Any mounted file system can be unmounted using the unmount command. One can even mount disk or disk partitions during the system boot using etc/fstab files. There files are referred as automounting devices. One can write information and create order of empty spaces in a file system. One can check and repair the file system using fsck command. There are various conditions to run this command which are as follows: failure in booting, encounter of input or output, and unexpected working of attached drives. However, superuser privilege is required to run this command. This command returns and exits the code after the execution which can be seen in fsck's manual. A virtual file system is created on the fly at the time of Linux system boot which immediately gets dissolved when the system shuts down. This virtual file is known as the proc file system(*procfs*) and is considered as the control and information hub for the kernel.

Journaling file system offers various advantages over static file system and provides more safety to the system. There is a storage device management tool known as LVM which enables the users with the power to pool and abstract physical layout of component storage which subsequently leads to hassle-free and flexible storage management. The concept of VFS allows kernel to keep various file system together, and further, without reinstallation of kernel, you can update the existing file system with new ones.

11.14 Review Exercise

1. Discuss, how a file system is important for an operating system, and highlight the key difference between ext3 and ext4 file system.
2. Why swap partition is required and how you can increase the size of it?
3. Briefly explain the file system layout in terms of blocks.
4. What is called for the unique entry for each file or directory in the file system? Explain briefly.
5. Write the procedure which you follow to extend an LVM partition.
6. How a user can view all the mounted partitions on the system?
7. Write steps:
    a. to view all disk partitions in Linux
    b. to print all partition tables in Linux
    c. to create a new partition in Linux
    d. to find out how much memory Linux is using
    e. to add a disk to a volume group
8. Discuss the backup of new LVM data structures.
9. What do the last two sections define in fstab file?
10. What are the steps to perform in order to increase the logical volume on the fly?
11. Discuss mount and unmount mechanism of a device.
12. How to migrate Ext2/Ext3 file systems to Ext4 on Linux?

13. How to manage and create LVM using vgcreate, lvcreate, and lvextend commands?

14. How the mounting, unmounting, and automounting of devices done? Write suitable commands.

15. Write a command that will look for files with an extension "c", and the occurrence of the string "Linux YOURSELF" in it.

16. Write the command to print the no. of errors in a file system.

17. Write down the key difference between perror() and strerror() functions.

18. Name the virtual file system which is created at the time of booting of Linux system. Explain /proc/PID in brief.

19. Write down the procedure to check and repair a file system.

20. Write down the various ways to kill a running process in Linux system.

References

1. Yogesh Babar and Technical Reviewer (Chris Negus). 2017. *Partitioning Disks with Parted*. Red Hat, Inc.
2. William Stallings.2005. *The Linux Operating System*. Prentice Hall.
3. Mark G. Sobell. 2012. *A Practical Guide to Fedora and Red Hat Enterprise Linux*. Prentice Hall.
4. J. Purcell. 1997. *Linux Complete: Command Reference*. Red Hat Software, Inc.
5. Wale Soyinka. 2012. *Linux Administration: A Beginner's Guide*, The McGraw-Hill.

12

Linux System Programming

In this chapter, we look at Linux programming procedure along with the functional description of a process, thread, and system call. The vi programming environment and the shell scripting concept are already explained in detail in Chapters 3 and 8, respectively. Now, this chapter presents additional programming features with compilation procedure of C and C++ source codes by using gcc/g++ compilers, respectively, in vim editor environment. The first part of this chapter covers program compilation, debug, and process-related details. The next part of this chapter discusses threads, device files, signals, system calls, and other related details concerned with Linux programming.

12.1 Getting Started

In the section, you will see brief details about vim and writing code of C or C++ programming languages with some standard common C library. The source code examples in this section are in C programming language. We assume that you are familiar with syntax and semantics of C programming language.

12.1.1 Editing with vim

An editor is a program that you can use to write and edit programming language source code and other texts. Many different editors are available for Linux operating system, but the most popular one with full-featured editor is certainly GNU vim. vim stands for "vi" improved that is the advanced version of vi editors, or also referred to as vi clones (see Chapter 4). It was written and is maintained by Bram Moolenaar. The latest stable version of vim is 8.1. For more details, you can visit vim.org Web page.

vim is a highly configurable text editor that incorporates numerous latest features that enable the editing of code in many new languages along with creating and changing any kind of text. For example, its context-specific language editing is started with C programming language and has grown up to incorporate C++, Java, C#, and many other programming languages.

vim provides and adds some essential features over the years in vi editor. Such features are considered to be essential and available in current text editors, such as ease of use, syntax highlighting and formatting, graphical interface, and customization. Presently, vim is widely in use and popular, particularly among Unix and its variants, like BSD and GNU/Linux variants. But, many distributions of GNU/Linux came with a default installation of vim editor. If the Linux system does not have vim editor, then you can simply download the vim source code, compile it, and install it. vim source code is available and it downloads from the vim home page.[*]

---

[*] Vim home page: http://www.vim.org.

You can start vim by typing vim in your terminal window with or without filename as given below:

```
# vim          //Start vim editor
```

OR

```
# vim file_name     //Start vim editor and open the given input file
```

You can see Chapter 4 for more details regarding vi editor operations. vim editor comprises all vi command line options along with some more command line options; some key additional options are listed in Table 12.1.

- **Opening a C or C++ source file**

  After starting vim editor, you can create a new source file, write the content, and save the content under the new source filename. If you want to create a C program source file, use a filename that ends in .c, such as *filename.c*. If you want to create a C++ source file, use a filename that ends in .cpp. With the below command, you can open c program file and, then, you can write the content as any ordinary word-processing program or text editor like notepad. When you finish using vim, you can exit from it.

  So, for example:

```
# vim Hello.c

int main ()
{
printf ("Hello, My name is Sanrit\n");
printf ("Hello, New World\n");
}
```

  After finishing the typing, you press <Esc> key, type :wq to save, and exit from vi editor as given below:

```
Press< Esc> key
:wq                         // Save the content and exit from vi editor
#
```

TABLE 12.1

Some Key Command Line Options of vim

Option	Description
-b	Start in binary mode
-C	Execute in vi compatibility mode
-g	Start GUI version of vim, i.e., gvim (GUI)
-c *command*	Command will be executed as an ex command
-m	Turn off the write option. Buffers will not be changeable
-f	Keep in the foreground in GUI mode
-n	Do not generate a swap file to avoid recovery of data
-p	Start a new tab for every given file on the command line
-R	Start in read-only mode, setting the read-only option
-o	Open all files in a separate window

12.1.2 Program Compilation with gcc/g++ Compiler

A compiler checks the syntax and semantics of a human-readable source code of a particular programming language and coverts into machine-readable object code, if it found all source codes in order. These machine codes may be ready for execution by a user/programmer. The compilers of choice on Linux systems are the GNU Compiler Collection, typically known as GCC.

GCC also includes compilers for C (gcc), C++ (g++), Objective-C, Objective-C++, Java (gcj), Fortran (gfortran), Ada (gnat), Go (gccgo), OpenMP, Cilk Plus, OpenAcc, and D, as well as libraries for these languages. GCC was originally written as the compiler for the GNU operating system including GNU/Linux. The GNU system was developed to provide 100% free software. The free software means that the users have got the right of 100% freedom on that software. At the time of writing this chapter, the current release version of GCC is 9.1, and to get more information about GCC, visit its official home page.* As mentioned in Section 4.7, this book primarily focuses on C and C++ programming language in vim editor.

- **Compiling a single source file**
  - In Linux, the compiler name of C programming language is gcc (GNU Cross-Compiler). You can use -c option to compile a C source file. For example, the below syntax at the command prompt helps to compile *hello.c* source file:

    ```
    #gcc -c hello.c
    ```

  - After compilation, the resulting object file is generated with the name of *hello.o*.
  - In the similar way, the compiler name of C++ programming language is g++. The operation of g++ compiler is very similar to gcc, for example,

```
#g++ -c Welcome.cpp
```

  - The -c option tells g++ to compile the program to an object file only. Without -c option, g++ will compile and link the program to produce an executable file. So, with -c option, the resulting object file is named *Welcome.o*.
  - You can see many more other options associated with gcc and g++. You can use the following command or visit the online available documentation. You just type the following command on shell prompt:

    ```
    #info gcc
    OR
    #man gcc
    ```

12.1.3 Automate Program Execution with GNU Make Utility

GNU Make is a utility that automates the generation of executables and other nonsource files of a program from the program's source files. Make utility uses a so-called makefile, which contains rules on how to build the executables. Make uses a file, known as makefile that comprises the rules to provide the knowledge of how to build your executable program. The *rule* in the makefile tells Make utility that how to execute a sequence of commands in order to build a *target* file from source files. When you write a program [1], you

---

* GCC home page: http://gcc.gnu.org.

must write a makefile for it, so that it is possible to use Make utility to build and install the program. For example,

Write a simple program to print **Hello New World** and save it as (hello.c) and convert into executable (hello.exe) via Make utility as follows.

```
<<< hello.c>>>

#include <stdio.h>

int main() {
  printf("Hello New World!\n");
  return 0;
}
```

Generate the following file named "makefile" (without any file extension), which comprises the rules to build the executable file and store it in the same directory where the source file is saved. Use "tab" to start the command, don't use spaces.

```
all: hello.exe

hello.exe: hello.o
        gcc -o hello.exe hello.o

hello.o: hello.c
        gcc -c hello.c

clean:
        rm hello.o hello.exe
```

Run the "Make" utility as follows:

```
#make
gcc -c hello.c
gcc -o hello.exe hello.o
```

Running Make utility without argument starts the target "all" in the makefile. A makefile comprises a set of rules. A rule consists of three parts: a target, a list of prerequisites, and a command as given below:

```
target: pre-req-1 pre-req-2 ...
        command
```

The target and prerequisites are separated by a colon (:). The command must be headed by a Tab, not by a Space.

In the above example, the rule "all" has a prerequisite "hello.exe". The Make utility could not locate the file "hello.exe" so it searches for a rule to generate it. The rule "hello.exe" has a prerequisite "hello.o". Again, it does not exist, so the Make again searches for a rule to generate it. The rule "hello.o" has a prerequisite "hello.c". The Make checks that "hello.c" exists and it is newer than the target, but the target does not exist. It runs the command "gcc -c hello.c". The rule "hello.exe" then runs its command "gcc -o hello.exe hello.o". In this way, Make utility automates the execution of hello.c program through makefile.

12.1.4 GNU Debugger (GDB)

The debugger is the program that allows you to test and debug another program or target source code to figure out why the target program is not working properly as the way it should behave. The GNU Debugger (GDB), also referred to as GDB, is the most widespread debugger used by most Linux programmers to debug C and C++ programs. With the help of GDB, you can check your code, set a breakpoint, and inspect the value of local variables. GDB can perform four key operations that help you to catch bugs in the program:

- Start your program, highlighting anything that might change its procedure.
- Write your program that stops on mentioned conditions.
- Evaluate the cause if your program has stopped working.
- Modify things in your program, so that you can test with modifying effects of one bug and move on to study about another.

GDB supports many programming languages; some are Ada, Assembly, C, C++, D, Go, Objective-C, OpenCL, and Rust. The latest release version of GDB is GDB 8.3. For more details, you can visit GNU home page of GDB [2].

12.2 File I/O

One of the most important topics in Linux is File I/O, as you knew that in Linux, every-thing is a file. The basic operation over a file is reading to or writing from files. The file I/O may be performed by the following key functions [3]:

- open()
- read()
- write()
- close()
- lseek().

A file must be opened before it can be read from or written to. The kernel keeps a per-process list of open files, known as the file table. This table is indexed with nonnegative integers, referred to as file descriptors (generally denoted as fds). System call is used to perform I/O to open files using a file descriptor. File descriptors are used to refer all types of open files, including FIFOs, pipes, sockets, devices, terminals, directories, and regular files. Every process has its own set of file descriptors. In the file table, every entry comprises the information about an open file, including memory details, inode, related metadata such as file position, and access modes. Table 12.2 lists the standard file descriptors.

In Linux, every process has a limit of maximum number of files that it can open. File descriptors start at 0 and go up to 1 less than the maximum defined value. By default, the maximum is 1,024, but it can be configured as high as required, that is 1048576 (which 1,024 *1,024). The negative numbers are not allowed as file descriptors, but -1is regularly

TABLE 12.2

List of Standard File Descriptors

File Descriptor	Description	POSIX Standard Name	*stdio* Stream
0	Standard input	STDIN_FILENO	stdin
1	Standard output	STDOUT_FILENO	stdout
2	Standard error	STDERR_FILENO	stderr

used to specify an error, if occurred during operation. The C library provides the preprocessor and defines STDIN_FILENO, STDOUT_FILENO, and STDERR_FILENO.

12.3 Processes

After the concept of files, processes are the most fundamental concepts in a Unix/Linux operating system as you knew that everything in Unix/Linux is a file. A file is preserved as a simple file when it stores on disk, but when the file is in execution mode, i.e., active, alive, running programs, it becomes a process. In another word, in passive mode, a file referred to as a regular file, but in active mode, a file referred to as a process. A running illustration of a program is called a process. The life of a process is finite; it gives birth to another process and dies also. Every executing program may be used to create many processes, or otherwise, many processes may be executing the same program.

Linux is a multitasking operating system that permits to run a huge number of processes for doing more work on a large system. Processes belong to Linux kernel, which is accountable, and manage these processes. Processes are executed by kernel and rapidly switched from one executing program to another.

When a system starts up, the kernel activates a few of its own activities as processes and starts a program, known as *init*. Further, init executes many shell script programs, located in /etc directory, referred to as init scripts, which activate and start all the required system services to boot the system. Many such system services are implemented as daemon programs that execute in the background without having any user intervention. The kernel maintains information about each executing process like process identification number (PID), parent process identification number (PPID), process state, priority, and other details. Every process has its own PID and PPID except the *init* process. The subsequent section briefs out all such details related to the processes.

12.3.1 Listing Processes, PID, PPID

In Linux, the process is simply an instance of a running program. A program can have more than one process. When a program starts execution, a process is said to be born and remained active or alive until the program is in execution.

Each process has its own name, generally the name of the program being executed, like if you execute *cat* command, then a process is to be created with a name cat. But the system identifies each process with its PID. These PIDs are 16-bit numbers and assigned with a unique number. The PIDs are assigned by kernel in ascending order, but *init* process is always allotted with PID 1. The PID is used to identify various characteristics of the process [4].

A process is always created by another process; hence, excluding the first process, every process has a parent. The processes are arranged sequentially by Linux when new processes are created. Except the init process, all other processes have a PPID. Due to this, all these processes are arranged in a treelike structure, in which init process keeps in the top as a root. So, you can say that all the processes of Linux are started by init process. Process characteristics are kept in a structured table, referred to as a process table, stored in memory, and maintained by the kernel. A process entry remains in the process table until the kernel kills the process accurately. The size of the process table is limited, that's why there are a finite number of processes that may execute on a system.

Listing processes

The ps command is used to list all the processes that are running on the system. The basic way to use the ps command is as follows:

```
# ps
PID           TTY          TIME          CMD
4046          pts/1        00:00:00      bash
4229          pts/1        00:00:00      ps
#
```

The basic ps command does not display all the important information of the running processes. By default, the ps command (without option) displays only those processes that are associated with current user and running on the current terminal. In the given above example, bash shell is running (the bash shell is also a program running on the system) ps command. The ps command itself is another process. The basic output of ps command displays the PID of each of the programs, the terminal (TTY) details from where they are executing, and the consumed CPU time by the respective process. Through the system call, you can also get the PID and PPID of a running process.

Linux systems support the GNU ps commands in three different formats of command line option parameters as follows:

- Unix-style parameters, which are headed by a dash,
- GNU long parameters, which are headed by a double dash,
- BSD-style parameters, which are not headed by a dash or double dash.

Some key ps command options are listed in Table 12.3 in Unix- and BSD-style formats.

TABLE 12.3

ps Command Options in Unix- and BSD-Style Format

Unix-style Parameters	BSD-style Parameters	Description
-l	l	Display long-listing information of processes
-e or –A	aux	All processes including user and system processes
-f	f	Full listing showing the PPID of each process
-u usr	U usr	Processes of user usronly
-a	-	Processes of all users excluding processes not associated with terminal
-o format	O format	Lists specific columns in formatto display along with the standard columns
-p pidlist	p pidlist	Shows processes with PIDs in the list pidlist
-t ttylist	t ttylist	Shows processes associated with a terminal listed in ttylist

Use -l option with ps command to get more detailed information of the process that produces long format outputs as mentioned below:

```
# ps -l
```

```
F S UID PID PPID C PRI NI ADDR SZ WCHAN TTY TIME CMD
```

```
Show the process information under above-mentioned columns that described
as follows.
```

```
F: System flags assigned to the process by the kernel
S: The state of the process (O= running on processor; S= sleeping;
R= runnable, waiting to run; Z= zombie, process terminated but parent
not available; T= process stopped)
UID: The user responsible for launching the process
PID: The process ID of the process
PPID: The PID of the parent process (if a process is started by another
process)
C: Processor utilization over the lifetime of the process
PRI: The priority of the process (higher numbers mean lower priority)
NI: The nice value, which is used for determining priorities
ADDR: The memory address of the process
SZ: Approximate amount of swap space required if the process was
    swapped out
WCHAN: Address of the kernel function where the process is sleeping
TTY: The terminal device from which the process was launched
TIME: The cumulative CPU time required to run the process
CMD: The name of the program that was started
```

```
#ps -a //Display all the processes running with all the terminals.
#ps -x //Display all the system processes, i.e., processes not
associated with any terminal.
#ps -axl // Display all the processes running on a system including
user and system processes in long format.
```

- **The Process-ID (PID)**: Each process is recognized by an exclusive integer number, referred to as PID. The PID is used to control a process during the execution. The *getpid()* system call is used to return the PID of the calling process or running program [5] as shown in the below program.

  ```
  #include <sys/types.h>
  #include <unistd.h>
  pid_t getpid (void);    // getpid( )system call returns the PID of
                     the calling process
  ```

  The *pid_t* is used as a data type of a PID return value of getpid() function that is an integer-type value. In Linux kernel 2.4 and earlier, the upper limit of PIDs was set to 32,767 by using the kernel constant variable PID_MAX, but this limit is customizable from Linux kernel 2.6 and onward kernels.

- **The Parent Process-ID (PPID)**: Every process has a parent (except init process); the process that created it and the details of PPID of the current process is also available in the process table. When several processes have the same PPID, it often makes sense to kill the parent process rather than killing all of its child processes separately. The *getppid()* system call is used to return the PPID of the calling process or running program [5] as shown in the below program.

```
#include <sys/types.h>
#include <unistd.h>
pid_t getppid (void);    // The getppid( )system call returns the
                         PID of the calling process parent.
```

The parent of any process may also be got by visiting the *ppid* field mentioned in the Linux process specific file, i.e., */proc/PID/status* .

12.3.2 Process State

In Linux, the process has a life cycle going through several states from its birth to death (kill/terminate or exit). Some processes exist in the process control table even after they are terminated, such processes are referred to as zombie processes (see Section 12.3.7). Broadly, process states are defined that what process is performing and what it is likely to do in the next slot of time [6]. Since the system runs many processes, you can see all these processes by using ps command that produces a long list. Let's look at the following command again:

```
# ps aux
```

In the output, a new column titled STAT has been added that discloses the current status of the process, where STAT is short for process state. Various process states are shown in Table 12.4.

TABLE 12.4

Process States

Process State Notation	Process State Meaning	Description
R	Running	The process is running or about to run
D	Uninterruptible sleep (waiting)	Process is waiting for input or output completion
S	Sleeping	The process is not running; usually waiting for an event to occur, such as a keystroke input
T	Stopped	A process that has been suspended or stopped
Z	Defunct or zombie process	A process is terminated, but it has not been cleaned up by its parent
<	A high-priority process	Process is running at high priority that will get more time on the CPU
N	A low-priority process, nice	A process with low priority (a nice process). It will get processor time only after higher priority processes have been completed its execution
+	Foreground process	The process is running in foreground process group
l	Multithread process	The process is multithreaded
s	Session leader process	The process is a session leader

12.3.3 Process Context

A process is an *executing* instance of a program. In Linux, multiple processes are executing in parallel and share various resources with their *parent* processes. When a process is interrupted from access to the processor, enough information about its current state and other information are stored in memory. When a process is again scheduled to run, it can resume its operation from the current state. This operational state data is referred to as its *context* and switching of a process or a task from processor to another process or task is known as a *process switch* or *context switch*. A process context also referred to as a context switch or a task switch.

A context switch is sometimes defined as the kernel suspending *execution of one process* on the CPU and resuming *execution of some other process* that had earlier been suspended. Context switches can happen only in *kernel mode* that is also known as privileged mode of the CPU. In this mode, only kernel processes execute and provide access to all memory locations as well as all other system resources. Process context is an important feature of *multitasking* operating systems. A multitasking operating system allows multiple processes to execute on a single CPU without interfering with each other and apparently parallel.

12.3.4 Creation of New Process: fork(), vfork(), execv(), system()

In Linux, it is very important to understand how a process is created and terminates or kills, as well as how a process can execute a new program. In Linux, the creation of a process is only created by another process. There are three basic steps involved in process creation, namely fork (), exec(), and wait() system calls [5,7]. This section describes various system calls used for process creation.

- **fork()**: The fork() system call is used to create a new process. The way of creating a new process is called forking. A successful call of fork( ) system call creates a new process, alike in almost all aspects of the calling process. The newly created process has a different PID and referred to as the child process, and the process that created it becomes its parent process. **In the child, a successful call of fork( ) returns 0. In the parent, fork( ) returns the PID of the child process.** The child process is an exact (almost) duplicate copy of the parent; the child obtains copies of the parent's data, stack, heap, and text segments. But due to different PID allocations to child and parent processes, they are different; otherwise, they have the same process images except for few essential differences; some key facts of fork() system call are listed here:
  - The fork () system call does not take any arguments, but returns a PID.
  - The *pid* of the child process is different from its parent process.
  - The child's parent *PID* is fixed to the *PID* of its parent process.
  - All the resource of the parent is reset to zero in the child process.

The fork() system call, in C language, is defined as;

```
#include <sys/types.h>
#include <unistd.h>
pid_t fork (void);
```

        In parent: it returns PID of child on success of fork() , or -1
        on error, and always returns 0 on successful creation of child

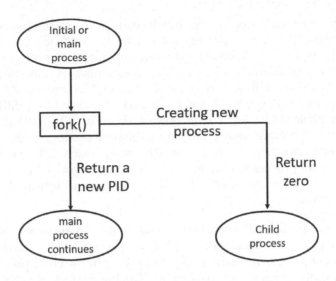

FIGURE 12.1
Creation of child process

- The two processes run the same program code, but they have distinct copies of the data, stack, and heap segments. The child process data, stack, and heap segments are first exact duplicate copies of the corresponding parts of the parent process memory. After calling the *fork()*, each process (child and parent) may update the variables in its own data, stack, and heap segments without disturbing the other process. The creation of the child process from the main process is shown in Figure 12.1.

- The following code is employed for child and parent processes after fork() call:

```
#include <sys/types.h>
#include <unistd.h>
                        /* Used in parent to store PID of child after
                        successful call of fork()*/

pid_t new_pid;
new_pid = fork();

switch (new_pid)
  {
        case -1:            /* fork() failed */
        /* Handle error */

        case 0:     /* Successful call of fork(), child comes here */
        /* Perform the execution mentioned to child */

        default:    /* After successful call of fork(), Parent comes here */
        /* Perform the execution mentioned to parent */

}
```

The data type of returned PID is *pid_t* that is defined in *sys/types.h* header file. Normally, the PID is an integer. Additionally, a process can use getpid() function to get the allotted PID to this process, and similarly, getppid() function gets the PPID of this process.

NOTE: Basically, there are three phases involved in process creation: fork(), exec(), and wait(). Fork system call creates a process but not in the condition to execute a new program. To enable this, the fork child is required to replace its own process image with the code and data of the new program image. This mechanism is done by exec(). There is a family of exec() functions built on a single system call which are execl(), execle(), execlp(), execlpe(), execv(), execve(), execvp(), and execvpe(). The exec file is either an interpreter script or an executable object file. For successful call of exec() function, no return value, the calling process is operationally changed by the new process. The subsequent section describes the execv() function. In the last phase, wait() system call is used to wait parent process until its child process is executing a new program and complete it. Thereafter, it picks up the exit status of the child. After the child process terminates, parent process *continues* its execution.

- **vfork():** vfork() is a special case of fork(). It is used to create new processes without copying the page table entries of the parent process. vfork() differs from fork() in that the parent process is momentarily blocked until the child process calls exec() function(or its family of functions) or exit(). During the parent process suspension, the child shares all memory with its parent, including the stack. After exec() or exit() function, the parent process continues its execution. The performance of the child process of vfork() is sensitive as compared to that of fork().

```
#include <sys/types.h>
#include <unistd.h>

pid_t vfork(void);
```

- **execv():** The exec() family of functions replaces the current process image with a new process image. The C language function of execv() is as follows:

```
#include <unistd.h>
int execv(const char *pathname, char *const argv[]);
```

The execv functions provide an array of pointers to null-terminated strings that signify the argument list accessible to the new program. By convention, the first must be pointed to the filename connected with the file being executed. The array of pointers should be completed by a NULL pointer. The execv functions never return the control to the original program except there is an execv() error.

- **System():** The **system()** function is similar to fork() to create a child process that executes the shell command specified as *command* using execl() as follows:

```
execl("/bin/bash", "bash", "-c", command, (char *) NULL);
```

During the execution of the command, **SIGCHLD** will be blocked, and **SIGINT** and **SIGQUIT** will be overlooked. **system()** function returns after the command has completed its execution.

The C language function of system() is as follows.

```
#include <stdlib.h>
int system(const char *command);
```

The system() function does not affect the wait status of any other children. If *command* is NULL, then system() returns a status indicating whether a shell is available on the system: a nonzero value if a shell is available, and 0 if shell is not available.

12.3.5 Terminating Process

Any process can be terminated in two ways: first by calling exit() function and secondly by using *kill* command. Kill is a built-in command that located in */bin/kill* directory. The kill command is very popular because it is used to terminate (or kill) a process manually through sending a signal to a process that terminates another process. The most powerful termination signal is SIGKILL, which ends a process immediately and may not be blocked or handled by a program [7,8]. The general syntax of kill command is as follows.

```
#kill -option pid...
OR
#kill [-signal] PID...
OR
#kill -kill PID...
```

where PID is the process ID of the terminating process and option is used to mention the sent signals or list out all available signals for the use of terminating a process. Signals can be classified into three ways:

- By number (e.g., -5),
- With SIG prefix (e.g., -SIGkill),
- Without SIG prefix (e.g., -kill).

To send a signal from a program, you must use the kill function. The first parameter is the target PID. The second parameter is the signal number. Use SIGTERM to simulate the default behavior of the kill command. The kill function is defined as follows:

```
kill (child_pid, SIGTERM);
OR
int kill(pid_t pid, int signal);

#Kill -l  To display all the available signals)

HUP INT QUIT ILL TRAP ABRT BUS FPE KILL USR1 SEGV USR2 PIPE ALRM TERM
STKFLT CHLD CONT STOP TSTP TTIN TTOU URG XCPU XFSZ VTALRM PROF WINCH IO
PWR SYS RTMIN ...............
#
```

Options available with kill command are listed in Table 12.5. Kill command sends the specified signal to the specified process. If no signal is specified, the TERMsignal is sent, having number 15. If necessary, you can use the KILL(9) signal to terminate a process forcefully because this signal cannot be caught. In Linux, the commonly used signals are listed in Table 12.9. Some examples of kill commands are given below.

```
#kill 4385  //This command terminates a process having PID 4385 by using
             the default signal sent by the kill command. The default
             signal number is 15(name SIGTERM). It may be possible that
             some programs may ignore this signal and continue the
             execution as usual.
```

TABLE 12.5

kill Command Options

Option	Meaning	Description
-s	--signal	The signal to be sent through name or abbreviated name or the signal number, preceded by a dash. For example, -SIGTERM, -TERM, or -15
pid...		Mention the list of PIDs of terminating processes that kill should signal. By default, kill command sends signal number 15 to remove any process. You can use *ps* command to know the PID of all the running processes. You can give more than one PIDs with kill command on command line
-p	--pid	Display the PID of the named processes and do not send any signals
-l,	--list[=*signal*]	Display the list of all the available signal names. The signals can be found in /usr/include/linux/signal.h
-L	--table	Display all the signal names along with their corresponding number in a tabular form

```
#kill -9 4385
OR
#kill -s KILL 4385   //This command terminates a process having PID 4385
                     by using SIGKILL signal sent by the kill command. The
                     SIGKILL signal number is 9 (name SIGKILL) that is
                     used for sure or forcefully terminate a process. The
                     SIGKILL signal cannot be ignored or caught by any
                     process.

#kill -9 $$
OR
#kill -s KILL 0      // You can kill all processes, including login shell.
                     $$ is an environmental variable that stores PID of
                     current shell.
```

12.3.6 Process Priorities

Linux is a multitasking operating system and kernel is accountable for assigning processor time to every running process on the system. Previously, in most OS implementation, the default model for processes scheduling for processor time is round-robin with time-sharing. With this model, every process will get the CPU to use for a fixed period or time slice. But in this model, CPU utilization is always a questionable concern. For high CPU utilization, process scheduling is one of the key factors. The process scheduling can be done through the process priority [5]. The scheduling of priority of a process is the quantity of CPU time the kernel allocates to the process comparative to the other processes. By default, all processes started with equal scheduling priority from the kernel on the Linux system.

The scheduling priority is an integer value, referred to as nice value, ranging from -20 (the highest priority) to +19 (the lowest priority). By default, all processes start with the same scheduling priority with nice value 0 in bash shell. The range of nice value is shown in Figure 12.2.

Sometimes, you want to change the priority of a process, either increasing or lowering its priority, depending upon the CPU processing power utilization. With the mechanism, you can give more processing time to needy processes. You can do this by using the nice command. With the help of nice command, you can set the scheduling priority of a command before starting its execution.

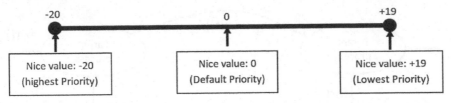

FIGURE 12.2
Range of nice value

```
nice - run a program with a modified scheduling priority. The syntax of
nice command is given as;

#nice -n <nice_value> < Command _names>

 # nice -n 8 ./hello.sh  // Execute the hello.sh script with low priority
                            as nice value increased from 0 to 8.

 # nice -n -18 ./hello.sh  // Execute the hello.sh script with high
                              priority as nice value decreased from 0 to
                              -18.
```

NOTE: If no nice value is provided with the nice command, then it sets a priority of 10 by default to the command arguments.

A command or program run without nice, then command run with default priority, i.e., with the nice value zero.

Only root can run a command or program with an increased or high priority.

Normal users can only run a command or program with a low priority.

You can see the changes done in process priority by using *ps* command as discussed in the above section. You can also use *getpriority( )* and *setpriority( )* system calls that allow you to give more control on process but they are a little bit more complex in operation.

12.3.7 Zombie and Daemon Process

This section describes the key factors of zombie and daemon processes.

- **Zombie process**: A process that has completed its execution but still has an entry in the process table and update to its parent process is referred to as a zombie process. The parent process has the responsibility to remove the entry of its zombie children from process table. When a child process completes its execution (i.e., dies), then its parent picks up the child's exit status and removes the child details from the process table and free the allocated resource to child. This is a normal and ideal situation, but if the parent process doesn't call wait() function to read the exit() status of child process, then the child process turns into a zombie process. In another word, due to not call of wait() system call by parent process, the child process has completed its execution and terminated but has not been cleaned up by its parent due to keeping the process table slot. Figure 12.3 shows the creation and termination of a zombie process.

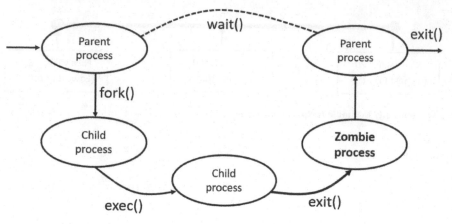

FIGURE 12.3
Creation and termination of a zombie process in Linux

Zombie process cannot be killed like a normal process with the SIGKILL signal because zombie processes are already dead. So, it will be removed from process table in two ways.

- In the first way, the exit() status of the zombie process may be read by the parent process by calling the wait() function. Thereafter, the zombie process is cleaned up from the system and released the allocated the PID and the process table entry of the zombie process. Now, the released PID can be reused by other processes.

- In the second way, you can send SIGCHLD signal to the parent process. This signal informs the parent process to execute the wait() system call and remove its child process that turned up into zombie state. He will **kill** the command and send the SIGCHLD signal along with the PID of the parent process as shown below:

```
#kill -s SIGCHLD pid
```

Zombie processes don't use any system resources but keep its own allocated PIDs. If the numbers of zombie processes are huge, then they exploit all the available PIDs. This situation creates hurdles to the other processes due to nonavailability of free PIDs. The zombie processes also specify as an operating system bug if their parent processes are not running or calling wait() system calls. If a number of zombie processes are less, then this is not a serious problem, but on the contrary, if numbers are very high, then this condition may generate some performance issues for the system.

The below lines of codes are creating the zombie process:

```
#include <stdlib.h>
#include <sys/types.h>
#include <unistd.h>

int main ()
{
int n;
```

```
char *message;
pid_t pid;
pid=fork();
switch(pid)
{
      case -1:
            perror("fork failed, no child creation");
            exit(1);
      case 0:
            message = "This is the child process";
            n = 4;
            break;
      default:
            message = "This is the parent process";
            n = 7;
            break;
}

for (;n>0;n++)
puts(message);
sleep(1)
}
```

- **Daemon process**: A daemon is an executing program on Linux OS that runs in the background, not connected to any terminal or not under the direct control of an interactive user. Daemon processes are generally started at boot time, run as root or some other special user mode that handles system-level processes or tasks, and terminate only when the system is shut down. In Linux, the processes are categorized into three types:
  - **Interactive process**: They are run interactively by a user at the command line (i.e., all-text mode execution of commands).
  - **Batch process**: They are submitted from a queue of processes and are not connected with the command line. They are well suitable for performing periodic tasks when system usage condition is short.
  - **Daemon process**: It must run as a child of *init* process (the first process started during the boot time of the system), and it must not be connected to user login sessions or a terminal.

Normally, all daemon process names are ended with the letter "d". Some examples are given as follows:

- **cron**: a daemon program that executes commands at a scheduled time.
- **httpd**: a HTTP server daemon that takes care of the Apache server for Web pages.
- **sshd**: the secure shell daemon that takes care of the ssh remote access connections.
- **inetd**: The Internet superserver daemon.

Linux operating system generally starts daemons at the boot time and terminated at system shutdown. The shell scripts kept in /etc/init.d directory are applied to start and stop daemon processes.

12.4 Threads

Nowadays, threads are a popular and widely used programming concept. They are like processes that permit an application program to perform more than one task simultaneously. A single process can have multiple threads that are independently executing the same program. All threads share the same global memory along with initialized data, uninitialized data, and heap segments.

Threads exist within a process as shown in Figure 12.4. When you start a new process, it creates a new thread for that process and this thread will generate more threads for the execution of the same process as required. Each thread performs a separate task of the currently running process. On a multiprocessor system, multiple threads may execute parallelly. If one thread is stopped for I/O request, then the other threads are still permitted to continue their execution. So, you can say that the thread enables the concurrent programming and true parallelism concept on multiple processor systems.

It is worth mentioning here the key difference between the fork system call and a new thread creation. When a process executes a fork system call, a new copy of the process is created with its own new PID and variables. This new process is executing completely independently with respect to the process that created it. On the other side, when you create a new thread in a process, the new thread is executing with its own stack, i.e., local variables, but shares global variables, signal handlers, file descriptors, and its current directory state with its created process.

12.4.1 Thread Creation

Generally, when a program starts its execution and becomes a process, it means that it starts with a default thread. In a simple way, you can say that every process has at least one thread. A process is recognized through a PID as same as a thread is identified by a thread ID. But PID and thread ID have some interesting features mentioned below:

- A PID is unique across the system, whereas a thread ID is unique only with respect to a single process.
- A PID is an integer value, whereas the thread ID is not essentially an integer value. Instead, it would be a well-defined structure.
- You can print the PID very easily, whereas a thread ID printing is not easy.

The above-mentioned points highlighted the key difference between a PID and thread ID, discussed more in Section 12.4.6. Once a program starts its execution, the resulting process

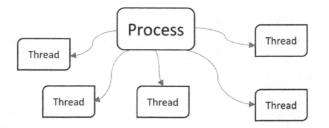

FIGURE 12.4
Threads of a process

will be comprised of a single thread, referred the *initial* or *main* thread. But a process may create many threads that are identified by unique thread IDs [5]. In C and C++ programming language, *pthread_t* data type is used to mention the thread IDs. New threads can be created using *pthread_create()* function. The clone() system call is also used to create the thread. Upon creation, every thread executes a thread function that is just an ordinary function and comprises the code that should be executed as a thread.

```
The pthread_create()function syntax is given as;
#include <pthread.h>
int pthread_create(pthread_t *thread, const pthread_attr_t *attr, void
*(*start)(void *), void *arg);
                //On Successful execution of function, returns 0; or
                a positive number on error.
```

The *pthread_create* function requires four arguments as described below:

1. A pointer *thread of *pthread_t*-type variable, which holds the thread ID of the newly created thread.
2. A pointer to a thread attribute object. This object comprises the control details of the thread, such as how the thread communicates with other threads of the program. A thread will be created with the default thread attribute properties, if input thread attribute is NULL, priority is one of the widely used thread attributes.
3. A pointer to the thread function of type *void* (*) (void*)*. This is an ordinary function pointer to keep the record that each thread starts with a function and that functions' address is passed here as the third argument, so that the kernel knows which function is to be started as the thread.
4. A thread argument value, *arg*, of type void*. It will hold the argument that you wish to pass to the thread function when the thread starts executing.

Every thread within a process is exclusively recognized by a thread ID. This thread ID is returned to the caller of *pthread_create()*, and a thread may get its own ID using pthread_self() function as given below:

```
#include <pthread.h>
pthread_t pthread_self(void);
        // Returns the thread ID of the calling thread
```

12.4.2 Thread Kill or Termination

A thread may terminate normally either by returning from its thread function or by calling the *pthread_exit* function. It is possible for a thread to request a termination of some other threads before its normal killing or termination; this is called canceling a thread. For canceling a thread, you can call *pthread_cancel* function and pass the thread ID of that thread which you want to be terminated or canceled before its normal termination. It can be finished using the following function syntax:

```
#include <pthread.h>
int pthread_cancel(pthread_t thread);
```

The return value of a canceled thread is given by PTHREAD_CANCELED variable. Generally, a thread could be a piece of code that must be executed completely or nothing in nature. For thread execution, it may allocate some resources, use them, and then deallocate them after execution. In case, thread is canceled in the middle of its execution, then it may be difficult to deallocate the resources, and thus, the resources will be blocked and unused for others. To avoid this situation, a thread in a supervised manner is canceled, i.e., when and whether a thread can be canceled. An executing thread may be canceled in the following manner:

- By using the return value with *RETURN* mentioning.
- By calling pthread_exit() function.
- By calling pthread_cancel() function.
- By calling exit() function in main thread

12.4.3 Thread Data Structure

Linux operating system has an exclusive way of thread implementation. Linux kernel implements all threads as standard processes. To the Linux kernel, there is no thought of a thread. Linux kernel does not provide any distinct scheduling concept or data structures to signify threads. Due to this, kernel recognizes that a thread is simply a process that shares certain resources with other running processes. Further, every thread has a unique *task_struct* and it looks to the kernel as a normal process, threads basically meant to share some resources with their creator.

The key factor of thread concept in operating systems is to provide a lighter, faster execution unit than the heavy process. Threads are simply meant in a manner of sharing resources between processes. In Linux, the kernel stores entire details of processes presently running on the system in a circular doubly linked list form referred to as *task list*. As mentioned earlier that a process descriptor of the type struct *task_struct*, mentioned in <linux/sched.h> that describes each process in the *task list*. The defined *task_struct* data structure, which comprises all the information about a specific executing process (nothing but an instance of task) is mentioned in Table 12.6.

12.4.4 Synchronization, Critical Section, and Semaphore

- **Synchronization**: Linux is a multitasking operating system that allows the execution of more than one process simultaneously. Every process performs some action in coordination with other processes. In such a situation, a strong synchronization mechanism plays a crucial role that provides a synchronized execution environment for cooperating processes to maintain data consistency requirements. The synchronization refers to one of the two separates but connected the concepts: process synchronization (synchronization of processes) and data synchronization (synchronization of data). For concurrent and cooperating executing processes, the process synchronization concept avoids simultaneous access of shared data to minimize the chance of data inconsistency or integrity. Process synchronization primitives are normally used to implement data synchronization. For example, parallelly updating a common part of a file or a shared memory region by various processes would lead to produce an inappropriate or incorrect result, if no

TABLE 12.6

Fields of Thread Data Structure

Data Structure Fields of `task_struct`	Description
volatile long state	The current state of a process, namely running, interruptible, uninterruptible, zombie, and stopped
struct thread_info	Comprises the low-level information of the process: flags, status, cpu, execution domain, etc.
unsigned long ptrace	Associated with debugging a process. The process is being traced by another process, such as a debugger
int prio	Nice value, dynamic priority of the process that is used by the scheduler to decide to execute sequence of next process to be run
cpumask_t cpus_allowed	Allowed CPU core where the process to run—or allowed to run
ptrace_children	Child process under that is being traced, again pointing to another struct (another structure)
ptrace_list	Other parent processes that are tracing this process (another structure)
struct mm_struct * mm	This mm is a pointer that points to another structure called mm_struct. This is the memory manager comprising information on the memory map of a process
int exit_code, exit_signal	These fields are used to comprise the exit value of a process. It will inform the parent process, how the child died. If the child is terminated by a signal, it will contain the number of the signal
pdeath_signal	The signal that is sent when the parent dies
pid_t pid	Contains the process identifier
struct task_struct * parent	Pointing to parent process
struct list_head children	List of processes created by this process
utime, stime, cutime, cstime	User time, system time, collective user time spent by process and its children, total system time spent by process and children
uid_t uid, euid, suid, fsuid	Owner id (uid) of the process, effective ID (euid), and other specific attributes
gid_t gid, egid, sgid, fsgid	Group id (gid) of the process, effective ID (egid), and other specific attributes
struct files_struct *files	Another structure having open file information
struct signal_struct *signal	Signal handler information
Tgid	Thread group identifier

synchronization mechanism is applied. Linux kernel provides a couple of synchronization facilities, such as signals, semaphores, file locks, pipes, and mutexes.

- **Critical section**: You can also understand the concept of process synchronization by deliberating the so-called critical-section problem. A critical section is a portion of code that must be executed under mutual exclusion mode [9]. Suppose that a group of cooperating processes are trying to access a shared code area, referred to as a critical section, then it is essential to protect a critical region to avoid collusion in a code/system. Therefore, to maintain the data integrity, only one process can execute its critical section at a given point of time. If any other process also wants to execute its critical section, it must wait until the first one finishes.

  Let's assume that a system is comprised of N processes {P0, P1, ..., Pn–1}. Every process has a piece of code, referred to as a critical section, in which the process may be altering common variables, data, and other contents. Hence, to maintain the data integrity, the system must ensure that only one process may allow is executing in its

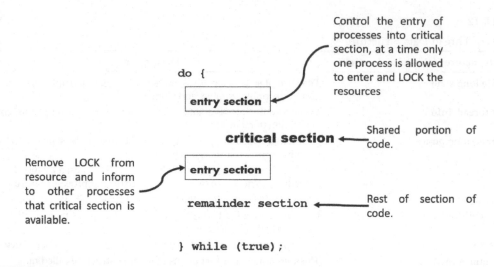

FIGURE 12.5
Critical section of code

critical section; then, no other process is permitted to execute in its critical section. It means, at the same time, no two processes can execute in their critical sections. This scenario is considered as a critical-section problem that provides a procedure or solution to cooperate among the group of executing processes. In critical-section problem, every process must request to take the permission to enter in its critical section. The critical-section code bounded with entry section and exit section is shown in Figure 12.5. Three key requirements, namely mutual exclusion, progress, and bounded waiting, must be satisfied to provide a solution to the critical-section problem. Only one process may enter into a critical area of code at a given point of time.

- **Semaphore**: Semaphore is another programming concept that can be used to solve synchronization problems among cooperating processes. The semaphore concept was offered by E. W. Dijkstra around 1965. A semaphore is simply a variable that is used to control access to a critical section by various concurrent running processes in a system, such as multiprocessing or multitasking operating system. In Linux, semaphores are acting as sleeping locks. It means when a multiple task/thread wants to share data and tries to acquire a semaphore that is unavailable, the semaphore places the task/thread onto a wait queue and places the task to sleep mode. When the semaphore becomes available, then one of the tasks/threads from the waiting queue (FIFO) will get invoked and acquire the semaphore.

    A semaphore S is an integer variable that takes only whole positive numbers through two standard atomic operation, namely wait () and signal (). In Linux programming, "wait ()" and "signal ()" operations have special meanings and originally notified as follows:

- P(semaphore variable) for wait(): Signify as "to test"

```
wait(S)
{
    While (s<=0)
            ;
    S- -;
}
```

- V(semaphore variable) for signal() : Signify as "to increment"
  ```
  Signal(s)
  {
      S++;
  }
  ```

For example, if two processes share the S semaphore variable. Once a process has executed P(S), it has acquired the semaphore and accessed the critical section. The second process is prohibited from accessing the critical section because, when it tries to execute P(S), it is informed to wait until the first process has left the critical section and executed V(S) to release the semaphore as given in the below pseudocode:

```
semaphore sv = 1;
loop forever {
P(sv);
critical code section;
V(sv);
noncritical code section;
}
```

A semaphore that can take only zero or one value is known as *binary* semaphore, or *mutex*, for *mut*ually *ex*clusive that is the most common form of semaphore. If a semaphore can take a range of many positive values, then it is known as *counting* semaphore.

The POSIX semaphores can be classified into two categories: unnamed semaphore and named semaphore. The unnamed semaphore does not associate with a name and might be shared between the processes or among the threads. If an unnamed semaphore is shared by more than one process, then it will allocate in process-shared memory and initialize. On the contrary, named semaphore has a name, and unrelated process may access the semaphore with the same name by calling *sem_opn()*. The semaphore functions are defined in <sys/sem.h> and <semaphore.h> header files. Some key functions that are associated with named semaphores [5] are listed in Table 12.7.

12.4.5 Demarcation between Process and Thread

This section briefly highlights some key differentiation between process and thread. These points may affect our choice, whether to develop an application as a set of processes or as a set of threads. Some remarkable features along with differences between thread and process are mentioned in Table 12.8.

TABLE 12.7

List of Functions for Named Semaphore

Function Name	Description
sem_open()	Opens or creates a semaphore
sem_post(sem)	Increment a semaphore's value
sem_wait(sem)	Decrement a semaphore's value
sem_getvalue()	Recovers a semaphore's current value
sem_close()	Removes the calling process's association with a semaphore that it formerly opened
sem_unlink()	Removes a semaphore name and sends the semaphore for deletion when all processes have closed it

TABLE 12.8

Differences between Process and Thread

Comparison Parameter	Process	Thread
ID	PID is unique across the system and it is an integer value	Thread ID is unique only with respect to a single process and not essentially an integer value. It would be a well-defined structure
Memory sharing	Processes are totally independent and don't share memory	Sharing memory between threads is easy because a thread may share some parts of memory with other threads
Creation time	Process creation is slow as compared to thread, i.e., requires more time	Thread creation is faster as compared to process, i.e., requires less time
Context switching time	Processes require more time for context switching as its heavier than thread	Threads require fewer time for context switching as its lighter than processes
Communication	Communication between processes needs more time as compared to threads	Communication between threads needs less time as compared to processes
Bug handling	Easy due to processes are more isolated from one another	Not easy, bug in one thread may affect other threads also due to sharing the same address space and other attributes
Resource consumption	Requires more resources than threads	Threads normally require less resources than processes
Blocked	If a process becomes blocked, but other processes may continue execution	If a user-level thread becomes blocked, then all its peer threads are also becoming blocked
Dependency	Distinct processes are autonomous of each other	Threads are slices of a process, hence dependent
Data and code sharing	Processes have their own data and code segments due to sharing of memory	A thread shares data segment, code segment, files, and other information with its peer threads
OS view	All the different processes are viewed or considered separately by the operating system	All user-level peer threads are viewed or considered as a single task by the operating system
Program execution	In a multiprocess application, different processes may execute different programs	In a multithreaded application, all threads must be executing the same program

The major differences between a process and a thread are given as follows:

12.5 Device File

Like the other operating system, Linux OS is also communicating with hardware devices through software segment components called device drivers. A device driver summarizes communication procedures of various devices from the operating system and permits the system to interact with the device via a standardized interface. In Linux, device drivers are parts of the kernel, it may be loaded either statically into the kernel or on demand as kernel modules.

Device drivers run as a part of the kernel, and it is not directly accessible to user processes. But Linux offers a mechanism by which processes can communicate with a device driver through file, referred to as a device file. These device files are kept in the file system that can open, read, and write into it through a program as a normal file. The device files are not an ordinary file; they do not represent data section on a disk-based file system [1]. The device file is one of the important file types, where the data reading or writing to a device file is the communication to the respective device driver.

- **Device type**

  In all distribution of Linux/Unix system, device files are categorized into two types:

  - **Character devices**: Character devices, regularly abbreviated as *cdevs*, are represented as hardware devices that are not addressable. It provides access to reads or writes data only as a stream, generally of characters (bytes). Some examples of character devices are keyboards, mice, printers, serial and parallel ports, tape drives, and terminal devices.

  - **Block devices**: Block devices, regularly abbreviated as *blkdevs*, are represented as hardware devices that are addressable in device-specified portions, referred to as blocks. It provides access to reads or writes data in fixed-size blocks. Unlike character device, a block device supports seeking and provides random access to data stored on the device. Some examples of block devices are hard drives, Blu-ray discs, and memory devices such as flash.

NOTE: It is important to mention here that there is another device type, known as network devices or occasionally referred to as Ethernet devices that are the most common types of network devices. It provides access to a network through physical connecter, generally 802.11 card. Network devices are not accessed through a device node, but with a special interface called as socket API.

- **Device numbers**:

  Linux recognizes every device by using two numbers, namely *major device number* and *minor device number*. The major device number identifies which driver the device links to. The communication from major device numbers to drivers is fixed and portion of the Linux kernel sources. It is possible that same major device numbers may communicate to two dissimilar drivers, one is a block device and another is a character device. But the minor device numbers differentiate distinct devices or components managed by a single driver. The minor device number is specific to the device driver.

- **Device directory**:

  GNU/Linux system has a directory, namely */dev* , under root (/) directory that comprises the complete detail entries of all devices including character and block devices of systems. The /dev directory has uniformed name entries corresponding to major and minor device numbers. To view the list of all devices known to the system by typing the below command:

```
#ls /dev       //Display the name of all devices
#ls -l /dev    // display the list of all devices along with details
```

12.6 Signals

A signal is a notification generated by Linux systems to a process that an event has occurred with some condition. With this notification, a process may take some action. One process can send a signal to another process that may be employed as a synchronization method, or as a form of passing information under interprocess communication (IPC) mechanism. It is also possible that a process sends a signal to itself. Sometimes, signals are defined as software interrupts, when it created by some error conditions, such as memory segment violations, floating-point processor errors, or illegal instruction statements. In all such cases, the programming interface is the same. Normally, the source of signal generation is kernel that delivered to a process. When a signal is communicated to a process, then process may perform any one of the subsequent default actions as per the signal behaviors as follows:

- **Ignore the signal:** rejected by the kernel and has no effect on the process.
- **Terminate (or Kill)** a process: default action is terminating the process. But, there are signals whose default action is to suspend or overlook the process. In maximum cases, every signal is associated with a default action.
- **Stopped** a process: suspended the execution of a process.
- **Resumed** a process execution: again, started the execution of previously stopped process.
- **Execute signal handler**: a signal handler is a function (program) that performs appropriate tasks in response to the delivery of a signal.

In Linux, kernel provides the standard signals that are numbered from 1 to 31. But, the Linux *signal(7)* manual page mentions more than 31 signal names. Signal names are declared by incorporating the header file *signal.h*. In C programming language, the signal is defined as a function with two input arguments, namely *sig* and *func*, [5] as given below:

```
#include <signal.h>
void (*signal(int sig, void (*func)(int)))(int);
```

To remove a process, you can use kill function that sends the specified signal, *sig*, to the mentioned process with given *pid* as given in the below code:

```
#include <sys/types.h>
#include <signal.h>
int kill(pid_t pid, int sig);
```

If you want to close your currently opened terminal, then compile the given below code:

```
#include<stdio.h>
#include<sys/types.h>
#include<signal.h>

int main()
{
    kill(getppid(),SIGKILL);
    sleep(3);
}
```

The above-mentioned code will take the *pid* of currently opened terminal and send SIGKILL (9) signal, i.e., sure killing of a process. After compiling the above code, the executable file is created as *a.out* and then run this file as follows:

TABLE 12.9

List of Signals

Signal Name	Signal Number	Description
SIGHUP	1	Hang-up (POSIX)
SIGINT	2	Terminal interrupt (ANSI)
SIGQUIT	3	Terminal quit (POSIX)
SIGILL	4	Illegal instruction (ANSI)
SIGTRAP	5	Trace trap (POSIX)
SIGIOT	6	IOT trap (4.2 BSD)
SIGBUS	7	BUS error (4.2 BSD)
SIGFPE	8	Floating point exception (ANSI)
SIGKILL	9	Kill (can't be caught or ignored) (POSIX)
SIGUSR1	10	User-defined signal 1 (POSIX)
SIGSEGV	11	Invalid memory segment access (ANSI)
SIGUSR2	12	User-defined signal 2 (POSIX)
SIGPIPE	13	Broken pipe, write on a pipe with no reader, (POSIX)
SIGALRM	14	Alarm clock (POSIX)
SIGTERM	15	Termination (ANSI)
SIGSTKFLT	16	Stack fault
SIGCHLD	17	Child process has stopped or exited, changed (POSIX)
SIGCONT	18	Resumed execution, if stopped (POSIX)
SIGSTOP	19	Stop executing (can't be caught or ignored) (POSIX)
SIGTSTP	20	Terminal stop signal (POSIX)
SIGTTIN	21	Background process trying to read, from TTY (POSIX)
SIGTTOU	22	Background process trying to write, to TTY (POSIX)
SIGURG	23	Urgent condition on socket (4.2 BSD)
SIGXCPU	24	CPU limit exceeded (4.2 BSD)
SIGXFSZ	25	File size limit exceeded (4.2 BSD)
SIGVTALRM	26	Virtual alarm clock (4.2 BSD)
SIGPROF	27	Profiling alarm clock (4.2 BSD)
SIGWINCH	28	Window size change (4.3 BSD, Sun)
SIGIO	29	I/O now possible (4.2 BSD)
SIGPWR	30	Power failure restart (System V)
SIGUNUSED/SIGSYS	31	Not defined on any architecture

```
# ./a.out   // press <enter Key > and execute, it will close the terminal,
    i.e., kill running terminal process
```

All signal names begin with "SIG" and Table 12.9 lists out all 31 signals.

12.7 Various System Calls

The system call is the fundamental interface between a user-space process (referred to as a user application) and hardware via Linux kernel. These system calls are built into the Linux kernel that invoked from Unix-C programs to access the hardware. Without system call interface, processes running in user-space cannot interact with the system. This interface obliges three key purposes as follows:

1 It offers an abstracted hardware interface for user space.
2 System calls confirm system security and stability.
3 A single common interface between user space and the rest of the system (including kernel) agrees for the virtualized system offered to processes.

In Linux, system calls are the only legal entry point into the kernel, excluding traps, and exceptions. Certainly, additional interfaces, such as device files or /proc, are eventually accessed through system calls.

- **Syscalls**
  System calls are generally referred to as syscalls in Linux that are characteristically accessed through function calls defined in the C library. Each of the system calls has a well-defined behavior, such as getppid() system call is defined to return an integer that is the PID of parent process of current processes. For example, some frequently used system calls are fork(), chmod(), chdir(), getpid(), kill(), mount(), pipe(),nice(), mkdir(), read(), write(), close(), exit(), fcntl(), perror(), sysinfo(), and so on.

- **System call numbers**
  In Linux, each system call is allocated a *syscall* number. This is a unique number assign to a specific system call. When a user-space process executes a system call, the *syscall* number recognizes which *syscall* was executed; the process does not use names to mention the *syscalls*. The Linux kernel saves a list of all registered system calls in the system call table, stored in *sys_call_table*. In kernel (5.0.21). The system call table for x86-64 architecture is defined in *arch/x86/entry/syscall_64.c*. This table allocates each valid *syscall* to a unique *syscall number*.

12.8 POSIX

The Portable Operating System Interface (POSIX) represents a set of standards specified by the IEEE Computer Society for keeping compatibility between the operating systems. POSIX standard was defined primarily for Unix-based operating systems in the 1980s to address the portability issue [10]. The standard was defined based on System V and BSD Unix. POSIX does not define the operating system; it defines the software (application program) compatibility with Unix and other operating systems. Although Windows operating systems may also run certain POSIX programs, like *Cygwin*, we just brief out of POSIX standard for Unix and Linux systems. The first POSIX standard was released in 1988. Formally, it was referred to as IEEE Standard 1003.1-1988 Portable Operating System Interface for Computer Environments.

Initially, POSIX was divided into several standards as given below:

- **POSIX.1**: Core Services (IEEE Std 1003.1-1988) like process, signals, file and directory operations pipes, C library (Standard ANSI C), I/O port interface and control, and others.

- **POSIX.1b**: Real-time extensions (IEEE Std 1003.1b-1993) like priority scheduling, real-time signals, clocks and timers, semaphores, message passing, shared memory, asynchronous and synchronous I/O, and memory locking interface.
- **POSIX.1c**: Thread extensions (IEEE Std 1003.1c-1995) like thread creation, control, and cleanup, thread scheduling, thread synchronization, signal handling.
- **POSIX.2**: Shell and utilities (IEEE Std 1003.2-1992), like command interpreter, utility programs.

After 1997, the Austin Group transported all the standards under a single umbrella. After that, the following POSIX standard has been released:

- POSIX.1-2001, referred to as IEEE Std 1003.1-2001
- POSIX.1-2004, referred to as IEEE Std 1003.1-2004
- POSIX.1-2008, referred to as IEEE Std 1003.1-2008
- POSIX.1-2017, referred to as IEEE Std 1003.1-2017.

Basically, the POSIX is created to enhance application portability. Any POSIX-compliant Unix/Linux system may be easily ported to another POSIX-compliant Unix/Linux system with the least alterations. Non-Unix/Linux systems may also be POSIX-compliant. The following operating systems certified as POSIX compliant operating systems with confirmation to follow one or more various POSIX standards are AIX, HP-UX, IRIX, Mac OS (since 10.5 Leopard), Solaris, QNX Neutrino, etc. On the contrary, the following operating systems followed most of the POSIX standards, but officially not certified as POSIX compatible operating systems, are Android, Linux (most distribution), FreeBSD, VxWork, OpenBSD, NetBSD, VMWare, MINIX3, etc.

12.9 Summary

This chapter elaborated a range of fundamental concepts related to Linux system programming. Understanding of these concepts must provide enough background to begin learning system programming on Linux or Unix. vim is an advanced vi editor available for Linux OS. GNU Make is a utility that automates the generation of executables and other nonsource files of a program from the program's source files. The GDB allows you to test and debug another program or target source code to check the proper functioning of the code. In the Linux system, everything is considered a file and when a file is in execution mode, it becomes a process. The Linux kernel manages the working and execution of different processors. A process is always created by another process, except the first process. Each process has a unique numeric identifier called PID and a PPID. Kernel mode is a privileged mode of the CPU in which the kernel runs and provides access to all memory locations and system resources, and only in this mode, context switches occur. Process termination can be done either by calling the exit() function or by using the kill command. After the execution of a process, the parent process does not call the wait() function, the process turns into a zombie process that does not use any resources but keeps its allocated PIDs leading to hurdles for other processes. Processes that permit an application program to perform more than one task simultaneously are called threads and the Linux kernel treats threads as regular processes. Process synchronization and data synchronization are the important concepts used in the Linux system. A critical section is a portion of code executed under mutual exclusion modes. Semaphores are also used to solve synchronization problem cooperating processes. A signal is a notification generated by Linux systems to a process that an event has occurred with some conditions. System calls allow processes to request services from the kernel that are inbuilt into the kernel. In the last, the compatibility issue between operating systems is addressed by POSIX, an IEEE standard.

12.10 Review Exercise

1. Define the term cross-compiler. List out the name of various programming languages that uses gcc as compiler.

2. How will you execute the makefile by just giving the "make" command?

3. On what fact the basis of shell program relies?

4. Generate a file "both.txt" which will be formed by a concatenation of "file_1st.txt" and "file_2nd.txt".

5. List out the key difference between process and thread in Linux.

6. Can the <cat filename> command give the following error message?

    --bash: cat: command not found.

7. How can we pass arguments to a script in Linux? And how to access these arguments from within the script?

8. Explain daemon characteristics and coding rules.

9. Explain the importance of File I/O in Linux and write an command to see various active processes of system?

10. Does Unix support preemptive multitasking? Explain.

11. Discuss the various system calls for process creation with its functional description?

12. What are the states associated with the thread? Why thread behavior is unpredictable?

13. Write a program for handling a user-defined signal *usr1*.

14. Write a command to pass any user-defined signal to a particular process using the kill command.

15. How to create a child process? Which system calls replace current process image with new process image?

16. Which command displays information about a selection of the active processes on the system?

17. Which signal is used to send to a process when its controlling terminal is closed?

18. Write a command to print all processes running as root.

19. Describe the usage and functionality of kill command. Further write a command to remove all active processes including login shell?

20. How signal is helpful in the synchronization of process?

21. What happen to the system when a thread is called? List the pros and cons of using the threads.

22. What precautions one needs to be taken to avoid deadlock (two threads are trying to read and write the same memory locations)?

23. What is a critical region? As a programmer, what necessary precautions you should take to avoid the deadlock.

24. List out various algorithms that are developed to avoid deadlock.

25. Look at the following algorithm:

```
a=True
b=False

while(a)                        //thread1
{
    Critical section
     b=!b
}

while(b)                        //thread2
  {
    Critical section
     a=!a
  }
```

Is this a deadlock-free algorithm?

26. Discuss the thread mechanism in both Linux and windows.

27. Discuss the device files in Linux? How it is useful in the communication with device drivers.

28. Describe the semaphore concept of programming to solve the synchronization problem.

29. What kind of data structure Linux provides to signify threads?

30. Discuss how program counter and stack pointer are managed while using threads in the system.

31. Discuss process priorities in Linux. Which process does not have any PPID?

32. If you want to kill all the processes with the same PPID then how will you kill all these processes in one operation?

33. How to identify a zombie process. Give examples.

34. Discuss how is semaphore helpful to maintain data integrity.

References

1. Mark Mitchell, Jeffrey Oldham, and Alex Samuel. 2001. *Advanced Linux Programming*. New Riders Publisher.
2. The GNU Project Debugger, GDB Documentation listing in the World Wide Web. https://www.gnu.org/software/gdb/.
3. Neil Matthew and Richard Stones. 2001. *Beginning Linux Programming*. Wrox Press Ltd.
4. Robert Love. 2007. *Linux System Programming*. O'Reilly Media.
5. Michael Kerrisk. 2010. *The Linux Programming Interface*. No Starch Press, Inc.
6. William Stallings. 2005. *The Linux Operating System*. Prentice Hall.
7. J. Purcell. 1997. *Linux Complete: Command Reference*. Red Hat Software, Inc.
8. Richard Blum and Christine Bresnahan. 2015. *Bible- Linux Command Line & Shell Scripting*. John Wiley & Sons.
9. Avi Silberschatz, Peter Baer Galvin and Greg Gagne. 2009. *Operating System Concept*. John Wiley & Sons Inc.
10. Wikimedia Foundation, Inc. The Portable Operating System Interface (POSIX) in the World Wide Web. https://en.wikipedia.org/wiki/POSIX. Last modified May 29, 2021.

13

Linux Inter-Process Communications

Inter-process communication (IPC) is the simplest way of communication among various processes of the system to communicate and transfer data with each other. There are various ways to communicate between parents and child processes, unrelated processes, and even processes of different systems. This chapter briefs various forms of IPC such as shared memory, message queue, pipe, and sockets, as well as procedure to synchronize their actions too.

13.1 Basic Concept

IPC is the transmission of data among processes. This involves synchronizing their action and managing shared data. Before describing various ways of IPC, you must understand the classification of various IPC services. For Linux programmer, the key nomenclature of IPC services is given as follows:

- **Communication services**: Related to communication of data between processes.
- **Synchronization services**: Related to synchronizing among various activities of processes.
- **Notification of signal services**: Primitive form of IPC or notification to the process.

(a) **Communication services**: Various communication services are further divided into two categories, data transfer and shared memory, as shown in Figure 13.1 that allows processes to exchange data with one another.
   - **Data communication mode**: This communication mode performs IPC through reading and writing operation of data. Here, one process writes data to buffer (IPC interface) and another process reads data from that buffer (IPC interface).
   - **Shared memory mode**: In shared memory communication, it permits processes to share information by keeping it in a common memory place that is shared among the communication processes.

(b) **Synchronization services**: To maintain the data integrity, synchronization mechanism is very important that allows various coordination processes to coordinate their actions to avoid simultaneous operation on a common or shared place or a similar part of memory or file. If no synchronization occurs, then concurrent updates may cause to alter data of a program to give undesirable results. The synchronization services may broadly offer through semaphore, file lock, thread-level sync, and some other synchronization mechanisms. Various categories of synchronization services are shown in Figure 13.2.

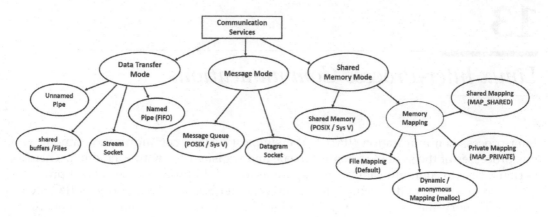

FIGURE 13.1
Linux communication services in IPC

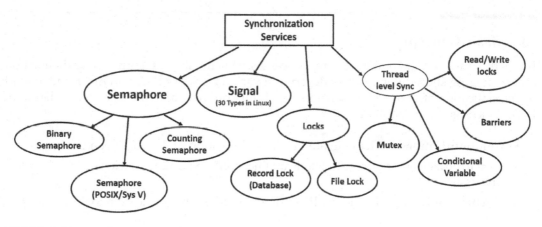

FIGURE 13.2
Linux synchronization services in IPC

(c) **Notification or signal services**: Signals are one of the oldest and inexpensive methods of IPC offered by Unix/Linux system. A signal is an event produced by the Linux system in reply to some condition, such as keyboard interrupt or error conditions (such as illegal instruction or memory segment error or floating-point processor errors). Every signal has a predefined message along with some action that a process may choose to act or ignore upon its receipt. Linux follows the Unix method of IPC in which a signal acts as a command that allows to control various processes, as discussed in Section 12.6.

The preceding chapter is already explained some key synchronization mechanisms under Section 12.4.4 and signal services under Section 12.6. Further, the following section described some key methods of communication services, such as shared memory, message queue, pipe, and sockets. Each communication service is explained along with its functional description of important system calls, data structure, header files, and other details in C programming codes [1–3].

13.2 Shared Memory

Shared memory is one of the basic and simplest ways of IPC, in which more processes (related or unrelated) may access the identical memory space. It is a very reliable and efficient way of communicating data between two executing processes. A process can make data available to other processes by storing it in the common shared memory region. The physical address of this memory region is shared among all the communication processes as shown in Figure 13.3. During communication, if any one of the processes updated the memory, then it can be viewed by all the other processes. The accessing speed of data from shared memory is almost like accessing data from nonshared memory, since the communication does not consist of system call or data transfer between user and kernel spaces.

In short, shared memory can offer very fast communication, but kernel does not synchronize shared memory access by various processes. Therefore, a programmer must apply suitable synchronization mechanism to avoid any inconsistency in data access. Some key points of shared memory are highlighted as follows:

The general points about shared memory are as follows:

- Data transfer is not facilitated by kernel user space. Data placed in shared memory is immediately visible to other processes.
- Provide very fast IPC as compared to pipes and message queue, where sender requires copy from user space to kernel space and receiver memory requires copy from kernel space to user space.
- Appropriate synchronization mechanism is required to prevent simultaneous updates or access to shared memory.
- Generally, semaphore is used as synchronization mechanism in shared memory IPC.

Linux offers three standard specifications for shared memory as shown in Figure 13.1. Those standard specifications for shared memory are as follows:

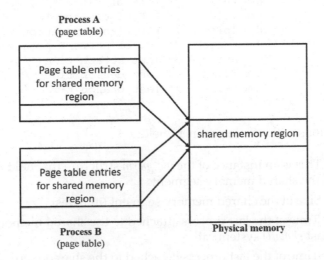

FIGURE 13.3
Physical memory layout of shared memory region among processes

- **System V shared memory**:
  - Sharing between unrelated processes
  - Widely used and original shared memory method
- **POXIS shared memory**:
  - Sharing between unrelated processes
  - Don't need to create a file, and avoid file system I/O overhead
  - To keep simpler and better than older Application Programming Interface (APIs)
- **Memory mapping**:
  - **Shared anonymous mappings**: Sharing between related processes via *fork ()* system call
  - **Shared file mappings**: Sharing between unrelated processes, supported by file in a file system

13.2.1 System V Shared Memory API

This section describes System V shared memory data structure along with various system calls. Shared memory permits two or more processes to share the same memory region, mentioned as a *segment*, of physical memory as shown in Figure 13.3. A segment may be formed by one process, and thereafter any number of processes can write to and read from it.

The kernel keeps a special internal data structure for each shared memory segment that presents within its addressing space. A shared memory segment is exclusively recognized by a positive integer, *shmid*, and has a related data structure of type *struct shmid_ds*, defined in *<sys/shm.h> as well as <linux/ipc.h>*. The structure of *shmid_ds* is given as follows:

```
struct shmid_ds
    {
        struct ipc_perm shm_perm;   /* operation permission*/
        size_t          shm_segsz;  /* size of segment(bytes) */
        pid_t           shm_cpid;   /* PID of creator */
        pid_t           shm_lpid;   /* PID, last operation */
        shmatt_t        shm_nattch; /* no. of current processes attaches */
        time_t          shm_atime;  /* last attach time */
        time_t          shm_dtime;  /* last detach time */
        time_t          shm_ctime;  /* last change time */
    };
```

Here are descriptions of the more pertinent fields:

- **shm_perm**: This is an instance of the *ipc_perm* structure that holds the access permissions on the shared memory segment
- **shm_segsz**: Size of the shared memory segment (in bytes)
- **shm_atime**: Time of the last process attached to the shared memory segment, i.e., time of the last *shmat()* system call
- **shm_dtime**: Time of the last process detached to the shared memory segment, i.e., time of the last *shmdt()* system call

- **shm_ctime**: Time of the last change to the *shmid_ds* structure (mode change, etc.) performed by the *shmctl()* system call.
- **shm_cpid**: The Process Identifier (PID) of the process that created the shared memory segment
- **shm_lpid**: The PID of the last process that operated on the segment by executing *shmat()* or *shmdt()* system call
- **shm_nattch**: Number of active processes presently attached to the shared memory segment

The system calls used to manage a shared memory segment are as follows:

- **SYSTEM CALL: shmget():-create a new shared memory segment**:
  This system call is used to create a new shared memory segment or obtain the identifier of an existing segment (which is created by another process). **The shmget() function is given as follows:**

  ```
  #include <sys/types.h>
  #include <sys/types.h>
  #include <sys/shm.h>
  int shmget(key_t key, size_t size, int shmflg);
  ```

  On successful execution, it returns segment identifier on success, otherwise –1 on error.
- **SYSTEM CALL: shmat() :- connect shared memory segment**:
  Initially, when a shared memory segment is created, then it is not accessible by any process. To allow access to the shared memory, you must connect or attach it to the address space of a process. Hence, by executing the *shmat()* system call, it attached the shared memory segment recognized by *shmid* to the virtual address space of the calling process. The *shmat()* function is given as follows:

  ```
  #include <sys/types.h>
  #include <sys/types.h>
  #include <sys/shm.h>
  void *shmat(int shmid, const void *shmaddr, int shmflg);
  ```

  On successful execution, it returns address at which shared memory is connected or attached, otherwise –1 on error.
- **SYSTEM CALL: shmdt() :- detach the shared memory segment**:
  If a process is no longer required a shared memory segment, then it should be detached by calling *shmdt()* function. The *shmdt()* function is given as follows:

  ```
  #include <sys/types.h>
  #include <sys/types.h>
  #include <sys/shm.h>
  int shmdt(const void *shmaddr);
  ```

The above function detached the shared memory segment located at the address mentioned by *shmaddr* from the address space of the calling process. On successful execution, it returns 0, otherwise –1 on error. It is noted here that after successful execution of *shmdt()*, the detached segment is not removed from the kernel, even the number of processes (*shm_nattch*) associated to *shmid_ds* structure is decremented by one. When this value becomes zero (0), then kernel will remove the segment automatically.

- **SYSTEM CALL: shmctl():- control the shared memory segment**:
  The shmctl() function performs various control operations specified by *cmd* on the shared memory segment whose identifier is given in *shmid*. The *buf* argument is a pointer to the segment data structure *shmid_ds*. The *shmctl()* function is given as follows:

    ```
    #include <sys/types.h>
    #include <sys/types.h>
    #include <sys/shm.h>
    int shmctl(int shmid, int cmd, struct shmid_ds *buf);
    ```

On successful execution, it returns 0, otherwise –1 on error.

13.2.2 POSIX Shared Memory APIs

The Portable Operating System Interface (POSIX) shared memory is supported on Linux since kernel 2.4 adheres to POSIX.1b standard. The POSIX shared memory API allows processes to communicate information by sharing a region of memory. POSIX discusses shared memory *objects*, while System V talks about shared memory *segments*. But both terminologies are used for referring memory regions that are shared between processes. A program using POSIX shared memory generally comprises the following steps:

- Create or open a shared memory object with *shm_open()*. A file descriptor will be returned on successful execution of *shm_open()*.
- Assign the shared memory object size with *ftruncate()*.
- Use *mmap()* and MAP_SHARED function to map the shared memory object into the current address space.
- Read/write the shared memory region.
- Unmap the shared memory region with *munmap()*.
- Close the shared memory object with *close()*.
- Delete the shared memory object with *shm_unlink()*.

The mentioned POSIX shared memory system calls are described in the subsequent section:

- **SYSTEM CALL: shm_open() :- create and open an object**:
  The *shm_open()* function creates and opens a new shared memory object with a specified name or opens an existing shared memory object. The *shm_open()* function is given as follows:

    ```
    #include <fcntl.h>
    #include <sys/stat.h>
    #include <sys/mman.h>
    int shm_open(const char *name, int oflag, mode_t mode);
    ```

TABLE 13.1

List of *oflag* behaviors in *shmopen()* System Call

Oflag	Description
O_RDONLY	Open the object for read access
O_RDWR	Open the object for read–write access
O_CREAT	Create the shared memory object if it does not exist
O_EXCL	If **O_CREAT** was also mentioned, and a shared memory object with the given *name* already exists, return an error
O_TRUNC	If the shared memory object already exists, truncate it to zero bytes

The *name* argument recognizes the shared memory object that to be created or opened. The *oflag* argument is a mask of bits that define the behavior of the created object, as defined in Table 13.1. On successful execution, it returns file descriptor on success, otherwise –1 on error.

- **SYSTEM CALL: ftruncate() :- set the size of the shared memory object:**
  The two system calls, namely truncate() and ftruncate (), are used to truncate a file to a specified length. **The ftruncate**() can also be used to set the size of a POSIX shared memory object. The **ftruncate** () function is given as follows:

```
#include <unistd.h>
#include <sys/types.h>
int ftruncate(int fd, off_t length);
```

The **ftruncate**() function set the size of shared memory object, as referenced by *fd* to be truncated to an exactly mentioned size, *length* in bytes. A newly created shared memory object has a length of zero. If the existing object was larger than this size, then the additional or extra data is lost. If previously, it was shorter, then it is extended, and this part reads as null bytes. In short, *ftruncate()* system call is used to expand or shrink the shared memory object as described. By using the system call *fstat()*, you can get the size of existing object. On successful execution of **ftruncate**() function, it returns 0, otherwise -1 on error.

- **SYSTEM CALL: mmap():- map the shared memory object:**
  The mmap() function maps the shared memory object into the virtual address space of the calling process. The starting address for the new mapping is mentioned in *addr*. The *length* argument specifies the length of the mapping that should be greater than the **mmap** () function as given as follows:

```
#include <sys/mman.h>

    void *mmap(void *addr, size_t length, int prot, int flags,
    int fd, off_t offset);
```

On successful execution of *mmap()* function, it returns a pointer to the mapped area, otherwise -1 on error. On the other hand, *munmap()* system call is used to unmap the shared memory object from the virtual address space of the calling process.

```
    int munmap(void *addr, size_t length);
```

On successful execution, it returns 0, otherwise -1 on error.

- **SYSTEM CALL: shm_unlink() :- remove a shared memory object name**:
  The *shm_unlink()* function removes the shared memory object, as mentioned by name. The POSIX shared memory objects have kernel persistence feature; it means objects remain in kernel until they are clearly deleted, or the system is restarted. If there is no further need of shared memory object, then it should be removed by using *shm_unlink()* function. The *shm_unlink()* function is given as follows:

  ```
  #include <sys/mman.h>
  int shm_unlink(const char *name);
  ```

On successful execution, it returns 0, otherwise -1 on error.

13.3 Message Queue

Message queue is one of the methods of the IPC among various processes. It permits processes to send and receive messages between them and access them over a common system, referred to as message queue. Message queue offers an asynchronous communication mechanism in which the sender and receiver do not need to communicate with each other before sending the message. The sender placed the message in message queue, and it is stored until the receiver retrieves it. Message queue is a type of one-way pipe, one or more processes to write messages in message queue, that will be read by one or more reading processes until an end-of-data condition occurs. Every IPC message has an explicit length, usually small (in bytes) and queue length (maximum number of pending messages).

Linux maintains a list of message queues that are exclusively recognized by a positive integer, *msqid*. Each element of message points to a data structure of type *struct msqid_ds*, defined in *<sys/msg.h>*. When message queues are created, a new *msqid_ds* data structure is allocated from system memory [1,2]. The message queue data structure, *msqid_ds*, is containing the following elements:

```
struct msqid_ds
  {
      struct ipc_perm msg_perm;
      msgqnum_t      msg_qnum;      /* no of messages on queue */
      msglen_t       msg_qbytes;    /* bytes max on a queue */
      pid_t          msg_lspid;     /* PID of last msgsnd() call */
      pid_t          msg_lrpid;     /* PID of last msgrcv() call */
      time_t         msg_stime;     /* last msgsnd() time */
      time_t         msg_rtime;     /* last msgrcv() time */
      time_t         msg_ctime;     /* last change time */
  };
```

- **msg_perm ipc_perm**: Structure that denotes the access permissions on the message queue.
- **msg_qnum**: Presently, number of messages on the message queue.
- **msg_qbytes**: Allowed maximum number of bytes of message text on the message queue.

- **msg_lspid**: Process ID of the process that performed the last *msgsnd()* system call.
- **msg_lrpid**: Process ID of the process that performed the last *msgrcv()* system call.
- **msg_stime**: Time of the last msgsnd() system call.
- **msg_rtime**: Time of the last msgrcv() system call.
- **msg_ctime**: Time of the last system call that changed a member of the *msqid_ds* structure.

It is also worth mentioning the key difference between message queue and shared memory, which are as follows:

1. In message queue, after receiving message by a process, it will not be no longer available for any other processes, whereas in shared memory, the data is available to access by various processes.
2. Message queue has inherent synchronization mechanism, but shared memory provides no inherent synchronization.
3. Shared memory seems to be faster as low overhead and high volume of data transferring as compared to message queue where message size is usually very small (in bytes).
4. Data integrity is high in message queue as compared to shared memory.

Basically, Linux offers two varieties of message queues: System V message queues and POXIS message queues. Both offer almost similar functionality, but system calls for the two are different. System V message queues came into existence in the 1980s while POSIX message queues were introduced in 1993 but both are still in the necessity of Unix/Linux-based system. The subsequent section describes the system calls of both message queues.

13.3.1 System V Message Queues

This section describes System V message queues that allow processes to exchange data in the form of messages. To perform IPC using message System V queues, the following steps are required along with the mentioned system calls:

Step 1: msgget() - Create a message queue or connect to an existing message queue

Step 2: msgsnd() - Write message into the message queue.

Step 3: msgrcv() - Read message from the message queue.

Step 4: msgctl() - Perform control operations on the message queue.

- **SYSTEM CALL:** *msgget()* –create a message queue or connect to an existing message queue:

   The msgget () system call creates a new message queue or connects to an existing queue through its identifier (*msgid*). The **msgget** () function is given as follows:

```
#include <sys/types.h> /* For portability */
#include <sys/msg.h>
#include <sys/ipc.h>
int msgget(key_t key, int msgflg);
```

The argument *key* is an integral value that is used to recognize the required message queue. The flag argument is used to specify zero or more flags that are used to specify whether the queue already exists or not. If the execution of **msgget** () function is successful, it returns message queue identifier, otherwise –1 on error.

- **SYSTEM CALL: msgsnd() –write message into message queue**:
  The **msgsnd**() system call is used to send or copy messages, pointed by *msgp*, to System V message queue, mentioned as *msqid*. The msgsz argument comprises the size of the message (in bytes), without the length of the message type.

- **Message buffer**: This is a particular data structure for message data for *msgsnd()* and *msgrcv()* calls. The *msgp* argument is a pointer to a caller-defined data structure *msgbuf* of the following form:

```
struct msgbuf {
    long mtype;        /* Message type must be > 0 */
    char mtext[1];     /* message text body */
};
```

There are two members in the *msgbuf* structure:

- *mtype:* The message type of a positive number must be greater than zero.
- *mtext:* The message data itself.

- **Kernel msg structure**
  The kernel keeps each message in the message queue in the given *msg* structure. It is defined in *linux/msg.h*. One *msg* structure for each message and the elements are as follows:

```
struct msg {
    struct msg *msg_next;   /* next message on queue */
    long msg_type;
    char *msg_spot;         /* message text address */
    short msg_ts;           /* message text size */
};
```

There are four members in the *msg* structure:

- *msg_next:* This is a pointer to the next message in the queue. All are stored as a single linked list within kernel addressing space.
- *msg_type:* This is the message type, as mentioned in *msgbuf* structure.
- *msg_spot:* A pointer to the start of the message body.
- *msg_ts:* The text size of message or message body.

The msgsnd() function is given as follows:

```
#include <sys/types.h>
#include <sys/ipc.h>
#include <sys/msg.h>
int msgsnd(int msqid, const void *msgp, size_t msgsz, int msgflg);
```

On successful execution, it returns 0, otherwise -1 on error.

- **SYSTEM CALL: msgrcv() – read message from the message queue**:
  The **msgrcv**() system call is used to receive messages from System V message queue specified by *msqid* and places it in the buffer pointed by *msgp* structure (used to hold the received message). The third argument *msgsz* specifies the maximum size of message text *mtext* (in bytes) for the structure pointed by the *msgp* argument. *The msgrcv()* function is given as follows:

```
#include <sys/types.h>
#include <sys/ipc.h>
#include <sys/msg.h>

ssize_t msgrcv(int msqid, void *msgp, size_t msgsz, long msgtyp,
                int msgflg);
```

On successful execution, it returns number of bytes copied into message buffer (*mtext* field), otherwise -1 on error.

- **SYSTEM CALL: msgctl() – read message from the message queue**:
  The **msgctl**() system call performs the control operation mentioned by *cmd* on the message queue that is identified with *msqid*. *The msgctl()* function is given as follows:

```
#include <sys/types.h>
#include <sys/ipc.h>
#include <sys/msg.h>
int msgctl ( int msgqid, int cmd, struct msqid_ds *buf );
```

On successful execution, it returns 0, otherwise -1 on error. The *cmd* argument specifies the following operations that to be operated on queue:

- **IPC_STAT**: Saves the copy of *msqid_ds* structure for a queue, in the address of the *buf* argument.

- **IPC_SET**: Sets particular fields of the *msqid_ds* data structure connected with message queue with the values provided in the *buf* argument.

- **IPC_RMID**: Immediately deletes the message queue object and its associated data structure (*msqid_ds)* from the kernel.

13.3.2 POSIX Message Queues

POSIX message queues permit processes to exchange data in the form of messages but provide similar functionality as compared to System V message queue. POSIX queues are like the named pipes. In POSIX queues, each message queue is recognized by a name, in the format of /*queue-name*; that is, a null-terminated string of maximum size to **NAME_MAX** (i.e., 255), in which the name string is comprised of an initial forward slash(/), followed by one or more characters (none of which are slashes) and ending with the null character. The message queue structures are found in the <*mqueue.h*> header file. POSIX message queues are priority-driven that are a comparatively recent addition to Linux kernel version 2.6.6 announced in 2004. POSIX message queues have kernel persistence: if not removed by *mq_unlink*(), it will exist until the system is shutdown.

The main system calls in the POSIX message queue are as follows:

- **mq_open()**: Opens an existing queue or creates a new message queue.
- **mq_send()**: Writes a message to queue.
- **mq_receive()**: Reads a message from queue.
- **mq_close()**: Closes a message queue.
- **mq_unlink()**: Removes a message queue name and cleans from kernel.

- **Message queue attributes:**
  Every message queue is related to set of attributes, out of which some may be set while creating or opening queue by using *mq_open()*. Further, there are two functions, namely *mq_getattr()* and mq_setattr(), that are used to retrieve and modify message queue attributes, respectively. The *mq_notify()* function allows a process to register for message notification from a queue. The *mq_open(), mq_getattr(),* and *mq_setattr()* functions also allow an argument that is a pointer to an *mq_attr* structure. The *mq_attr* structure is defined in *<mqueue.h>* header file and given as follows:

```
struct mq_attr {
        long mq_flags;      /* Message queue description flags*/
        long mq_maxmsg;     /* Maximum number of messages on queue*/
        long mq_msgsize;    /* Maximum message size (in bytes)*/
        long mq_curmsgs;    /* Number of messages currently in queue */
};
```

- **SYSTEM CALL: mq_open()** - creates **a new message queue or** opens **an existing queue**:
  The mq_open() function creates a new POSIX message queue or opens an existing queue. The queue is identified by *name. The mq_open()* function is given as follows:

```
#include <fcntl.h>
#include <sys/stat.h>      /* For mode constants */
#include <mqueue.h>

 mqd_t mq_open(const char *name, int oflag);
 mqd_t mq_open(const char *name, int oflag, mode_t mode,
                  struct mq_attr *attr);
```

On successful execution, it returns a message queue descriptor, otherwise (mqd_t) –1 on error. The *oflag* argument mentioned flags that control the operation of the *mq_open()*. The detail definition flags are included in *<fcntl.h>* header file. The associated operation with each flag is described in Table 13.2.

- **SYSTEM CALL: mq_send()** - **sends a message to a queue:**
  The **mq_send**() function incorporates the message, pointed by *msg_ptr* to the message queue, named by descriptor *mqdes*. The *msg_len* argument denotes the size of the message pointed by *msg_ptr*. This size must be less than or equal to the *mq_msgsize* attribute, else *mq_send()* gives *error of EMSGSIZE* . The message size

TABLE 13.2

oflag Operation Description

Oflag	Description
O_CREAT	Create the message queue if it does not exist
O_EXCL	If O_CREAT was mentioned in *oflag*, and a queue exists with the given name, then it fails with the error EEXIST
O_RDONLY	Open the queue for reading/receiving only
O_WRONLY	Open the queue for sending/writing only
O_RDWR	Open the queue for both reading and writing
O_NONBLOCK	Open the queue in nonblocking mode

of zero length is acceptable. The *msg_prio* argument is a nonnegative integer that mentioned the priority of each message. *The mq_send() function is given as follows:*

```
#include <mqueue.h>
int mq_send(mqd_t mqdes, const char *msg_ptr, size_t msg_len,
unsigned int msg_prio);
```

On successful execution, it returns 0, otherwise -1 on error.

- **SYSTEM CALL: mq_receive() - receives a message from a message queue**:
  The **mq_receive**() function deletes the oldest message with the highest priority from the message queue as mentioned by the message queue descriptor *mqdes* and copies it in the buffer pointed by *msg_ptr*. The *msg_len* argument denotes the available space in the buffer (in bytes) pointed by *msg_ptr*. The *msg_len* must be greater than or equal to the *mq_msgsize* attribute of the queue, otherwise *mq_receive()* fails with the error EMSGSIZE. If *msg_prio* is not NULL, then the priority of the received message is placed into the location pointed by *msg_prio*. The *mq_receive()* function is given as follows:

```
#include <mqueue.h>
ssize_t mq_receive(mqd_t mqdes, char *msg_ptr, size_t msg_len,
unsigned int *msg_prio);
```

On successful execution, it returns number of bytes in received message, otherwise -1 on error.

- **SYSTEM CALL: mq_close() - Closing a message queue**:
  The *mq_close()* function closes the message queue descriptor *mqdes*.

```
#include <mqueue.h>
int mq_close(mqd_t mqdes);
```

On successful execution, it returns 0, otherwise -1 on error.

- **SYSTEM CALL: mq_unlink() - removes a message queue**:
  The *mq_unlink()* removes the mentioned message queue by *name*. The message queue name is removed immediately.

```
#include <mqueue.h>
int mq_unlink(const char *name);
```

On successful execution, it returns 0, otherwise -1 on error.

13.4 Pipes: Named and Unnamed

A pipe is another oldest method of IPC that allows unidirectional communication or inter-action among processes. Pipes came into the existence in Unix in the early 1970s. In other words, a pipe is a method of connecting two or more processes in which the standard output of one process is connected to the standard input of another process and so on. This concept is extensively used on the Unix/Linux command line (in the shell).

Basically, the pipe connects two or more processes, in which it allows to flow data from one process to another, i.e., the output of one process is the input of another process till the last process in the list of processes. Every user of Linux must be familiar with the use of pipes in communicating commands on shell. This shell feature is known as pipelines. The pipeline can be implemented by using a vertical bar character (|), referred to as unnamed pipe.

The basic syntax of pipelining is given as follows:

```
$command_1 | command_2 | command_3 | command_4 ...
```

In the above syntax, pipe operator is placed between two commands to form a connection between them, in which the output of one command (command_1) becomes the input for another command (command_2). In a similar way, the output of command_2 becomes the input of next command (command_3) and so on. Therefore, if you want to communicate *N* different commands, then you need *N-1* pipe operators. In pipelining, the output of the last command is the output of all N commands that are connected through pipe operators. For example,

```
$ ls | wc | lpr
```

The above example combines three commands *ls*, *wc*, and *lpr*. In order to execute the above command, the shell creates two pipes and executes three commands in which the output of *ls* command is the input of *wc* command, and the output of *wc* command is the input of lpr command, and the output of *lpr* command is the final output of all three communicating commands. The *lpr* command prints total number of lines, words, and characters in the current directory.

Some key facts regarding pipes are given as follows:

- **Pipelining acts as filters:**
  You can create pipelining which can also be used to perform complex task on given data. Therefore, it has the possibility of combining numerous commands together into a pipeline, to create customized filters, for example:

  ```
  #ls /etc /usr/bin | sort > file1
  ```

- **Allow sequentially movement of data:**
  A pipe is a byte stream, which indicates that there is no concept of messages or message boundaries when using a pipe. The data passes through the pipe in sequentially byte stream form, i.e., bytes are read from a pipe, exactly in the same sequence as they were written into pipe. It is not possible to randomly access the data in a pipe.

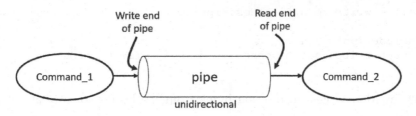

FIGURE 13.4
Pipe layout

- **Pipes are unidirectional**:
  In pipes, data can move only in one direction from one command to another command. One end of the pipe is used for writing data, known as input end of pipe, and the other end is used for reading data, known as output end of pipe, as shown in Figure 13.4.

13.4.1 Creating pipes in C

The *pipe()* system call creates a pipe, a unidirectional data channel that can be used to set up an IPC channel between related processes. The *pipe()* system call returns two open file descriptors in the array file descriptor *fd[2]* which are referring to the ends of the pipe. The array file descriptor *fd[0]* refers to the read end and *fd[1]* refers to the write end of the pipe. The *write()* system call is used to send or write the data into pipe and *read()* system call is used to retrieve or read the data from the pipe as shown in Figure 13.5. The c syntax of *pipe()* system call is given as follows:

```
#include <unistd.h>
int pipe(int fd[2]);
```

On successful execution, it returns 0, otherwise -1 on error.

```
fd[0])→ for read() ;   fd[1])→ for write()
  read () - read from a file descriptor
       #include <unistd.h>
       ssize_t read(int fd, void *buf, size_t count);
```

The *read()* function tries to read up to *count* bytes from file descriptor *fd* into the buffer starting at *buf.*

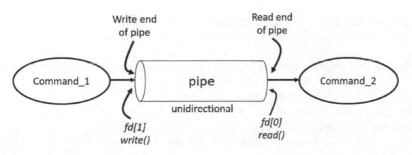

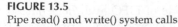

FIGURE 13.5
Pipe read() and write() system calls

```
write() - write to a file descriptor

   #include <unistd.h>
   ssize_t write(int fd, const void *buf, size_t count);
```

The *write()* function writes up to *count* bytes from the buffer starting at *buf* to the file descriptor *fd*.

13.4.2 named pipes : FIFO

Linux also supports *named* pipes, which is known as first-in, first-out (FIFO). It is one of the methods of IPC between unrelated processes (e.g., a client and server). A FIFO is an extension to the traditional pipe concept that operates on a FIFO principle. The main difference with traditional pipe (referred to as unnamed pipe) is that a FIFO has a name within the file system and opened in the similar manner as regular file. Once FIFO has been created, you can use the same input/output system calls as used with pipes and other files like read(), write(), and close().

Any process may create and open a FIFO, subject to file permission to permit it. The *mkfifo()* function creates a new FIFO, named pipe, with the given name *filepathname*, which exists as a device-specific file in the file system [4]. The C function of *mkfifo()* is given as follows:

```
#include <sys/types.h>
#include <sys/stat.h>
int mkfifo(const char *filepathname, mode_t mode);
```

On successful execution, it returns 0, otherwise -1 on error. You can also create a FIFO from the shell by using the *mkfifo* command given as follows:

```
$ mkfifo [ -m mode] filepathname
```

In the above command syntax, the *filepathname* is the name of the FIFO that to be created, and -m option is used to mention a permission *mode* that is similar to *chmod* command. The *mode* argument specifies the permissions of newly created FIFO.

The read end of FIFO is opened, using *open()* system call, with the *O_RDONLY* flag, and the write end is opened using the *O_WRONLY* flag. The C function of *open()* is given as follows:

```
        #include <sys/types.h>
        #include <sys/stat.h>
        #include <fcntl.h>

        int open(const char *pathname, int flags);
        int open(const char *pathname, int flags, mode_t mode);
```

For example, we can create a named pipe in */home/myself* by calling *mkfifo* command given as follows:

```
$mkfifo -m /home/myself
$ ls -l /home/yourself
prw-r--r-- 1 sanrit sanrit 0 2019-02-18 16:04 /home/myself,
```

In the above commands execution, -m is the permission mode of FIFO and */home/myself* is the pathname of the FIFO which created. The output of *ls -l* is produced as prw-r--r-- 1 sanrit sanrit 0 2019-02-18 16:04 /home/yourself, in which **"p"** in the permission column indicates the type of file that is named pipe along with permission mode of it.

To show the working of named pipe, two terminal consoles are to be opened. In one console, we will give the following command:

```
$mkfifo /home/myself
$ cat < /home/myself
```

After that as soon as enter key button is pressed, the terminal will hang. This state is referred to as blocked pipe. Subsequently, we will open another terminal (second console) and type the following command:

```
$ cat > /home/myself
Hello, I am a new user!!
```

And enter text or anything into this console (second). It will appear in the first terminal (first console) as a write to the pipe. And this state is called as nonblocking pipe state. The display is given as follows:

```
$ cat < /home/yourself
Hello, I am a new user!!
```

which is similar to client–server architecture, where one terminal is acting as client and another terminal as server.

13.5 Sockets

Now, this section describes another method of IPC, referred to as socket, which was announced in Berkeley versions of Unix, BSD 4.2 in 1983. Hence, in other words, sockets are a method of IPC, that is, an extension of the pipe concept with some key difference. Sockets allow data to be communicated between processes that allow client/server systems to be established either on the same host (computer), i.e., locally on a single machine, or on different hosts connected over a network. Linux functions, such as printing, and network utilities, such as telnet, ping, rlogin, and ftp, generally use sockets to communicate. The socket mechanism may implement numerous clients attached to a single server.

We are going to provide a general overview of various system calls involved in socket communication [5]. In a general client–server situation, processes that communicate using sockets are as follows:

- Every process creates a socket through which communication occurred. Both communication processes, known as client process and server process, required one socket each for communication.
- A socket is created by using the *socket ()* system call, which creates an endpoint (client and server) for communication and returns a file descriptor that refers to that endpoint.
- A socket is characterized by three attributes. The names of such attributes are domain, type, and protocol.

- **Connection-oriented client–server communication**
- **Client**
  - Create socket for communication
  - Establish connection
  - Knowledge of address and port of the server
  - Active socket for communication
  - Close connection
- **Server**
  - Create socket for communication
  - Assign port to a socket
  - Listen for communication
  - Respond to clients, if available
  - Accept and communicate
  - Close connection
- **Socket creation in C: socket():**
  - The socket () system call is used to create socket, and the c function of *socket()* is given below:

```
#include <sys/types.h>
#include <sys/socket.h>
int sockfd = socket(int domain, int type, int protocol);
```

  - On the successful execution, it returns an integer file descriptor that is used to refer the socket, otherwise -1 on error.
  - The *socket()* function has three attributes: domain, type, and protocol. The subsequent section describes each attribute in brief.

- **Domain (socket communication domain):**
  Domain attribute specifies the network medium that socket used for communication. The basic features of socket domain are given below:
  - The technique of identifying a socket, i.e., the format of socket *"address."*
  - The range of communication (means either between processes on the same system or between processes on different systems connected via a network).

  Almost all Linux distributions support the following domains:
  - The UNIX(AF_UNIX) domain permits communication between the processes on the same host, i.e., AF_UNIX, AF_LOCAL for local communication and file addresses.
  - The IPv4(AF_INET) domain permits communication between the processes running on systems connected through an Internet Protocol version 4 (IPv4) network. AF_INET is common socket domain that is used on many Unix local area networks and, of course, the Internet itself.

- The IPv6(AF_INET6) domain permits communication between the processes running on systems connected through an Internet Protocol version 6 (IPv6) network (IPv6 is the successor of IPv4) Currently, It is also most extensively used protocol.

- **Socket type**:

    The implementation of each socket offers at least two categories of sockets, namely stream and datagram. These socket categories are supported in both fields: Unix/Linux and computer networking (Internet).

    - **Stream sockets (SOCK_STREAM)**: It offers a reliable, bidirectional, and byte stream communication channel. It is also mentioned as connection-oriented and usually uses service of the Transmission Control Protocol (TCP). They are implemented in the AF_INET domain by TCP/IP connections. But they are also the normal type in the AF_UNIX domain. This section will concentrate on SOCK_STREAM sockets only. TCP/IP stands for Transmission Control Protocol/Internet Protocol. TCP offers sequencing, flow control, and retransmission to confirm large data transfer smoothly. IP is the low-level protocol for packets that offer routing through the network from one computer to another.

    - **Datagram sockets (SOCK_DGRAM)**: It offers an unreliable and connectionless communication channel to transfer data in the form of messages called datagrams. With datagram sockets, message boundaries are preserved and use user datagram protocol (UDP) to transmit messages. Due to not setting up and maintaining a connection, messages may be lost, duplicated, or arrived out of sequence at the receiver end. Datagram sockets are also known as connectionless sockets and implemented in the AF_INET domain by UDP/IP connections (sequenced and unreliable service).

- **Protocol**: It specifies protocol type as given below, but usually set to 0 to select the system's default protocol for the given combination of socket family and type.

    - **IPPROTO_TCP**: TCP transport protocol
    - **IPPROTO_UDP**: UDP transport protocol

System calls for socket IPC:

The main system calls for socket IPC are given as follows:

- **The *socket()* system call**: creates a new socket for connection.
- **The *bind()* system call**: binds a socket to an address.
- **The *listen()* system call**: permits a stream socket to accept incoming connections from other sockets.
- **The *accept()* system call**: accepts a connection from a listening stream socket.
- **The *connect()* system call**: set up a connection with another socket.
- **The *send()* System call** : transmits data over connection
- **The *receive()* system call**: receives data over connection
- **The *close()* system call**: closes the connection

Figure 13.6 demonstrates the use of socket IPC system calls over TCP communication, known as stream sockets.

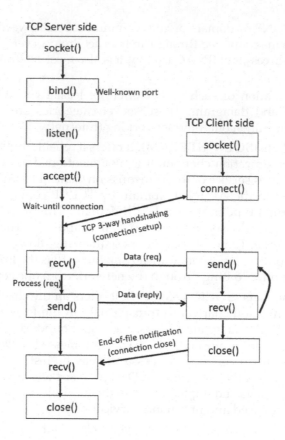

FIGURE 13.6
Overview of socket IPC system calls over TCP communication

- **Socket address structures: struct sockaddr**
  - The socket API specifies a generic address structure, struct *sockaddr*, to identify the information of various domain-specific address structures, which are used as arguments in socket system calls. The *sockaddr* structure is basically defined as follows:

```
struct sockaddr
{
    unsigned short sa_family; /* Address family (for ex. AF_INET) */
    char sa_data[14];         /* Family-specific address information */
}
```

  - The below structure is a particular form of the *sockaddr* used for TCP/IP addresses:

```
struct sockaddr_in
{
unsigned short sin_family;  /* Internet protocol (AF_INET) */
unsigned short sin_port;    /* Address port (16 bits) */
struct in_addr sin_addr;    /* Internet address (32 bits) */
char sin_zero[8];           /* Not used */
}
```

sockaddr_in may be casted to a *sockaddr*. This structure assists as a template for all of the domain-specific address structures.

- **Dealing with IP addresses:**
  The four bytes of an IP address create a single 32-bit value. An AF_INET socket is fully described by its domain, IP address, and port number. The IP address structure, *in_addr*, is given as follows:

```
struct in_addr
{
 unsigned long s_addr;   /* Internet address (32 bits or 4 bytes) */
}
```

- **Assign address to socket: bind():**
  After creating a socket by using *socket()* system call, then you must attach an address to this socket by using the *bind()* system call to make it available for use by other processes. The *bind()* system call is defined as follows:

```
#include <sys/types.h>
#include <sys/socket.h>
int bind(int sockfd, const struct sockaddr *addr, socklen_t addrlen);
```

  - The *sockfd* argument is a file descriptor obtained from a *socket()* system call.
  - The *addr* argument is a pointer to a struct *sockaddr* that contains information about IP address and port of the machine; for TCP/IP server, Internet address is usually set to INADDR_ANY, i.e., choose any incoming interface.
  - The *addrlen* argument specifies the size of the address structure, i.e., struct *sockaddr*
  - It returns 0 on success, otherwise –1 on error

- **Listening for incoming connections: listen():**
  The *listen()* system call is used to accept incoming connections from other sockets. A server program must create a queue to store pending requests, and this can be done using *listen ()* system call:

```
#include <sys/types.h>
#include <sys/socket.h>
int listen(int sockfd, int backlog);
```

  - Waits for incoming connections.
  - *sockfd* is the socket file descriptor returned by *socket()* call.
  - *backlog* is the integer number of active connections that can "wait" for a connection.
  - It used by the server only to get new sockets.
  - It returns 0 on success, otherwise -1 on error.

- **Accepting a connection: accept():**
  After creating and naming the socket, the server program uses the *accept()* system call to accept an incoming connection on the listening stream socket denoted by the file descriptor *sockfd*. A file descriptor for the connected socket for an incoming client is returned as *sockfd* after the successful call of *accept()* function:

```
#include <sys/types.h>
#include <sys/socket.h>
int accept(int sockfd, struct sockaddr *addr, socklen_t *addrlen);
```

- *sockfd* is the listening socket descriptor.
- The *addr* argument is a pointer to a struct *sockaddr* that contains information about IP address and port of destination machine.
- The *addrle*n argument specifies the size of the address structure, i.e., struct *sockaddr*
- It returns socket file descriptor of connected client socket on success, otherwise -1 on error
- At this point, connection is established between client and server and ready to exchange data.

- **Set up connection with a peer socket: connect()**
  The *connect()* system call to establish a connection between client programs and servers. It connects the active remote socket mentioned by the file descriptor *sockfd* to the listening socket. The *connect()* system call is defined as follows:

```
#include <sys/types.h>
#include <sys/socket.h>
 int connect(int sockfd, const struct sockaddr *addr, socklen_t
addrlen);
```

- It connects to a remote host referred to as *sockfd*.
- The *addr* argument is a pointer to a struct *sockaddr* that contains information about IP address and port of destination machine.
- The *addrle*n argument specifies the size of the address structure, i.e., struct *sockaddr*.
- It returns 0 if successful connect, otherwise -1 on error, which means no connection and wish to reattempt the connection

- **Socket-specific I/O system calls: send() and recv():**
  The two functions, **send() and recv(),** are for communication over stream sockets or connected datagram sockets. Here, we explained exchanging data with stream socket only.

  - **Transmit data: send():**
  The *send()* function is used to transmit data over network which is defined as follows:

```
#include <sys/types.h>
#include <sys/socket.h>
 ssize_t send(int sockfd, const void *msg, size_t length, int flags);
```

  - *sockfd* (received from *accept()* system call) is the socket file descriptor to which you want to transmit data.
  - *msg* is a pointer to the data to be transmitted.
  - *length* is the size of data (in bytes) to transmit.

– *flags*, a bit mask that modifies the behavior of the I/O operation, usually set to 0.

– On successful execution, *send()* returns the number of bytes actually transmitted, and it may be less than the number you requested to send it or -1 on error.

- **Receive data: recv():**

The *recv()* function is used to receive data over network which is defined as follows:

```
#include <sys/types.h>
#include <sys/socket.h>
ssize_t recv(int sockfd, void *msg, size_t length, int flags);
```

- *sockfd* is the socket file descriptor to which you want to receive data.
- *msg* is a pointer to the data to be received.
- *length* is the size of data (in bytes) to receive.
- *flags*, a bit mask that modifies the behavior of the I/O operation, usually set to 0.
- On successful execution, *recv()* returns the number of bytes actually received or -1 on error
- If *recv()* returns 0, the remote side has closed current connection.

- **Connection termination: close():**

The usual way of terminating a stream socket connection is by calling *close()* function that closes connection corresponding to the socket descriptor, *sockfd*. You should always close the socket at both ends after communication is over. The *close()* system call is defined as follows:

```
#include <sys/types.h>
#include <sys/socket.h>
int close(int sockfd);
```

- *sockfd* is the socket file descriptor to which you want to close connection.
- Return 0 if successful, otherwise -1 on error.
- Frees up the port used by the socket.

13.6 Summary

This chapter provided insight of various methods of IPC facilities that are used to communicate processes (and threads) with others and transfer data with each other. There are various ways to communicate between parents and child processes, unrelated processes, and even processes of different systems. The IPC services are classified into communication, synchronization, and signal. Some key IPC services, message queues, shared memory, pipes, FIFOs, and sockets are described with their system call in C language. Shared memory is one of the basic and simplest ways of IPC, in which two or more processes (related or unrelated) may access the same memory region. There is no binding of kernel interference in the case of data exchange through shared memory. After placing the data into shared memory, it is immediately visible to other processes. Linux offers three standard specifications for shared memory: System V shared memory, POXIS shared memory, and memory mapping.

Message queue is another method of IPC with the inherent feature of synchronization mechanism, but shared memory provides no inherent synchronization. On the other hand, shared memory seems to be faster than message queue. Pipes were the oldest method of IPC under the Unix system. A pipe, created by *pipe()* system call, is a unidirectional and byte stream data flow. It acts as filter that can be used for communication among related processes. On the command line, a pipe can be represented by using vertical bar (|) between two commands that can also referred to as unnamed pipe. Linux also supports *named* pipes, known as FIFOs that operate in exactly the same way as pipes, except that they are created using *mkfifo()*, have a name in the file system and can be opened by any process with appropriate permissions. Socket, another IPC mechanism, allows communication between applications and processes of same system or on separate system connected via network. In wider prospect, generally processes use one of two socket types: stream or datagram. Stream sockets (SOCK_STREAM) offer reliable, bidirectional, and byte stream communication channel between two processes. On the other hand, datagram sockets (SOCK_DGRAM) provide unreliable, connectionless, and message-oriented communication. In a stream socket communication, you can create a stream socket server by using *socket()* system call. The other system calls which required to step up communication between client and server are *bind(), listen(), accept(), connect(), send(), recv(), and close()*.

13.7 Review Exercises

1. Highlight various synchronization services along with their classifications.
2. Write down the name and syntax of various system calls used for message queues?
3. Predict the output of the following program code?

```
main()
{
fork();
printf("hello World!");
}
```

4. Write a program of full-duplex communication and function syntax for opening an existing message queue.
5. How can we sort a file and print unique values using pipes?
6. What does mkfifo() create?
7. Which system call is used to create System V message queue?
8. Which is the fastest IPC method, justify with a suitable reason?
9. Discuss the persistence level of shared memory segments?
10. Which structure keeps the information about shared memory in the kernel?
11. Write a command for knowing the number of active processes attached to shared memory segment.

12. One process requires M resources to complete a job. What should be the minimum number of resources available for N processes so that at least one process can continue to execute without blocking/waiting?

13. Which call is used to set the resource count of semaphore?

14. Discuss various types of sockets? Explain AF_INET socket.

15. What is the difference between a TCP socket and a UDP socket? Why would you use one or the other?

16. Write down the basic structure of socket address and how to create a socket.

17. Discuss and pictorially represent each and every socket IPC system call that takes place during TCP communications.

18. What happened, if return data type of pipe() is not mentioned? explain, why pipe is very helpful in filtering process, justify with suitable example.

19. Is it possible to reallocate shared memory?

20. Where the shared memory is created, in user space or kernel space?

21. How shared memory is accessed if we have two processors trying to acquire the same region?

22. How would you set a socket to nonblocking mode? In what situations would you do that?

23. Explain the following system calls:
    a. Bind
    b. Listen
    c. Accept
    d. Connect
    e. Send and receive
    f. Close

24. Why the pipes are called unidirectional? Create the pipe which acts as a filter.

25. Write a brief description of the following flags:
    a. O_RDONLY
    b. O_RDWR
    c. O_CREAT
    d. O_EXCL
    e. O_TRUNC

26. Write a brief description of the following fields of System V shared memory data structure:
    a. shm_perm
    b. shm_segz
    c. shm_dtime
    d. shm_ctime
    e. shm_cpid
    f. shm_lpid
    g. shm_nattch

References

1. Michael Kerrisk. Linux man pages in the World Wide Web. http://man7.org/linux/man-pages/.
2. Michael Kerrisk. 2010. *The Linux Programming Interface*. No Starch Press, Inc.
3. J. Purcell. 1997. *Linux Complete: Command Reference*. Red Hat Software, Inc.
4. Mark Mitchell, Jeffrey Oldham, and Alex Samuel. 2001. *Advanced Linux Programming*. New Riders Publisher.
5. Neil Matthew and Richard Stones. 2001. *Beginning Linux Programming*. Wrox Press Ltd.

14

X Window System Overview and Programming

This chapter describes the very important aspect of Linux environment, i.e., the X Window System. All the previous chapters were dedicated to various other aspects of the Linux system, i.e., the kernels, architectures, shells, shell scripting, etc. X Window is a graphical window system that Linux and many other Unix systems use. It is equivalent to the window environment in Windows, OS/2, or Mac OS. The basic idea is to create an X Window System by MIT to achieve a uniform environment for graphical programs. X Window System offers a fundamental set of graphical operations which may be used by user interface applications like window managers, file managers, desktops, etc. These sets of operations help window managers to create widgets for operating windows, like scroll bars, resize boxes, and close boxes. X Window System is available for most Linux distributions. Hence, this chapter provides an overview of X Window System, starting with the introduction of X Window System and its customization, and the latter sections will cover various topics, ranging from X client, X server, X protocol, and Xlib to desktop environments and upgrading X window tools.

14.1 X window System and Its Customization

In today's era, graphical user interface (GUI) is an essential portion of all running operating systems, widely referred to as a desktop environment. The standard GUI is the X Window System, also known as X or X11, which was initially intended for Unix-based frameworks. The X Window System is used to provide a GUI for most Linux and Unix systems that was originally implemented in 1984 by the Massachusetts Institute of Technology (MIT) Laboratory in cooperation with Digital Equipment Corporation (DCE) as part of the project "Athena." In 1987, MIT released the X Window System as X11 version. X11 is the version most frequently seen in Linux distributions. MIT implemented the X Window System but later it was handed over to the X.Org Foundation for further development.

Initially, the Linux distributions used the X Window System from the XFree86 project.* It has freely redistributed open source implementation of X Window System that runs on Linux, Unix, BSD series, Mac OS, OS/2 Solaris(x86 series), etc. Because of some licensing matters, various key Linux providers (including Red Hat, SUSE, and Slackware) changed their X Window System from XFree86 to X from X.Org [1]. The X Window System is currently managed by the X.Org group on behalf of TOG (The Open Group). TOG is a consortium of over hundred companies, including Fujitsu, HCL, IBM, Oracle, Philips, Intel, and many others [2]. The X.Org Foundation, non-profitable organization, maintains the existing X Window System code and provides free official Window System update as well as the latest releases to the general public free of cost. You may visit the X.Org official home

---

* The XFree86 Project, Inc : https://www.xfree86.org/.

page to get more about it [1]. Presently, X Window System is used on most Linux systems under the GNU Public License.

Nowadays, you have come across or are familiar with either Microsoft Windows, X, or Apple's macOS. When you are configuring X by yourself, you can select the video driver, monitor settings, mouse configuration, and other basic graphical features required to get your display working perfectly. To run X Window System, you must install X window server software. Almost, all currently available video cards can run the X Window System. X is not designed for any specific desktop interface, but every window environment is programmed differently to each system.

To maintain the portability and work on different hardware and software platforms, the X Window System was created and implemented through its four components; X client, X server, X protocol (communication link), and X window manager [3]. All these components interact with each other to provide GUI. Basically, all the components of X interact or perform an execution in a client/server style. It may be possible that both X servers and X clients are in the same system or different systems across a network so that computation is processed independently from display rendering, whereas in the X server control hardware, they do not describe display look and not offer any tools to manipulate clients. The key responsibility of X server is making numerous shapes and colors on the screen.

14.1.1 X Client, X Server, and X-Protocol

The client–server architecture, network design of X Window System, is a modularized system that divides work among four separate modules, but closely between two separate and linked programs, known as servers and clients. The X Window System comprises four modules that are mentioned as follows:

1. X client - the application
2. X server - interacting with the user
3. X protocol - client–server interaction
4. X window manager - basically, it is used to manage display task

In client–server model, X server is a server, on which X client's application program runs. In X client–server design, the server is on a local or nearby machine and the client is on a remote or neighborhood or local system, which is opposite of the ordinary client–server model as shown in Figure 14.1. The basic correspondence between X server and X client occurs with the help of X protocol. This is a fundamental communication between both (client and server) through messages about designs UI activities. X protocol is a bundle set–based protocol, which takes the help of TCP/IP protocol when X client and X server are on different PC. However, when X client and X server are both on the same PC, the X protocol is executed by utilizing UDS (Unix space attachments), which works as a library in Linux OS (or other Unix OS) for local communication.

- **X client:**
  In client–server design, X is the application program that requests services from X server. The X server then returns information to the X client to display on the video output device/system. Due to client–server model, the X Window System has an advantage to set up X client locally or remotely. The user programs, CAD

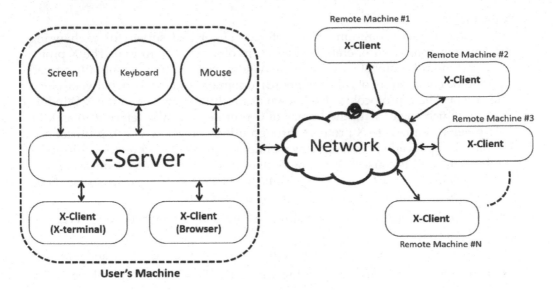

FIGURE 14.1
Layout of X client and X server structure of a local and remote system.

tool, spreadsheet, or Photoshop, are the example of X client that act as a graphical output system. Some other terms which also refer to X client are as follows:

- Web browser application (Firefox)
- Screen (complete desktop display)
- Cursor pointer (used to represent the position of the mouse cursor)
- Window (display frame use another application)
- Open office (office application), mail program (KMail), a terminal emulator

- **X server:**

   X server or X display server is a program in the X Window System that provides direct hardware support to the entire GUI. The most common version of X server used in Linux is XFree86. X server runs on a local machine or computer system (which is accessed by the user directly) and handles all graphical display hardware, such as display cards and screens, and input devices (keyboard and mouse) of that local computer. The X server responds to the request from the X client applications to draw graphics on the display screen or take input from the keyboard and mouse over screen display. Simply, X servers manage hardware, but they don't define the look of the display and there is no tool available to manipulate the client. X server is responsible for rendering various shapes and colors on screen. The *XFree86* or X.Org provides the latest updates and package of X server regularly, and with this support, Linux has the ability to support most hardware for the entire X Window System. The X server performs the following tasks:

   - Grants permission to multiple clients to access the display output
   - Translates network messages received from the client
   - Transfers user input data to the client by sending network messages
   - Preserves complex data resources, including windows, cursor, fonts, and graphic contexts

- **X protocol:**

  X protocol offers communication between client and server for exchanging information about GUI operation in the form of a message format. The X protocol is a standard protocol used by the X Window System that is mutually agreed upon some common set of rules (standard) to connect and transmit between client and server. The X protocol is also known as machine language of the X Window System, which comes with the library of C routine generally referred to as Xlib. Xlib gives you access to X protocol through utilities routine. The X protocol distributes the application processing by mentioning client–server relationship at the application level because X Window System has separated the application part from the hardware part, which permits the hardware system to disassociate from the software. X protocol provides the following advantages:

  - Various languages and the operating system easily supported by X server and X client.
  - No performance degradation in the application while using locally and remotely.
  - In fact, local and remote computing are identical for user and developer.

14.1.2 Xlib

Xlib, also referred to as libX11, is a low-level library written in C language to provide an API for X protocol in X Window System that came into existence around 1985. Xlib is a standard library of X client task that provides methods for using X protocol to interact with X server. It handles low-level details of using X window protocol like drawing, lines, symbols, shapes, and event handling (mouse movement and keyboard-input reaction) and sets up the connection between client and server over the appropriate communication channel. Although Xlib is very convenient and its capabilities are basic, it doesn't provide a function that displays menus, buttons, scroll bars, etc.; if you want such functions, you must write them by yourself. XCB (X Protocol C-Language Binding), started in 2001 by Bart Massey, library is an attempt to replace Xlib that is written in C language and distributed under the MIT license.

14.2 X Toolkits

Xlib is the lowest level of X system software architecture. It is so powerful that many applications can be written directly using Xlib alone, but it is rather a cumbersome or, we can say, time-consuming task to write complex GUI programs using only Xlib. But not to worry, there are various higher-level subroutine libraries to take care of this, known as *X toolkits*.

X toolkit (high-level languages of X) is a set of libraries used by X client in X window server. With the help of this toolkit, a user can create windows, menus, buttons, frames, and many more GUIs. It can also be used to create any window object called widgets. Nowadays, numerous X toolkits are available as shown in Figure 14.2. The best-known tools are *Xt Intrinsics library suite* that comprises X and *two commercial products:* Sun's OpenLook (XView) and Motif. Some of these tools are Qt, GTK+, OpenLook, Xt, and Motif, which are popular and described as follows;

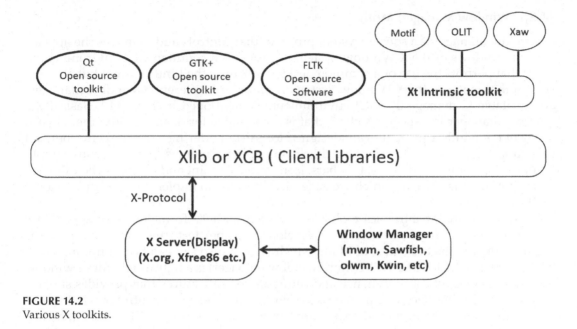

FIGURE 14.2
Various X toolkits.

- Xt is a free library written on the upper of X to act as an interface layer that streamlines the application programming. Xt supports widgets but does not provide any widgets. On the other hand, you can get the particular widget by using Xaw (X Athena Widgets) and Motif.
- OpenLook is a free toolkit offered by Sun that implements a different look and feel. It's built on the upper layer of the library called Xview, which is alike to Xt.
- Motif is built on top of Xt. It defines a programming pattern that any programmer may follow whether they are really using the Motif toolkit or not and see the subsequent section for more details.
- Qt is a library of KDE desktop environment, which is part of many Linux distributions.
- GTK+ is a multiplatform toolkit of GNOME system to create GUIs, which comes with many Linux distributions.

Motif is an industry-standard toolkit developed under Motif development project to get a common view and feel like the Unix desktop. The source code of Motif is freely available to open source developers, but a commercial license for Motif may also be taken for any Unix or Linux system that permits business agreement and gets the job easier. Motif is built on top of the Xt toolkit. It has two major parts: the first part is comprised of files, which describe constants used in Xt functions and the other part is comprised of a library of convenient functions to simplify the creation of elements like dialogs and menus. Motif also describes a programming approach that can be followed by any programmer whether they are really using it or not, and the subsequent section will give more details on it.

14.2.1 Window Manager: Motif

A window manager is a special system program that controls and manages the appearance of windows of the X Window System. It provides a common GUI environment to users that permits the use of a common desktop, not only the Linux system but also any system that may use the X Window System. A variety of different X window managers are available to manage the GUI environment, as mentioned in Table 14.1. Basically, a window manager is a special X client that is accountable for dealing with other clients. Many of the window managers are designed to assist in offering a desktop environment that supports essential functionality, like overlapping windows, popup menus, graphics hardware, pointing devices, title bars, icons, mouse (point and click), keyboard(click and type), and many more, which are generally written and implemented using a widget toolkit.

Extension to the Xt Intrinsics toolkit is provided with X. To make GUI's appearance attractive, many software houses have developed custom features. One such toolkit is Motif. Widget is the basic unit of Motif, which we can also say is a basic building block for GUI. As per Table 14.1, MWM (Motif window manager) is a lightweight Motif window manager, having robust compliance and configuration of features that provides support to common user interface, common desktop environment, element control of window elements (like size, icon, normal display, input/output, etc.), X resource database (/home/app-defaults/ and runtime), X session manager protocol, X edited resource protocol (edit widget data), image decoration, and virtual desktop support.

MWM first came into existence in the early 1990s along with the Motif toolkit. MWM is not a full desktop environment, simply a window manager. Therefore, it only manages windows and other features like configuration, programs, and sound provided by other programs. Motif and MWM were created by the Open Software Foundation (OSF). Sometimes, Motif was also referred to as OSF/Motif, but presently owned by TOG. The latest version of Motif is Motif 2.3.4 and available under LGPL license as open source software; you can get more details about Motif's (MWM) latest release, documents, and other details on the official home page [4].

The basic syntax of MWM is as follows:

```
mwm [options]
```

TABLE 14.1

List of Some Keys Available in Window Managers

Window manager	Description
awesome	Dynamic window manager(dwm), developed in the C and Lua programming languages
kwin	Window manager of KDE, default window manager of KDE plasma, KWin 5.0 is the first release based on KDE frameworks 5 and Qt 5
mwm	Motif window manager, originally developed by the Open Software Foundation and toolkit for the Common Desktop Environment
Olwm	The OpenLook window manager
twm	Tab window manager, written in C language and managed by X consortium
FLWM	Fast light window manager is a stacking window manager written in C++, default window manager for Tiny Core Linux
Fvwm	Popular in Linux, it supports virtual desktops and has configuration files that allow it to emulate other window managers, developed by Robert Nation

TABLE 14.2

Motif Window Manager Options

Option (with Argument)	Description
-display **Display**	It specifies the display to use
-xrm **Resource string**	This option specifies a resource string to use
-multiscreen	It manages all screens on the display. By default, it manages single screen only
-name **name**	**mwm** get resources using the specified name as in **name*resource**
-screens **name [name [...]]**	Mentions the resource names to use for the screens managed by **mwm**. If **mwm** is managing a single screen, only the first name is used from the list. If **mwm** is managing multiple screens, then the names are assigned to the screens from the list as written in a sequence or order, i.e., starting with screen 0. Screen 0 gets the *first name*, screen 1 gets the *second name*, screen 2 gets the *third name*, and so on

The associated options are listed in Table 14.2.

It is advisable for most GUIs to be developed in Motif to look and behave in the same manner. Default action is provided to each widget by the Motif to enforce such features. Motif style guide is where one can find information about the Motif GUI design. The general behavior of every widget is characterized as a major aspect of Motif (Xm) library. The Xt library characterizes certain base classes of widgets which structure a typical formation for almost all Xt-based widget sets. There are two wide Motif widget classes: the primitive widget class contains real GUI parts like catches and content widgets and, the second, the manager widget class characterizes widgets that hold various widgets.

14.3 Creating and Managing A Window

After the first two sections, this section will introduce the concepts of creating and managing X windows. Initially, this section starts with basic window concepts and later on presents window operations.

14.3.1 Basic Window Concept

The X Window System supports one or more screens covering overlapping windows or sub-windows. A screen is a physical monitor and hardware that is used to display the information, maybe in text or graphics. A window is rectangular in shape, meant for controlling GUI [5]. Through this, one or more processes can be executed and displayed inside that visual area as shown in Figure 14.3. Creation and management of a window is the responsibility of X client application. X client controls and manages many general aspects of the window system, like look and feel, which includes appearance and functionalities. Some of the functions of a window may include minimization, maximization, and closing of it. The positioning and coordination of a window defines its appearance. These graphical properties are included in X display.

In other terms, top-level GUIs in the X Window System are referred to as a "window." The term "window" is also used for windows that reside within another window, called sub-windows or child windows of a parent window. While using windows, we can understand graphical elements, like buttons, menus, icons, etc. A window can only be created as a sub-window of a parent window. In this way, you can make and arrange all the windows

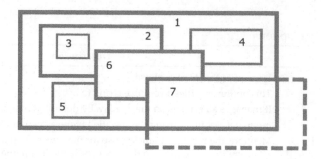

FIGURE 14.3
Imaginable placement of some windows: 1 is the main or root window, which covers the whole screen; 2, 4, and 5 are the sub-windows of 1; 3 is the sub-windows of 2; 6 and 7 are top-level windows. Some parts of window 7 are not visible due to outside with main window.

in a tree or in a hierarchy. The root of this window tree is called root window or main window or output screen, which is automatically created by the server. You can see and set the size of the root window as large as the output device screen. The top-level windows are precisely the direct sub-windows of the main or root window that are arranged in well-defined orders. Simply, you can say that screen may comprise N numbers of sub-windows arranged in a strict hierarchy. If X server creates many screen windows, then it is possible that the root window is partially or completely covered by sub-windows. All created windows by X server have parents, except root windows. Each sub-window or child window may have its own children window. Each application carries at least one screen. There is usually at least one window for each application program, but an application program may generate a randomly deep tree child window on each screen. X offers graphics, text, and raster operations for windows.

14.3.2 Window Operation

All data about windows, fonts, etc., are stored in the server. The most important window operations are pointing and clicking on some icon inside the window or the whole window itself. And other operations include double click, right click, selections, cut buffers, dragging and dropping of window, scrolling window, and many more. These are the various methods used in X Window System to permit a user to transfer data from a window to another. One very popular and easy way to transfer data between windows is drag-and-drop operation. In this, the user needs to select something in a window, then click on the selection, and drag it into another window for data transfer. Correspondingly, selections and cut operation are also applied, where a user needs to select some text or data in a window, then make a click to select cut operation and paste it in another window for data transfer.

In window operation, the client program connected with the selection of text in a window offers a protocol for transmitting data to the demanding application window. On the other side, the cut operation is a passive method, in which the user picks some text from a window and its content is shifted to a temporary buffer, known as cut buffer, where it remains alive until it is pasted into another window. If there are two windows, assume that they are controlled by two separate applications. In this case, data transfer involves two separate clients associated with a similar X server to cooperate with clients. The key protocol of X Window System comprises some types of requests and events that are particular to selection and then exchange. However, the transfer is mostly done using the usual client-to-client event transmitting and window features which are not restricted to selection transfer. Between two clients, the transferred data may be of various types, generally text, but may also be a number, list of objects, PixelMap, and many others.

14.4 Starting and Stopping X

For Linux, X is the standard GUI like other GUIs like Microsoft Windows and Mac OS. *X.Org Foundation*, an open source organization, provides X Window System tools and related technologies. Therefore, you need to install the required programs and components on your system that permit X to start. These categories are as follows:

- Basic XFree86 program
- X servers
- Window managers
- Fonts

Most Linux users run XFree86, a freely available software system compatible with X. Currently, the latest version of X Window System is 11 with revision 7, which is commonly referred to as X11R7. The X11R7 X servers are nowadays developed and maintained by the XFree86 project organization. In Linux, all supporting libraries are installed in */usr/X11R7/ lib* directory, whereas similarly the X Window System (clients and servers) applications are installed in */usr/X11R7/bin* directory. After Linux installation, generally, you need to require to configure or reconfigure the X Window System with the help of various configuration tools and files, including X. These configuration tools and files are xinit, .xinitrc, XF86Setup, and XF86Config.

After you log in on the Linux command line interface, you can start X by typing *startx* command as follows:

```
# startx
```

The system screen should show X's screen with other graphical features in a few seconds, i.e., its start X Window System with window manager and desktop. The *startx* command uses the *xinit* command to run the X Window System. The *xinit* command first searches for the X Window System initialization script, known as *.xinitrc* located in the user's home directory. If the user's home directory could not find the *.xinitrc* script, then *xinit* uses */etc/ X11/xinit/xinitrc* as its initialization script. Both files, *.xinitrc* and */etc/X11/xinit/xinitrc*, comprise commands to configure the user X Window System server and run any preliminary X commands, like starting up the window manager.

After starting X, the X Window System uses numerous configuration files and commands to configure X Window System. The user must be very careful that some configuration files belong to the system and should not be altered. Each user may have their own group of configuration files, like as *.xinitrc*, *.xsession*, and *.Xresources*, which comprise the personalized configuration information. The *XF86setup* configuration tool is a GUI program that starts a default X Window System, which permits configuration of X Window System. Similarly, the *XF86setup* is a program or command that is used to configure the

X Window System at the command line; here the user must be logged in as a root user. The configuration file for fonts is *fontconfig* that is located at */usr/share/X11/fonts*. The XFS font server takes care of the X Window System fonts, which are configured with the */etc/X11/fs/config* configuration file.

To stop X, the user must click on exit or quit or log out from the menu bar. Upon exit, X will return to a text-based console of Linux shell prompt.

14.5 X Architecture and Application

X architecture is based on the client–server model but in an inverted way from the general client–server model. Here, the local system contains X server, which can communicate to a remote system that contains different applications which act as different clients. A workstation equipped with the ability of bit-mapped graphics is the core requirement of X. Both color and monochrome variants are equally good. Most of the time when you are studying X, then you will come across a term called "display." A combination of keyboard, one or more screens, and a pointing device (which could either be a mouse or a trackball, or a graphics tablet) are referred to as "display."

14.5.1 Client and Server Architecture

The concept of client–server is inherent when we talk about X as shown in Figure 14.4. There could be multiple systems involved in the execution of application, and it may be possible that the system on which the application is executing is different from that of a display. Managing so many devices is a cumbersome task and X finds it difficult too, and this is when the concept of client-server was introduced. If we run a program on a single workstation using X, it makes the practicality of the client–server system transparent to the user, so the user needs not to worry about the same. But, some notions of the concept

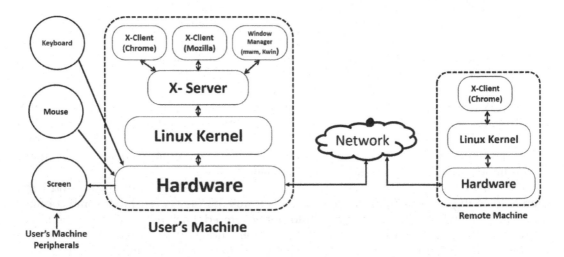

FIGURE 14.4
Layout of X client–server architecture.

are necessary to fully comprehend the working of X. Server is referred to any program that controls each display. The learner sometimes finds it difficult to digest that servers are mostly seen as something working in a remote location as a file server rather than controlling the display. In X, the server acts as a local program responsible to control the display. There is a possibility that the same display is shared within two or more systems, in that particular case the system acts as a true display server. The user programs are what we call the clients, and the server is responsible to act as a go between the clients or applications and the resources of the local system. It could either run on a local or remote system [6]. The following tasks are performed by the server:

1. Providing access to multiple clients.
2. Interpretation of network messages received from the clients.
3. Receiving requests for displaying two-dimensional graphics.
4. Maintaining local resources which include windows, cursors, fonts, and graphics.

Once connection is founded between client-server, four kinds of packets are swapped among client and server across the channel, which are as follows:

1. **Request:** The client demands/requests data from server or requests it to do an action.
2. **Reply:** The server replies to a request, it's not necessary to create replies for all requests.
3. **Event:** The server transmits an event to the client, e.g., touchscreen or mouse input, or keyboard input or window (resized, moved, or visible), or any other operations.
4. **Error:** If the server receives a request which is invalid, then it can return an error packet.

The key application of X is to provide the basic framework for creating a GUI environment, which is responsible for drawing and moving windows on the display device and interacting with a mouse, keyboard, or touchscreen. Some developers have people who try to create an alternative or replacement for X. Some currently available alternatives for X are listed as follows:

- **macOS**, a mobile version of iOS, implements its own window system, which is identified as Quartz.
- **Android**, a mobile OS which runs on the Linux kernel, uses its own window system for drawing the user interface, referred to as SurfaceFlinger, and 3D rendering is handled by EGL. EGL stands for "embedded system graphics library."

14.6 The X Programming Model

This section will introduce you to the concepts of X programming model. Connectors are paths via which client and server are connected with each other. But, how to provide a path, the answer is by means of a low-level C language interface library known as Xlib. It is the lowest level, which comes in the system software hierarchy or the architecture. It is so powerful that many applications can be written directly using Xlib alone, but it is rather a cumbersome or, we can say, time-consuming task to write complex GUI programs using only Xlib. But, not to worry, there are various higher-level subroutine libraries, called toolkits to take care of this. It is important to note that X is not confined to the arena of a single language, operating system, or user interface. It is quite common to link calls to X from most programming languages. There are only X protocol messages which are required by X to generate and receive messages. These protocol messages are readily available as C libraries in Xlib. In hierarchy, there are usually two more levels of toolkits above Xlib as shown in Figure 14.2 that are Xt Intrinsics library and widget sets (like as Motif). The application program in X generally consists of two parts. The first one is GUI written using one or more of Xlib, Xt, or Motif and the second part is the algorithmic or functional part of the application where the input from the interface and other processing tasks are defined [6]. Figure 14.5 explains the software architecture of X based on Xlib library.

- **Xt Intrinsics library (X toolkit)**
- There is a range of user interface features or application environments, such as menus, buttons, or scroll bars which are implemented using the X toolkit. Xt library is part of X toolkit that enables the programmers to build and use new widgets. With proper usage of widgets, it simplifies the X programming process that is useful to preserve the look and feel of the application which makes it easier to use. Some key features of Xt are as follows:
- It is a powerful library and built upon Xlib (lowest level interface to X).
- Well matched with other X applications.
- Wide support to all popular user interface conventions.

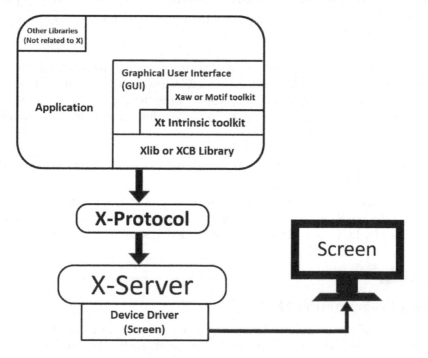

FIGURE 14.5
Software architecture of Xlib-based application.

- Written in C language, extensively available and understandable to all programmers.
- Offering an object-oriented layer to support the user interface abstraction, referred to as a widget.
- **Widget:** A widget is a refillable and configurable piece of code that runs independently of the application except through prearranged interaction. A group of widgets, commonly referred to as widget sets, offer a consistent appearance, which is similar to *look* and *feel* of the application interface.

14.6.1 Xlib Overview

Xlib is a low-level library, written in C language, interface to the X Window System protocol. Xlib library is used to write a client program using Xt Intrinsics library and set of widgets for X Window System and provides functions for interacting with an X server. Xlib performs a very crucial task of translating C data structures and procedures into the special form of X protocol messages, which are then sent off. It is not only responsible for this forward procedure, but the converse of the same is also true. The interface between the client (application) and the network is also handled by Xlib. It also includes functions which provide linking to a specific display server, creating windows, drawing graphics, setting foreground/background colors, acting on events, and many more. Xlib does not offer support buttons, menus, scroll bar, etc., which are referred to as widgets. Such widgets are supported by other libraries, like GTK+, Qt, Xt (Intrinsics), Motif, Xaw, etc. The XCB library is trying to offer an alternative option to Xlib.

Some key points of Xlib are given as follows:

- Lowest level C language library, interface to X.
- Offers full access to the capabilities of X protocol.
- Handles the interface between an application and network.
- Helps in application optimization that offers improved network usage.

14.6.2 Xlib Usage

Xlib functions are used to provide a major operation in X Window System application. All Xlib functions begin with a capital "X" character [6, 7]. Some key functions are highlighted under four major categories, which are given as follows:

- **Connection with X display:** Open a connection to the X server that controls the display; some functions are **XOpenDisplay()**, **XDefaultScreen()**, **XCloseDisplay()**, etc.
  - The **XOpenDisplay()** function is as follows:

    ```
    Display *XOpenDisplay(display_name);
    ```

    - **display_name:** Mentions the hardware display name, which controls the display and the communication domain that is to be used.
  - The **XDefaultScreen()** is as follows:

    ```
    int XDefaultScreen( *display);
    ```

    - **display:** Mentions the connection to the X server

- The **XCloseDisplay()**is used to close a display or disconnect it from the X server:

  ```
  XCloseDisplay( *display);
  ```

  - **display**: Mentions the connection to the X server to disconnect.
- **Server operations:** The functions which make requests to the server for operations; some functions are XCreateWindow(), XCreateColormap(), XGetWindowProperty(), etc.
  - The XCreateWindow() is used to create an unmapped window and set its window attributes. The function description is as follows:

    ```
    Window XCreateWindow( *display, parent, y, height, border_width,
    depth, class, *visual, valuemask, *attributes);
    ```

    - **display**: Mentions the connection to the server
    - **parent**: Mentions the parent window
    - *x*
    - **y**: Specify the x and y coordinates.
    - **width**
    - **height**: Specify the width and height.
    - **border_width**: Specifies the width in pixel of the created window border.
    - **depth**: Mentions the window depth.
    - **class**: Specifies the class of created windows.
    - **visual**: Mentions the visual type.
    - **valuemask**: Mentions which window attributes are defined in the attribute's argument.
    - **attributes**: Mentions the structure from which the values are to be taken as mentioned by valuemask.
  - The XCreateColormap() is used to create a colormap for a screen. The function description is as follows:

    ```
    Colormap XCreateColormap(*display, w, *visual, alloc);
    ```

    - **display**: Mentions the connection to the server
    - **w**: Specifies the window
    - **visual**: Mentions a visual type that is compatible with the screen.
    - **alloc**: Mentions the allocated colormap entries.
  - The XGetWindowProperty() is used to get the properties of a given window. The function description is as follows:

    ```
    int XGetWindowProperty( display, w, property, long_offset, long_
    length, delete, req_type, actual_type_return, actual_format_
    return, nitems_return, bytes_after_return, prop_return);
    ```

    The above function will return the actual type of window properties on the various parameters as mentioned in the form of arguments.

- **Event operations:** The Xlib library maintains all events of X Window System through various functions. The X Window System may broadly divide all events into seven categories, which are as follows:
  - *Mouse Event*
  - *Keyboard Event*
  - *Expose Event*
  - *Colormap Notification Event*
  - *Window State Notification Events*
  - *Window Structure control Events*
  - *Inter-client Communication Events*

Some event operation functions are XEventsQueued(), XPending (), XsendEvent, XcreatImage(), XSaveContext(), etc. All generated events are stored in an event queue before processing by the X client.

- The XEventsQueued() is used to check the number of events in an event queue. The function description is as follows:

```
int XEventsQueued( *display, mode);
```

  - **display**: Mentions the connection to the server.
  - **mode**: Mentions the mode.

- The XPending() is used to return the number of events that are pending. The function description is as follows:

```
int XPending( *display);
```

  - **display**: Mentions the connection to server.

- **Errors handling:** The Xlib library detects an error whenever it is occurred and calls an error handler. Your program can provide an error handle, if not the error is printed, and your program terminates. Xlib provides two default error handlers.

  - **Fatal error handler**: It handles fatal I/O errors, which refer to an error that happens when Xlib tries and fails to send a request to or read replies from a display server due to a machine crash or broken connection or any other causes. The *XIOErrorHandler()* function is used to handle fatel I/O errors.

  - **Protocol error handler**: It handles protocol errors from the X server due to some errors in parameters sent to the X server. It invokes *XSetErrorHandler()* function.

  - Both error handlers by default display a message using *XGetErrorText()* and then exit.

  - The XSetErrorHandler() is used to handle protocol error. The function description is as follows:

```
int *XSetErrorHandler( *handler);
```

  - **Handler**: Mentions the programs supplied to an error handler.

- The XGetErrorText is used to obtain textual descriptions of specified error code. The function description is as follows:

```
XGetErrorText( *display, code, *buffer_return, length);
```

- **display**: Mentions the connection to server.
- **code**: Mentions the error code, for which you wish to get a description.
- **buffer_return**: Returns the error report.
- **length**: Mentions the buffer size.

After successful execution of many Xlib functions, it retune an integer resource ID, mentioned as XID, that permits you to describe the objects kept on X server. Basically, this type of XID is utilized as generic resource IDs, which may be the type of font, window, colormap, cursor, pixmap, and GContext, as specified in the header file <X11/X.h>. Some Xlib standard header files are listed in Table 14.3.

14.6.3 Event-Driven Applications

The programming concept of window-enabled graphical system is totally different from the traditional programming concept. Generally, event-driven programming concept is used to develop graphical-based application which is controlled by various events responded by the X client. An event is a kind of data generated by X server and communicated to X client on the basis of device activity with the help of Xlib function. X client must be ready to reply to such events. The possibility may include user's input, like a mouse click, mouse movement, keypress, and others, and interaction with other programs like opening an icon click to open a new window, minimization of window, switching from one application to another application area, and others. Further, the users' moves and resize of window on the screen, click on the button or scroll bar, or switch keyboard focus from one window to another. Simply, we can say that every window application must be capable to respond to anyone of the above different events at any point of time.

X server generates specific event categories through its type. For every event type, a corresponding constant name and structure is declared in <X11/Xlib.h> library, which is used while referring to an event type [7].

TABLE 14.3

List of Xlib Standard Header Files

Header File Name	Description
<X11/Xlib.h>	Xlib main header file, generally all Xlib functions are declared by using this header file
<X11/X.h>	Comprises types and constants for the X protocol
<X11/Xcms.h>	Comprises symbols for color management services
<X11/Xutil.h>	Describes numerous functions, types, and symbols, utilized for inter-client communication and application service functions
<X11/Xresource.h>	Comprises functions related to resource manager facilities
<X11/Xatom.h>	Has declaration of all predefined atoms
<X11/cursorfont.h>	Comprises cursor symbols for the standard cursor font
<X11/keysym.h>	Comprises all standard KeySym values
<X11/Xlibint.h>	Comprises all functions, types, and symbols used for extensions
<X11/Xproto.h>	Declaration of types and symbols for basic X protocol
<X11/X10.h>	Declaration of all functions, types, and symbols which are used for the X10 well-matching functions

TABLE 14.4

List of Some Major Event Categories

Event category	Event type
Keyboard events	KeyPress, KeyRelease
Pointer events	ButtonPress, ButtonRelease, MotionNotify
Input focus events	FocusIn, FocusOut
Exposure events	Expose, GraphicsExpose, NoExpose
Colormap state notification event	ColormapNotify
Client communication events	ClientMessage, PropertyNotify, SelectionClear, SelectionNotify, SelectionRequest

The basic syntax and member of event structure is defined as follows:

```
typedef struct{

        int type;
        unsigned long serial;   /* No of last request processed by server*/
        Bool send_event;        /* Check send event status*/
        Display *display        /* Display the event*/
        Window *window;

}XAnyEvent;
```

Table 14.4 highlighted some major event categories and their associated event type.

As per Table 14.4, there are different categories of an event which may occur at any time and in any order that require proper handling. All these generated events are placed on a queue as they are generated and processed by the client in the same order. Handling these events is the key difference between event-driven programming for window application and traditional procedural programming. For example, if an application wants to open a file, then the user has the option to select a particular file from the system but in the same window, the user also has the option to cancel this operation by clicking on the cancel button. Such type of interaction gives full control to the user over the application. To get such kind of flexibility, the programmer must follow a well-defined programming model, in which the application continuously checks what the user wishes and accomplishes the suitable task to fulfill a request. The programming model which handles various tasks is also referred to as an event, i.e., request of pointing on the screen by mouse, pressing a key on the keyboard, or performing various operations on application (opening and closing a file, printing a document, menu navigation, or any other action) is known as event-driven programming model to make an event-driven application for the X Window System.

14.7 Desktop Environment

The desktop environment integrates the application and helps to unify the user interface. Also, it provides services which are not included in the default setup. A desktop environment provides flexibility in customization of window manager. The key objective of window manager is to design the foundation for uniformity and standardization of the appearance and strategies on programming methodologies and developing applications

[8]. It combines various X clients to offer a common GUI environment and development platform to the user. The two most popular desktops used with X are KDE and GNOME.

14.7.1 KDE

The K desktop environment, KDE, is a freely available desktop that includes KWM (KDE window manager). KDE uses Qt toolkit to create various user interface controls. KDE is a vast improvement over a standard X setup. It installs quite easily. It looks very nice and works pretty much like you'd expect, borrowing its shortcut keys from Windows rather than emacs as many other Linux programs do. KDE offers a wide range of accessories and excellent applications for everyday tasks including:

- Graphical applications like Kfract and Kview (an image viewer).
- Games like Kpoker, Kmines, and Ktetris.
- Multimedia applications like Kmedia, a media player, and Kmix, a sound mixer.
- Open office suite like KOffice.
- Network applications like Kmail—a mail client, Knu—a network utility, etc.

New KDE accessories and applications along with updates are available at regular intervals, and you may visit KDE official page to get more details on the latest announcements and updates [9]. Further, you may also see Section 9.9 for more details.

14.7.2 GNOME

It is a freely available desktop environment from the GNU Project. Under GNU Project, many free programs and functions are written for Linux-based systems. In contrast with KDE, GNOME is an open source software that comes with Red Hat Linux. GNOME uses GTK+ toolkit, which is a multiplatform toolkit used for creating GUI. GTK+ comprises GLib, GDK, and GTK libraries that have a C-based object-oriented architecture. GNOME offers various desktop tools in line with KDE offering that includes:

- GNU image manipulation program (GIMP)
- Games like Gnome Mines, FreeCell, Gnobots, and Gnometris
- Multimedia applications like Audio Mixer
- Network applications like Mailman, Talk, Synchronize (help to synchronize files on numerous systems).
- Text editor as gEdit and spreadsheet as Gnumeric
- Well-known browser application Netscape Navigator

GNOME developers, like KDE developers, always make an effort to provide excellent graphical environment and applications for Linux system. You may visit GNOME official page to get more details on the latest releases and announcements [10]. See Section 9.9 for more details. Some proprietary desktop systems are also offered for Linux system like Common Desktop Environment (CDE). It follows all requirements of Unix/Linux community and is like what you would get on an HP/UX or Solaris desktop. CDE comes under the proprietary software category, therefore no free version is offered. However, this environment seems comfortable to convert from another Unix rather than Windows.

TABLE 14.5

List of Major Open Source Communities which handle X Window Tools

Name of open source projects & communities	Web link	Description
X.Org project	https://www.x.org/releases	Free and open source community for the implementation of X Window System along with *freedesktop.org* project community. They work together for the development of X window server platform and offer the latest releases and tools
XFree86 Project	http://www.xfree86.org/	Free open source project of X Window System since 1992. You can see the current versions and tools of X under release section at the given web link.
GTK (GNOME Project)	https://www.gtk.org/	GTK, a set of widgets, is a toolkit of GNOME that provides GUI for the X Window System, written in C language, and works in cross platform, and supports a wide range of languages. Presently, GNOME 3 is the default desktop environment of GNOME that is offered in many Linux distributions, including RedHat, Fedora, Debian, Ubuntu, SUSE Linux, Kali Linux, CentOS, Oracle Linux, Solaris, etc.
Qt toolkits (KDE Project)	https://kde.org/	Qt is an open source toolkit for developing GUI for the X Window System, also known as multiplatform toolkit. Plasma desktop is the default desktop environment in many Linux distributions, like openSUSE, Kubuntu, PCLinuxOS, KaOS, KDE Neon, OpenMandriva, etc.

14.8 Upgrading X Window Tools

As discussed in earlier sections, X Window System (X11), also known as X, X11, or X Windows, is an open source, cross-platform, client–server computer software system that offers GUI in a network environment. The X.Org project provides an open source implementation of X Window System, being carried out in coordination with freedesktop.org community; you may visit its home page for further details. * Other open source organizations are also contributing to the development of X Window System applications and tools. But, currently, most of the work of X11 is carried out and managed by four major nonprofitable communities, which are described in Table 14.5. These open source platform projects provide new releases and updates of various X tools. You can visit the respective project and community website to get complete information and procedure upgradation of its respective release and tool.

14.9 Summary

GUI provides an interactive way to communicate with the system. Nowadays, GUI is an essential part of all running operating systems generally referred to as desktop environment. The standard GUI is provided by X Window System also known as X or X11, which

---

* Free Desktop Community: https://www.freedesktop.org/wiki/.

is currently managed by the X.Org foundation. The X Window System follows client–server architecture, in reverse from general client–server paradigm, that comprises X line, X server, X protocol, and X window manager. The X protocol provides communication between client and server that details are given in the Xlib library. Xlib comprises other library tools like GTK+, Qt, Xt Intrinsics, Motif toolkit, etc. Motif is a window manager that controls and manages the appearance of windows of X Window System. A basic window is defined as rectangular in shape for controlling a GUI. The creation and management of window is the responsibility of the X client system. Window data are stored in X server and perform various operations like pointing and clicking on an icon, double click, right click, selection, drag and drop of window, scrolling the window, and many more. By using the command *startx* on command prompt, you can start the X Window System along with window manager and desktop. X programming model concept deals with the development of X Window System protocol. The Xlib library is used to write client programs, using Xt Intrinsics library and set of widgets. Xlib offers various functions to provide key operations in the X Window System, like connecting with X server, server operations, event operation, and error handling. All X Window System functions begin with X. The two most popular desktop environments used with X are KDE and GNOME.

14.10 Review Exercise

1. How X Window System helps Linux popularity? List out promptly GUI tools for Linux?
2. Justify this statement "X Window is a graphical-rendering framework."
3. Which command is used to start and stop the X? Discuss some applications of X Window System.
4. How X server, X client, X protocol, and X manager are important with respect to X Window System?
5. Describe various low-level libraries used as an interface for application programming.
6. If Xlib is present for providing application programming interface, then why X toolkit is required?
7. List out the names of various tools and procedures that can be used for configuring X under Linux?
8. How to set up a default GUI which starts automatically when you load Linux? Explain with suitable examples.
9. Which toolkit of X Window System helps to make the interface interactive?
10. How X architecture is an inverted way of general client–server architecture, justify with an example?
11. List out the name of a toolkit which enables programmers to build new widgets other than Motif, and by using that toolkit write a program to create a new widget.
12. Why Motif is called a third-party toolkit in X Window System for building new widgets?

13. Describe the importance of a desktop environment. How it helps to provide the services which are not included in the default setup?

14. List out various desktop tools offered by GNOME and KDE. Describe the procedure to set up one such tool to configure a desktop for your own purposes.

References

1. The X.Org Foundation, The X.Org documentation in the World Wide Web. https://www.x.org/wiki/Documentation/.
2. The Open Group, Members of the Open Group in the World Wide Web. https://www.open-group.org/our-members.
3. Neil Matthew and Richard Stones. 2001. *Beginning Linux Programming*. Wrox Press Ltd.
4. The Motif development project, The ICS motif Zone in the World Wide Web. https://motif.ics.com.
5. Richard Petersen. 2008. *Linux: The Complete Reference*. The McGraw-Hill.
6. Adrian Nye and Tim O'Reilly. 1990. *The X Window System*. O'Reilly & Associate, Inc.
7. James Gettys, Robert W. Scheifler et al. 2002. *Xlib - C Language X Interface: X Consortium Standard*. The Open Group (Open Source Document).
8. Trevor Kay. 2002. *Linux+ Certification_Bible*. Hungry Minds Inc.
9. The K Desktop Environment (KDE) Software Open Communities, The KDE Announcements in the World Wide Web. https://kde.org/announcements/.
10. The GNOME Project, The GNOME's technologies in the World Wide Web. https://www.gnome.org/technologies/.

Section III

Case Studies

15

Linux Distributions (Linux Distro)

Initially, every new Linux user who comes from other operating systems (OSs), such as Windows and Mac OS, feels that Linux is a single unit referred to as Linux OS. But in fact, Linux is only a kernel, not an OS. Therefore, in true spirit, Linux OS is a Linux distribution, and more exactly a GNU/Linux distribution. The Linux distribution concept is very popular and frequently referred to as a Linux distro. A rich number of Linux distros are currently available. Therefore, the task of selecting the right Linux distro for your requirements may simply become an interesting task. This section carefully selected some Linux distros, which are very popular in user and developer community. You can find the complete details of all prominent Linux distributions by visiting the *DistroWatch* webpage [1]. This webpage provides a complete list of Linux distributions along with an extensive technical description of each one.

15.1 Getting Started with Various Linux Distributions

A Linux distribution (often abbreviated as distro) is an OS, developed with the help of a set of software packages, which is centered upon the Linux kernel and package management system. It also includes the processes needed to install these components to attain a running system. Today, some Linux distributions may integrate with different versions of kernel to support different hardware architectures. Simply you can say that Linux kernel along with GNU software packages/tools makes a way for a Linux-based OS. It was a difficult task to build up an entire OS out from scratch, and it is better to produce with the support of available software packages (open source or commercial) and Linux kernel according to the needs and purposes.

In 1992, the first Linux distributions were produced, but the year 1993 witnessed the birth of two great Linux distributions: Slackware and Debian. Slackware was created on the Softlanding Linux System (SLS), and Debian was developed by Ian Murdock, due to different views or unhappiness with SLS distribution. Presently, a huge number of different Linux distributions exist due to the Linux kernel itself and the GNU open-source project. Nowadays, there are around 600 Linux distributions presented but approximately four to five hundred of these are currently in active development and use [2,3]. The top 20 popular Linux distros in 2019 are listed in Table 15.1 based on hit per day index.

TABLE 15.1

The Top 20 Popular Linux Distros in 2019

Rank	Name of Linux Distro	Source Link	HPD (Hit Per Day Index-2019)
1	MX Linux	https://mxlinux.org/	4618
2	Manjaro	https://manjaro.org/	2877
3	Mint	http://linuxmint.com/	2133
4	Debian	http://www.debian.org/	1499
5	Ubuntu	https://www.ubuntu.com/	1463
6	elementary	http://elementary.io/	1396
7	Solus	https://getsol.us/	1095
8	Fedora	https://getfedora.org/	1000
9	Zorin	https://www.zorinos.com/	871
10	Deepin	https://deepin.org/	829
11	openSUSE	http://www.opensuse.org/	788
12	antiX	https://antixlinux.com/	731
13	KDE neon	https://neon.kde.org/	675
14	CentOS	http://www.centos.org/	652
15	ArcoLinux	https://arcolinux.com/	624
16	Arch	http://www.archlinux.org/	596
17	PCLinuxOS	http://www.pclinuxos.com/	594
18	ReactOS	http://www.reactos.org/	526
19	Pop!_OS	https://system76.com/pop	516
20	Kali	http://www.kali.org/	485

15.1.1 Why So Many Distros?

Why there are many Linux distributions and what are the key factors to be taken into account when preferring a Linux distribution? There are many factors for the growth of Linux distribution, but some key reasons are low cost, availability of the source code, alternative of proprietary versions, and developed to perform general to specific task depending on the requirement. Most Linux distributions install software packages including the kernel and other core OS software packages in a prearranged configuration. Broadly, Linux distributions may be categorized into two parts, namely general-purpose OS and specific task-oriented OS.

When a user wants an OS for their own general daily use rather than to perform any specific task, they must choose an OS comprising all tools ranging from editing documents, painting, listing music, accessing Internet to editing videos, making movies, etc., like the big contenders of Linux, namely MS Windows and Apple OS, but in Linux distribution category, Some popular Linux distros are Red Hat, Fedora, Debian, Ubuntu, Knoppix, etc. On the other hand, if a user wants to use Linux to perform some specific task that is unavailable in general Linux distributions, then it required some necessary or dedicated features, which may require intense customization and installation of specific software packages in general Linux distributions. Such customization is not an easy practice in general Linux distributions. Therefore, developers started writing codes to develop such Linux distros, which focus to perform on a set of specific tasks or one task. Further, they removed all such software packages and features which are not required that make the OS lighter and lead to better performance. Some specific task-oriented Linux distributions are Android (mobile), Chrome OS (Chrome Books), Kali (Security), SteamOS (Gaming), CoreOS (automated and Open Cloud Services), etc.

A massive number of Linux distributions existed in active development, which are being used in a wide variety of commercially as well as open-source community-based distributions, including desktops, laptops, tablets to servers, data centers, and embedded devices/ systems. Some key commercial Linux distributions are Fedora (Red Hat), openSUSE (SUSE), Ubuntu (Canonical Ltd.), etc., and open-source Linux distributions are Debian, Slackware, Gentoo, etc. [4].

Linux distributions may be categorized into the following terms:

- Commercial or non-commercial (open source).
- Designed for enterprise users, home users, or powerful computing-oriented users.
- Designed to support numerous types of hardware or specific hardware architecture.
- Designed for data centers, cloud servers, desktops, laptops, mobile, and embedded devices or systems.
- Designed for general-purpose or event-specific system functionalities like networking devices including gateway, routers, firewalls, gaming, music composing, or other computer-specific functionalities.
- Mainly designed for security, reliability, portability, completeness, open source (free), accessing source code, multiprocessing, multitasking, etc.

15.2 Red Hat Enterprise Linux /Fedora

After the birth of Linux in 1991, the first five years of Linux history signifies a great start in era of Linux distribution. The four key distributions were coming into existence, namely Red Hat, Debian, SUSE, and Slackware. In 1994, Marc Ewing developed his own Linux distribution, referred to as Red Hat Linux. In late 2003, Red Hat discontinued Red Hat Linux and offered two Linux distributions: Red Hat Enterprise Linux (RHEL) and Red Hat Professional Workstation.

Initially, Red Hat came to understand that there are two different types of people showing their interests in its distribution. One type of people belongs to business type who want secure software that comes with support and update, and they are ready to pay considerable money for such support and services. The other type of people belongs to those users who want to use Red Hat and advanced features of Linux for their own use and experiment, but they don't want to pay any money for software and support. Hence, Red Hat has decided to cater to both types of people and offered multiple distributions. Such distribution world makes both types of users happy and a win-win situation for Red Hat to generate revenue from its own efforts in the form of providing support and services. Some of the key relevant Linux distros that came between 1995 and 2005 based on Red Hat (RH) are Mandriva, SCO, Fedora, and CentOS.

In 2003, Red Hat developed another Linux distribution, referred to as RHEL for the commercial organization or market. RHEL is released in server versions for x86, x86-64, Itanium, PowerPC, ARM64, and IBM System z architectures, and a desktop version for x86 and x86_64 processors. Today, RHEL offers supports and functions to software and technologies in various fields of applications like automation, cloud, middleware, containers, open stack, big data, Internet of things (IoT), mobile, data center, application development, virtualization, management, and microservices. Linux plays the most important part of various Red Hat offerings. Linux is not merely an OS for desktops or servers, but it is the backbone and base for the latest information technology stack. The Red Hat Academy Program provides support and training in various Linux certification courses including the popular course, Red Hat Certification Program, which is focused to provide technical skill on the RHEL platform. Red Hat was the first to adopt the RPM package management system [5].

Further, as everyone is aware that Linux is free, therefore, the subscription for RHEL server is offered at no fee for development purposes. But for this, every developer must sign up or register for the Red Hat Developer Program and accept licensing terms to prevent production use. This program was started on March 31, 2016, to provide technical skill to all Linux lovers as free of cost. Many other programs are offered as academic editions of the desktop and server area. All such programs are offered to schools and students, offered comparatively at a very low cost with Red Hat technical support as an option. Further, the Red Hat technical support may also be offered in web mode, which can be preferred by many customers and can be purchased separately. In Red Hat Enterprise Linux 5, some key new editions replace former Red Hat Enterprise Linux AS/ES/WS/Desktop, which are given below:

1. Red Hat Enterprise Linux (former ES) (limited to two CPUs).
2. Red Hat Enterprise Linux Desktop with Multi-OS option.
3. Red Hat Enterprise Linux Advanced Platform (former AS).
4. Red Hat Enterprise Linux Desktop (former Desktop).
5. Red Hat Enterprise Linux Desktop with Workstation option (former WS).
6. Red Hat Enterprise Linux Desktop with Workstation and Multi-OS option.

Red Hat builds, supports, and contributes to many free software projects. It has bought numerous proprietary software products with code through corporate mergers and purchases; thereafter, it has issued such software under open-source licenses. Now, the latest release of Red Hat Enterprise Linux 8.1 is available. You can visit the following weblinks to get more resources, update, and support related to RHEL:

Distribution Name	Red Hat (Enterprise) Linux
Home Page:	http://www.redhat.com/
Mailing Lists:	http://www.redhat.com/mailing-lists/
Documentation:	https://access.redhat.com/documentation/en/red-hat-enterprise-linux/,
Red Hat Knowledge Base:	https://access.redhat.com/knowledgebase
Download Mirrors:	https://access.redhat.com/site/downloads/
Bug Tracker:	https://bugzilla.redhat.com/
Wiki:	https://en.wikipedia.org/wiki/Red_Hat

- **Fedora**:
    - Fedora is always among the top five most popular Linux distributions and was founded on September 22, 2003, by Red Hat and Warren Togami. Sometime in 2002, Warren Togami began Fedora Linux as an undergraduate project at the University of Hawaii. This was done to provide non-Red Hat users a single repository with well-tested third-party software packages that are easy to use, develop, and find. The repository so created would help to set up a ground for global volunteer communities to collaborate and contribute to the repository. Red Hat first offered an enterprise Linux support subscription for Red Hat Linux 6.1. It was not a separate product, but the subscription level was branded as Red Hat 6.2E. Subsequently, Red Hat started creating a separate product with commercial service-level agreements and longer life cycle based on Red Hat Linux.
    - Thereafter in 2003, just after the release of Red Hat Linux 9, Red Hat decided to split Red Hat Linux into RHEL to retain the Red Hat trademark for its commercial products and another OS, managed by the community, known as Fedora, to promote free and open-source software. It is worth mentioning here that Red Hat is not using Fedora as the community edition of its distribution, rather its control and promotion by Fedora project which is sponsored by Red Hat. Fedora project is a community-supported free software project that aims to promote the speedy progress of free and open-source software, content, and fast innovation using open community support and forums. Fedora is also promptly referred to as Fedora Linux, a volunteer project that provided extra software for the Red Hat Linux distribution. The GNOME desktop environment and GNOME Shell are the default desktop and default interface in Fedora, respectively. Other desktop environments, including KDE, Xfce, LXDE, MATE, and Cinnamon, are also available. Fedora project also distributes specific sets of software packages, offering different desktop environments or targeting specific packages such as gaming, security, design, scientific computing, and robotics.
    - Both the Fedora Linux distribution and RHEL are open-source technologies. Fedora is built by the community for the benefit to all Linux users. RHEL is developed by Red Hat with the explicit intent of being used as an enterprise IT platform. In short, Fedora is a bleeding-edge distribution with the latest and greatest features of the repository, whereas RHEL is Red Hat's commercial server for Linux; also, there is another open-source version of RHEL, called CentOS. At the time of writing this chapter, the latest release of Fedora 31 is available [6]. You can visit the following weblinks to get more resources, update, and support related to Fedora:

Distribution	Fedora Project
Home Page:	https://getfedora.org/
Mailing Lists:	http://fedoraproject.org/wiki/Communicate
Documentation:	http://docs.fedoraproject.org/ http://fedoraproject.org/wiki/Docs
Download Mirrors:	https://getfedora.org/ https://admin.fedoraproject.org/mirrormanager/mirrors http://torrent.fedoraproject.org/
Bug Tracker:	https://bugzilla.redhat.com/
Wiki:	https://en.wikipedia.org/wiki/The_Fedora_Project

15.3 Debian GNU/Linux:

Debian is a GNU/Linux distribution that is one of the most famous and oldest Linux distributions. In the year 1993, Debian was created and founded by Ian Murdock, an American software engineer, who devised the idea of Debian project when he was a student of Purdue University. The Debian project was named after the name of his then-girlfriend Debra Lynn (**Deb**) and his name (**ian**). Later, he married her and then got divorced in January 2008. In the same time period, Slackware has made a significant presence in the field of open-source software. Debian generates many distributions, in fact, almost half of the currently active Linux distribution based on Debian, in which some popular distributions are Ubuntu, Knoppix, Xandros, Kali etc.

Debian GNU/Linux is a free and open-source OS built on the Linux kernel making it one of the earliest OSs on this kernel. Debian GNU/Linux is also known as Debian. It is based on a graphical user interface. It comes with thousands of software applications that are easy to install, run, and develop with the help of its GNU capabilities. Debian initially had a semi-official logo that was replaced by the new "swirl" logo designed by Raul Silva as a part of a contest in 1999. It is one of the prominent distributions, which is completely an independent and community-driven distribution. Debian is basically a general-purpose distribution with a distinctive feature for all kind of work; that's why the same distribution image is used for both environments: desktop and server. Ian Murdock was its first chief of the Debian project as founder from 1993 to 1996. H had clear objectives, which he conveyed in the *Debian Manifesto*. The free OS that he wanted would have two main features: quality and non-commercial distribution. He aimed at developing Debian with the greatest care and value of the Linux kernel. It also wanted that it would be appropriately realistic to compete with major commercial distributions for its greatest use.

Debian project continues to be one of the key leaders in Linux development and distribution. The development work is carried out over the Internet by a team of volunteers under the guidance of the Debian Project Leader. All the development and release follow three foundational documents, namely the Debian Social Contract, the Debian Constitution, and the Debian Free Software Guidelines. Debian works for its users. Hence, "**Debian Social Contract" Version 1.1 ratified on April 26, 2004. Supersedes Version 1.0 ratified on July 5, 1997**, summarizes the commitments of Debian project toward users, which are as follows:

- Debian will remain 100% free.
- We will give back to the free software community.
- We will not hide problems.
- Our priorities are our users and free software.
- Works that do not meet our free software standards.

The **Debian Free Software Guidelines (DFSG)** follows all the principles of free software with the following guidelines as per **"Debian Social Contract" Version 1.1**:

1. Free Redistribution.
2. Source Code.

3. Derived Works.

4. Integrity of The Author's Source Code.

5. No Discrimination Against Persons or Groups.

6. No Discrimination Against Fields of Endeavor.

7. Distribution of License.

8. License Must Not Be Specific to Debian.

9. License Must Not Contaminate Other Software.

10. Example Licenses (The *GPL*, *BSD*, and *Artistic* licenses are examples of licenses that we consider free).

Debian was the first Linux distribution to include a package management system for easy installation and removal of software. It was also the first Linux distribution that could be upgraded without requiring reinstallation. Now, at the time of writing this chapter, the latest release of Debian 10 is available [7]. You can visit the following weblinks to get more resources, update, and support related to Debian:

Distribution	Debian Project
Home Page:	http://www.debian.org/
Mailing Lists:	http://lists.debian.org/
User Forums:	http://forums.debian.net/
Documentation:	http://www.debian.org/doc/
	https://debian-handbook.info/browse/stable/
	(Debian Administrator's Handbook)
Download Mirrors	http://www.debian.org/distrib/ftplist
	https://www.debian.org/CD/ (Debian on CD)
Bug Tracker	http://bugs.debian.org/
Wiki:	https://en.wikipedia.org/wiki/Debian

15.4 Ubuntu Linux:

Ubuntu is the youngest and most famous member in the family of Linux distributions. Ubuntu is a free and open-source Linux distribution based on Debian. The Ubuntu was first announced in October 2004 and developed by Canonical Ltd., which is UK-based company to provide the commercial support and other related project services for Ubuntu. Mark Shuttleworth, South African entrepreneur, is the founder and CEO of Canonical Ltd., and he has created the first project of Ubuntu.

It is a complete desktop Linux OS, which is fast, secure, freely available, and simple with the support from professional and volunteer community. Due to its popularity, Ubuntu functions in millions of PCs across the world to compete with the other major commercial OS arenas like Microsoft Windows and Mac OS. Ubuntu is officially released in four editions: Ubuntu Desktop, Ubuntu Server, Ubuntu IoT, and Ubuntu cloud. Ubuntu is a popular OS for cloud computing, with support for OpenStack. All the editions can run on the computer alone, or in a virtual machine. Simply we can say that Ubuntu is available on every platform, desktop, server markets, mobile devices, the cloud, and the IoT. Ubuntu is readily available and easy to use. It is also very popular among beginners who are passionate about coding on Linux platform.

Ubuntu includes a Linux kernel that can use multiple processors, which allows you to use Ubuntu in more advanced computing environments with greater demands on CPU power. The latest details of Ubuntu editions are as follows:

- **Ubuntu Desktop**: For desktop, Linux OS with the latest release is 19.10 ("Eoan Ermine"), and the latest long-term support (LTS) release is 18.04 LTS ("Bionic Beaver"). The desktop edition can replace or run with your current OS, maybe Windows or Mac OS.

- **Ubuntu Server**: This version is installed in cloud and data center servers that come with five years of free upgrades, like ARM-based server systems, IBM Power server (for mobile, social, cloud, big data, data analytics, and machine learning).

- **Ubuntu IoT**: Available for IoT board or devices like Raspberry Pi 2, 3, or 4, Raspberry Pi Compute Module 3, Intel NUC, KVM, Intel Joule, Qualcomm DragonBoard 410c, Samsung ARTIK 5 or 10, UP2 IoT Grove, Intel IEI TANK 870.

- **Ubuntu cloud**: Available as images for the most important public clouds such as Amazon AWS, Google Cloud, or MS Azure.

- **Ubuntu Phone (Mobile)**: It is not openly available to download from the website and install on tablets or phones, rather it comes with factory-loaded smartphones and tablets as an original equipment manufacturer (OEM) OS.

Ubuntu is moving toward to take over the everyplace of the IT world. Now, at the time of writing this chapter, the latest release of Ubuntu 19.10 is available [8]. You can visit the following weblinks to get more resources, update and support related to Ubuntu:

Distribution	Ubuntu
Home Page:	https://www.ubuntu.com/
Mailing Lists:	https://lists.ubuntu.com/mailman/listinfo/
User Forums:	https://ubuntuforums.org/
Documentation:	https://askubuntu.com/
	https://help.ubuntu.com/
	https://wiki.ubuntu.com/UserDocumentation
Download Mirrors:	https://www.ubuntu.com/download/ •
Bug Tracker:	https://bugs.launchpad.net/
Wiki:	https://en.wikipedia.org/wiki/Ubuntu

Summary of Red Hat, Fedora, Debian, and Ubuntu key features:

Feature	Red Hat	Fedora	Debian	Ubuntu
Home page	http://www.redhat.com/	https://getfedora.org/	http://www.debian.org/	https://www.ubuntu.com/
Birth Year	1994	2003	1993	2004
Latest Release	Red Hat Enterprise Linux 8.1	Fedora 31	Debian 10	Ubuntu 19.10
Latest Release Date	11-05-2019	10-29-2019	07-07-2019	10-17-2019
Architecture	aarch64, i386, ia64, IBM Z, ppc, ppc64el, s390, s390x, x86_64	aarch64, armhfp, i686, x86_64	aarch64, armel, armhf, i386, i686, mips, mipsel, ppc64el, s390x, x86_64	armhf, i686, ARM64, PowerPC, ppc64el, s390x, x86_64
Desktop	GNOME, KDE	Awesome, Cinnamon, Deepin, Enlightenment, GNOME, KDE Plasma, LXDE, LXQt, MATE, Openbox, Pantheon, ratpoison, Xfce	AfterStep, Awesome, Blackbox, Cinnamon, Fluxbox, flwm, FVWM, GNOME, i3, IceWM, ion, JWM, KDE, LXDE, LXQt, MATE, Openbox, pekwm, ratpoison, WMaker, XBMC, Xfce	GNOME, Unity
Category	Desktop, Server	Desktop, Server, Live Medium, Workstation	Desktop, Live Medium, Server	Desktop, Server, Cloud, IoT
Product	Red Hat Enterprise Linux, Red Hat Directory Server, Fedora, Red Hat Virtualization, Red Hat Storage Server, CloudForms, etc.	389 Directory Server, Fedora OS, Cloud	Debian-Derived Distributions like Grml, Linux Mint, Knoppix Ubuntu	Kubuntu, Xubuntu, Lubuntu, Ubuntu MATE, Ubuntu Budgie, Ubuntu Kylin, Ubuntu Studio
Founder	Bob Young Marc Ewing	Warren Togami, Red Hat	Ian Murdock	Mark Shuttleworth

15.5 Ethical aspects of using Linux:

Linux is an open-source Unix-like OS based on the Linux kernel. A huge range of Linux-like OSs are available in the form of Linux distribution to cater to one particular feature or the range of features as per the usability. Linux comes with prebuilt and installed compilers and interpreters. So the task of the programmer becomes much easy, whereas the commercial OS does not offer any such functionality. Any package required by the programmer can be installed with simple command line tools: *apt-get install <package name>*.

Linux contains various command line tools like awk and grep, which are very helpful for creating command line tools. More or less, it is programmer's playground. Programmers love Linux for its speed and security. It also has inbuilt package manager. The other thing that made Linux so popular is that it is completely open source, and you have complete control of almost everything. There are also open-source tools that can integrate with Linux to help you in your task. Also, Linux requires lesser hardware requirements as compared to Windows or MAC OS, which is possible for Linux to revive older computer systems. However, that does not mean that every Linux distribution can run on 256MB of RAM with outdated processors, but yes there are certain distros that can even work on such outdated systems as "puppy Linux."

Linux kernel is superior as compared to commercial OS terminal and has various command line tools available for programmers. Other OSs offer updates only when they receive issues and some major problems, and on the other hand in the case of Linux, due to its large community, updates are produced more frequently. Linux is distinguished from other popular OSs in two most important ways. First, Linux is a cross-platform OS that runs on various computer platforms or models in contrast with other known OS like Microsoft Windows to run on an architecture. Second, Linux is free, no need to pay to get an OS image copy for desktop or server. Linux and many Linux applications are distributed in source code form. The whole source code of Linux is available under General Public License (GPL), which allows people to borrow and change code as they need, provided they properly credit the original authors. Because of that, everyone gets to pitch in and make the software more usable, all authors and contributors get credit, and software is available to use for free by everyone. Such a thing attracts both types of people, users and programmers, to use/contribute to Linux by having minimum legal liabilities. Because of such freedom, Linux is being continuously enhanced and updated. To prove this, Linux was the first OS to support Intel's Itanium 64-bit CPU. In Linux family, every day, large numbers of people, including users, learners, programmers, code reviewers, and testers, are joining to make Linux more user-friendly and useful.

References

1. Linux Distributions, The Major Distributions: An overview of major Linux distributions and FreeBSD in the World Wide Web. https://distrowatch.com/dwres.php?resource=major.
2. Andreas Lundqvist, Donjan Rodic, Mohammed A. Mustafa, Konimex, Fabio Loli, The GNU/Linux Distributions Timeline (Version 19.04) in the World Wide Web. https://en.wikipedia.org/wiki/Linux_distribution#/media/File:Linux_Distribution_Timeline_Dec._2020.svg.

3. Jose Dieguez Castro. 2016. *Introducing Linux Distros*. Apress.
4. Christopher Negus. 2008. *Linux Bible*. Wiley Publishing, Inc.
5. The Red Hat Enterprise Linux, The Red Hat Linux in the World Wide Web. http://www.redhat.com/.
6. Fedora, The Fedora Linux in the World Wide Web. https://getfedora.org/.
7. Debian, The Debian Linux in the World Wide Web. http://www.debian.org/
8. Ubuntu, The Ubuntu Linux images and documents in the World Wide Web. https://www.ubuntu.com/.

Index

Note: **Bold** page numbers refer to tables and *Italic* page numbers refer to figures.